Fundamental Physics

SECOND EDITION

Jay Orear

PROFESSOR OF PHYSICS, CORNELL UNIVERSITY

JOHN WILEY & SONS, INC., NEW YORK, LONDON, SYDNEY

The line drawings are by Herschel Wartik and Frank McCarthy.
The airbrush drawings are by John Yue of the Illustration Department
of John Wiley & Sons, Inc.

SECOND PRINTING, JULY, 1967

Library of Congress Catalog Card Number: 66-26754

Printed in the United States of America

Enrico Fermi, 1901-1954

Note to the student

This book contains a new feature which is a kind of built-in "teaching machine". At the bottom of each page (whenever there is room) is a "test your understanding" question, and on the next page is the answer. To gain the full value from these page-by-page questions, the reader should stop and answer each one before going on or looking at the answer. These questions are relatively easy and many of them can be answered without using paper and pencil. Whenever one of these questions is missed, it is probably because the reader did not understand fully what he had just finished reading. This is a common problem in the study of physics. It can be easy to read, but the understanding requires more than reading or rote learning—it requires thought, and there is no easy way to avoid the necessary thought. So if you miss one of these questions, let it be a warning that you should not go ahead without another reading and thinking through of that section. Remember, we become wise only when we are fully aware of what we do not understand.

Further help of this nature can be obtained from the *Programmed Manual for Students of Fundamental Physics* also published by John Wiley & Sons, Inc. This is a teaching machine in the form of a programmed manual which covers more thoroughly nearly all the material of the textbook.

Preface to second edition

This second edition is not the kind of revision in which the problems are renumbered and the physics brought up to date by adding a few pages at the end. It is a serious attempt to rewrite the entire book in order to increase the amount of explanation without much increase in subject matter. Few new concepts have been added; there are, however, many more examples, figures, and problems. Over 100 of the illustrations and about 100 of the problems are new. Many of the new examples and explanations are based on five years of feedback from students and teachers. Other sources are the *Programmed Student Manual* and conferences and study groups sponsored by the Commission on College Physics.

A major innovation is the thorough and unified treatment of quantum mechanics running through Chapters 12 to 16. I think of these chapters as a serious introductory course in quantum mechanics at the pre-calculus mathematical level. I know of no other introductory physics text at this level which develops the concepts and applications of quantum mechanics so thoroughly and so quantitatively, and I am convinced that a surprising amount of quantum mechanics can be taught without calculus. In most of the situations in physics in which both calculus and noncalculus explanations exist, I prefer the noncalculus explanation—even for erudite graduate students. The noncalculus explanation usually gives more physical insight and is easier to visualize.

It has been my experience that college freshmen can master some of the basic concepts of quantum mechanics as well as special relativity, and that they derive special enjoyment and excitement from these subjects. Consequently, even more emphasis is placed on quantum mechanics and relativity in this revised edition. Perhaps this is the new trend—these subjects are now taken seriously in a few of the more recent introductory textbooks. But I believe this book gives a more "grown-up" quantum mechanical explanation of atomic structure, theory of metals, nuclear structure, diffraction scattering, semiconductors, hybridization, lasers, radioactivity, etc. Other features of the new edition are a quantitative treatment of cosmology and Mach's principle, and the relativistic foundation of magnetic force.

Back in 1960 when the first edition was published, it was

considered daring to involve an introductory course so deeply with relativity, quantum mechanics, and its applications to atomic, solid state, nuclear, and high energy physics. Other innovations of the first edition were the format, the use of two colors throughout, extensive use of bubble chamber photos, quantitative drawings of electron clouds, Gaussian units in electricity, and a "cheat-proof" programmed students manual. Since then new books have appeared from the high school level on up incorporating some of the above innovations. Additional subject matter pioneering has been done in connection with the recent Cal Tech and Berkeley introductory courses for physics majors. I have found some of this new material appropriate for nonscience majors and have incorporated it with suitable modifications in this new edition.

A new feature starting with this edition is a kind of page-by-page built-in teaching machine. Experience with the Programmed Manual which accompanies this text has pointed out that one of its major advantages is to let the student know whether or not he really understood what he just got through reading in the text. It takes an exceptionally wise person to know just what he does and does not understand. Students should be urged to answer the questions at the bottom of each page as they come to them before looking at the answers.

In this new edition a serious attempt has been made to keep up-to-date with the exponentially increasing advance of science. I must confess that physics has progressed farther in the last six years than I had anticipated. The pace seems to increase with time. Take for example the number of elementary particles. When the first edition was completed the number of elementary particles had appeared to settle down to 30. But in the past five years the number has increased from 30 to about 200 with no end in sight. Fortunately the newly discovered particles have revealed new patterns or symmetries which make even the original 30 easier to understand. But then recent violations have been found of other more cherished symmetries or conservation laws which should have been obeyed by the elementary particles. These and other new developments such as quasars and lasers have been incorporated into whatever spot in the book is most appropriate. Even the classical physics in this book is "modern". All the physics is put in the context of 1966, not 1900 as most books, or 1925 as in some of the so-called modern physics books.

I am grateful to students, teaching assistants, and faculty at Cornell University for much of the new material. I wish to thank in particular Professors Alan Bearden, Robert Sproull, Phil Morrison, and R. Rajaraman. I also wish to acknowledge stimulating discussions with Professor Matt Sands concerning the new Cal Tech course developed by Feynman, Leighton, and Sands which cannot help but influence future physics textbook writers.

Jay Orear

September 1966

Preface to first edition

Just what is physical reality? What makes the universe tick? What are the "secrets" of nature? These questions exemplify the spirit and motivation of this book. It is primarily designed for use as a text in a one-year course at the college level for students who have had little previous training in mathematics and science, but it should also be helpful to science majors learning college physics for the first time.

In this textbook the main emphasis is on the first principles or fundamental laws of nature upon which all science is based. This area of study is properly a part of the vast collection of topics called physics. Thus this is another college physics textbook, but it deals lightly with the topics of more applied nature which are traditionally also called physics. For example, little will be said about machines, rotational dynamics, photometry, optical instruments, a-c theory, calorimetry, elasticity, acoustics, etc. By restricting the book to "fundamental physics" as opposed to "applied physics" there will be fewer topics, with the accompanying advantage that they can now be treated with greater depth. The main exception to this rule is Chapter 10 which goes deeply enough into electronics to explain the basic principles of radio and television in order to give a feeling for the vast technological implications of our understanding of the basic laws of nature. This chapter may be skipped by those who dislike teaching engineering in a physics course.

By its very nature, fundamental physics is deeply philosophical. It is a continual struggle of man's mind against nature, with nature full of shocking surprises to which man must then adjust. There is a temptation to present this battle of man's mind vs. nature to the liberal arts student using a philosophical and historical approach. Such an approach often makes the mistake of teaching about science rather than teaching the science itself, leaving the student without any real understanding of the physical world in which he lives. The "educated man" should be exposed to both approaches. Ideally he should learn about science and its relation to the other disciplines in his philosophy courses. On the other hand, there is no harm in presenting some of this "humanistic approach" in the introductory physics course as long as primary emphasis is placed on the teaching of the science itself. Consequently, this book gives some attention to the impact of science on our culture. The methods of science, the "art" of scientific discovery, the social responsibility of scientists, along with philosophical, social, and political relations of science to our culture are discussed, but they play a secondary role to presenting the subject matter of the science itself. The excitement and cultural value of the history of physics are made use of where appropriate to the main goal. The acid test of whether a student has grasped science itself is whether he can successfully solve problems that require some thought. My experience with much of the material and problems in this book in college courses at Columbia and Cornell Universities indicates that it is possible for nonscience liberal arts students to master the fundamental principles as evidenced by their ability to apply them in problem solving.

The order of presentation is mainly determined by what I feel to be pedagogically preferable. This tends toward development of logical sequences. Sometimes the "modern" physics manages to get presented before the "classical" and "preclassical" versions are discussed. Usually I have attempted to avoid the sequence: teach a topic according to the "old" physics, tell the student it is now wrong, then try to get the student to unlearn it and relearn it the new way. When a new topic is encountered, the final version is usually given first. Then from this viewpoint the student can observe and understand more fully the various "old" ideas. The sequence of topics was not determined by tradition, but by the requirements of a rather tight logical development. For example, the general phenomenon of barrier penetration is first introduced in the text in a discussion on field emission of electrons from a metal. This discussion depends on the previous presentations of Fermi energy, metallic binding, and potential diagrams, which depend in turn on the presentations of electron waves in a box, atomic electron clouds, and electric potential. Going back further in the logical chain of development are wave-particle duality, wave interference, electricity, energy, mechanics, etc. Hence most of the pages of this book belong to a central chain of logical development. Because of the compact, unified, theoretical structure of physics, my chapters do not correspond to the old, compartmentalized series of "independent" topics found in traditional textbooks. I was unable to tie the chapters up into neat, self-contained, airtight packages. To present classical physics in such a manner may be easier and more satisfying to some, but that just is not the way nature happens to be. I have attempted to present the true product: the imperfections and limitations of present-day physics are openly admitted.

I have also tried to face up to the fact that our present-day description of most physical phenomena requires use of the quantum theory. With the recent development of quantum electrodynamics enormous progress has been made in our understanding of the physical world and the structure of matter. If at all possible, students at the college level should be given a glimpse of our present understanding. Some physics teachers may feel that such topics as electron clouds, quantum theory of chemical binding, Fermi energy, time dilation, nuclear structure, cosmology, and conservation of parity are too difficult and abstract for college freshmen. However, it has been my experience in teaching this kind of course for the last six years that liberal arts students have more trouble in understanding Newton's Third Law than they do in understanding Fermi energy or charge conjugation invariance. Not only do they finish the course with a good physical feeling for modern physics, but they also express a strong preference for it over classical physics.

In conclusion, I wish to thank my colleagues at Cornell for their encouragement and particularly Professors Robert Sproull, Phillip Morrison, John DeWire, and Richard Feynman for their many helpful suggestions. Many people kindly contributed visual material for this book which would have been impossible for any one man to produce. I have drawn upon the visual material of the Physical Science Study Committee of Educational Services, Inc. and am grateful to them and Professor Francis Friedman for their help and cooperation. Mr. Kim Choy of Cornell was of considerable help in calculating the hydrogen electron clouds and the deuterium nuclear cloud. I am grateful to Mr. Francis Schrag and the many other students who contributed a great deal to the preliminary edition of this textbook while in use at Cornell for the last two years.

My greatest debt is to Enrico Fermi, who not only taught me much of the physics I know, but also how to approach it. As a teacher, Fermi was well known for his great ability to make the most difficult topics seem beautifully simple in a clear, direct way with little mathematics, but much physical insight. The goal I have been aiming at is to try to present the spirit and excitement of physics in the way that Fermi might have done.

Jay Orear

September 1960

Contents

* May be omitted in a fast paced or shorter course.

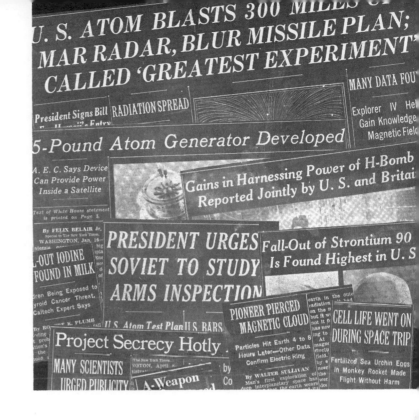

Introduction

Chapter 1

Introduction

1-1 What Is Physics?

Physics is what physicists do late at night

There is no clear definition of what is physics—or which topics belong in physics and which do not. A study of the table of contents and the index to this book should give an idea of some of the topics which the author believes belong to physics.

One of the goals of physics is to find out the "rules" of the universe in which we live. Some of these rules have come as a shocking surprise to scientists and philosophers. In fact, some of the discoveries were so shocking and so contrary to common sense that they were slow to gain acceptance, even by Nobel prize winners. For example, Einstein presented the theories of relativity and the photoelectric effect in 1905, but he was not awarded a Nobel prize for the latter until 1921. He never did receive a Nobel prize for his work in relativity. Apparently some of the old prize winners who help choose the new ones found the theory of relativity too radical for them.

By now physicists have become suspicious of common sense and what seems obvious. Common sense is a product of man's mind. There is no reason why Mother Nature should oblige. For example, now we even mistrust $2 + 2 = 4$ when applied to the physical world. We have learned that 20 billion centimeters per second plus 20 billion centimeters per second is not 40 billion centimeters per second, but it is 27.3 billion centimeters per second! We now know that one must add velocities in a way that seems peculiar to common sense. The resultant velocity is always less than the sum of its components. If the velocities are small compared to the speed of light, the effect is still there, but it is a very small effect. As an example of this, consider a river boat traveling downstream at 20 mph (miles per hour) with respect to the water. If the water velocity is 10 mph, then the resultant velocity of the boat with respect to the shore should be exactly the sum of the two or 30 mph. However, according to the laws of physics, the velocity of the boat will be slightly less than the sum of the two velocities. The correct answer would then be $v_{boat} = 29.9999999999999866$ mph. We can see why it took so long to discover this effect. In Chapter 11 we will study

Fig. 1-1. Boat sailing downstream. In (*a*) the boat is viewed by a person floating in the water. In (*b*) the boat is viewed from the shore.

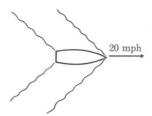

(*a*) Boat in still water

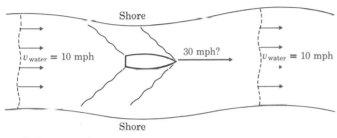

(*b*) Same boat in stream flowing at 10 mph

Fig. 1-2. Electron gun shooting beam of electrons through holes *A* and *B*.

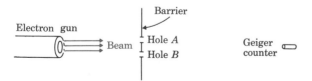

the law of physics to be used here. It is the Einstein addition of velocities which says

$$v_{\text{resultant}} = \frac{v_1 + v_2}{1 + \dfrac{v_1 v_2}{c^2}}$$

where $c = 670$ million mph is the velocity of light. In our example $v_1 = 10$ mph and $v_2 = 20$ mph. We see that for everyday practical problems $v_1 v_2/c^2$ is so much smaller than one that the old classical law

$$v_{\text{resultant}} = v_1 + v_2$$

is quite adequate.

We shall consider one more example that violates common sense. In this example $2 + 2$ can be zero or, if you prefer, we can make $2 + 2 = 8$. We shoot a beam of electrons from an electron gun at an opaque barrier that has two holes, A and B, in it. We put a small geiger counter far behind the barrier, and we plug up hole B. Then the counter counts 2 electrons per second. Now we open hole B and close hole A. Again we get 2 counts per second. We now open up both holes together. Here we get no counts at all! Not only is the whole less than the sum of its parts as in the Einstein addition of velocities, but here it is even less than either of the parts by itself. If we prefer to count 8 electrons per second, we just move the geiger counter slightly in the vertical direction, and we can find a place where the whole is twice as big as the sum of its parts. At first encounter all this may appear hard to accept, but it is true, and such phenomena have been observed in the laboratory (see Fig. 12-8, page 308). This particular phenomenon is due to the wave nature of matter. In Chapter 12 we shall learn that all particles have certain wave properties and that such phenomena then follow naturally. Figure 1-3 shows how the electron waves would look as they pass through holes A and B together.

The preceding examples of the river boat and the electron beam illustrate applications of relativity and quantum theory, respectively. Many of the more basic physical phenomena are applications of relativity and quantum theory. Consequently a significant part of this book must be devoted to relativity and quantum theory.

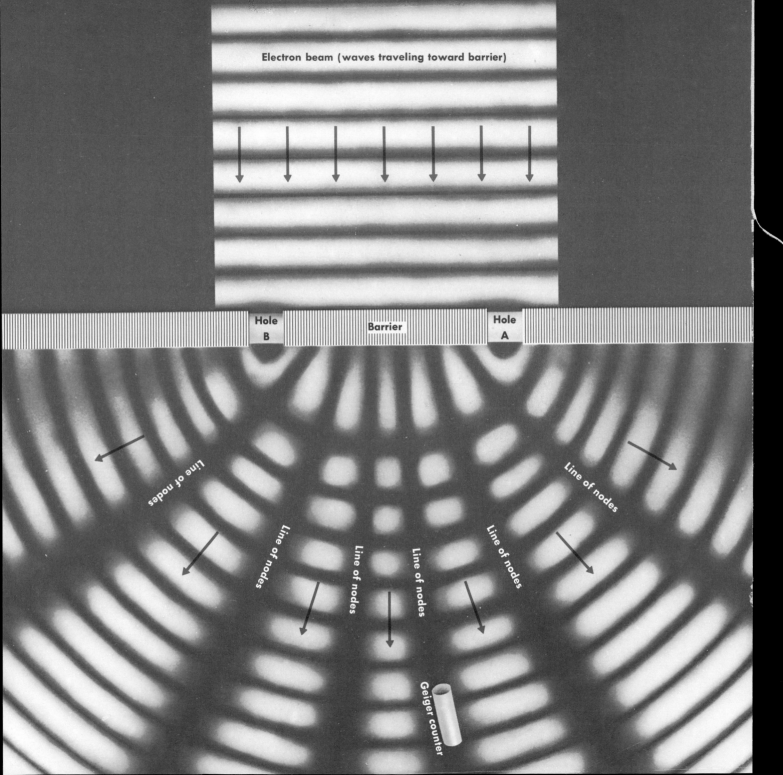

TABLE 1-1

Physical quantity	Units
Area	L^2
Volume	L^3
Velocity	L/T
Acceleration	L/T^2
Density	M/L^3
Momentum	ML/T
Force	ML/T^2
Energy	ML^2/T^2

Fig. 1-3. A beam of light waves passing through two holes. This is what the light intensity would be at a given instant of time. Black corresponds to zero intensity. According to the modern wave theory of matter, electron waves would behave in a similar fashion.

1-2 Units

L, T, and M

Much of physics deals with measurements of physical quantities such as length, time, frequency, velocity, area, volume, mass, density, charge, temperature, and energy. Many of these quantities are interrelated. For example, velocity is length divided by time. Density is mass divided by volume, and volume is a length times a second length times a third length. Most of the physical quantities are related to length, time, and mass. Some of these relationships are shown in Table 1-1. We will study these physical quantities as they appear later in the book.

Length

The definitions of length, area, and volume are given in Euclidean geometry. There are several standard units of length in use today, such as the meter, inch, foot, mile, and centimeter. Except for the United States and parts of the British Commonwealth, nearly all the nations of the world now use the metric system. Even England is now committed to changing over to the metric system. Although the English system of units is still the official system in the United States, American scientists use the metric system almost exclusively. In addition to using some English units, we will make use of the metric system throughout this book.

The meter was originally defined in terms of the distance from the north pole to the equator. This distance is close to 10,000 kilometers or 10^7 meters. The standard meter of the world is the distance between two scratches on a platinum-alloy bar that is kept at the International Bureau of Weights and Measures in France. However, Mother Nature provides us with a unit of length much more accurate than the distance between two scratches on a piece of metal. This is the wavelength of light from any sharp spectral line. The standard meter in France has been calibrated in terms of the number of wavelengths of light of a certain spectral line. The United States inch is defined in terms of the standard international meter so that one inch equals exactly 2.54 centimeters (100 centimeters = 1 meter).

TABLE 1-2 ORDERS OF MAGNITUDE OF DISTANCES

Distances in centimeters	
	Distance to farthest photographed galaxy
10^{25}	
	Distance to Great Nebula in Andromeda (nearest galaxy)
	Radius of our galaxy
10^{20}	
	One light year (distance light travels in one year)
	Size of solar system
10^{15}	
	Distance from earth to sun
	Radius of the sun
10^{10}	
	Radius of the earth
	Height of Mt. Everest
10^{5}	One km or $\frac{6}{10}$ of a mile
	One m or one yard
10^{0}	One cm or $\frac{4}{10}$ of an inch
	One mm
	Thickness of a hair
	Diameter of red blood corpuscle
	Wavelength of light
10^{-5}	
	Size of organic molecules
	Diameter of hydrogen atom
10^{-10}	
	Diameter of a uranium nucleus
	Diameter of an elementary particle
10^{-15}	

Other conversions are given in the Appendix at the end of the book. In order to solve everyday problems it will often be necessary to convert English units into metric units before proceeding.* Table 1-2 shows various lengths encountered in physics from the smallest (diameter of an elementary particle) to the largest (distance to the farthest observed galaxy).

Time

Time is a physical concept and thus its definition is related to certain laws of physics. For example, the laws of physics say that to very great accuracy the period of rotation of the earth must be constant. This fact can then be used to define a basic unit of time, called the mean solar day. Also the laws of physics say that the period of oscillation of a vibrating slab of crystal in a crystal oscillator should remain constant if the temperature and other external conditions are kept constant. So an electronic crystal oscillator can be made into a very accurate clock. The same is true of the vibrational frequency of atoms in a molecule. In fact, atomic clocks that count up these vibrations are the most accurate of all. The basic unit of time used in both the English and metric systems is the second which is

$$\tfrac{1}{60} \times \tfrac{1}{60} \times \tfrac{1}{24} = \tfrac{1}{86,400}$$

of a mean solar day. (We will use the abbreviation sec for second.) Table 1-3 lists various time intervals encountered in physics from the time it takes light to travel across an elementary particle to the age of the earth.

When we base a concept such as time on the laws of physics, we cannot be sure that these laws are absolutely correct. For example, suppose the speed of light is slowly increasing with time. This would then cause a change in some of our standards of length and time. So far there is no experimental evidence that any of the universal physical constants are changing with time, but this does not rule out the possibility of a very slow change beyond the accuracy of present measurements. In the course of this book we shall see that it

*We shall use the abbreviations cm for centimeter, m for meter, km for kilometer, in. for inch, ft for foot, and mi for mile.

TABLE 1-3 ORDERS OF MAGNITUDE OF TIMES

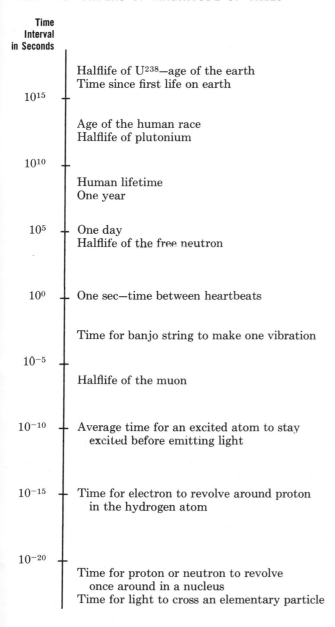

Time Interval in Seconds

10^{15} — Halflife of U^{238}—age of the earth
Time since first life on earth

Age of the human race
Halflife of plutonium

10^{10} —

Human lifetime
One year

10^5 — One day
Halflife of the free neutron

10^0 — One sec—time between heartbeats

Time for banjo string to make one vibration

10^{-5} —

Halflife of the muon

10^{-10} — Average time for an excited atom to stay
excited before emitting light

10^{-15} — Time for electron to revolve around proton
in the hydrogen atom

10^{-20} —

Time for proton or neutron to revolve
once around in a nucleus
Time for light to cross an elementary particle

is not uncommon for a "sacred" law of physics to be overthrown by new experimental data. We must learn to be openminded about our existing "laws" of physics and be prepared to modify them if experimental evidence should ever appear against them.

If we ask deeper questions concerning the nature of time, we discover that physicists and philosophers do not completely understand it. Our present understanding of the physical world is deep, but not as deep as we would like. The question whether we can ever achieve the ultimate truth is an age-old question. Unfortunately physics is not in a position to settle this classic question.

Mass

Mass is also a physical concept and must be defined in terms of certain laws of physics. In Chapter 3 we give the modern definition of mass in terms of the law of conservation of momentum. In the metric system the unit of mass was originally defined as that amount of mass contained in 1 cc (cubic centimeter) of water (at a specified temperature and pressure). This amount of mass is called the gram. Thus the density of water is conveniently one gram per cubic centimeter. In the English system the unit of mass is the pound. One kilogram (10^3 grams) = 2.204 pounds of mass.* Table 1-4 lists some of the masses encountered in physics.

In physics the quantities such as force and energy are usually measured either in meters, kilograms, and seconds, or in centimeters, grams, and seconds. The former is called the MKS system of units and the latter the CGS system of units. Both of these metric systems are referred to in this book. When working problems it is very important to convert all units to MKS or else all to CGS. *Never* use mixed units.

1-3 Mathematics in Physics

$$10^a \times 10^b = 10^{a+b}$$

In physics quantities can be calculated and measured with very great accuracy. This quantitative approach requires

---

*We shall use the abbreviations gm for gram, kg for kilogram, and lb for pound.

TABLE 1-4 ORDERS OF MAGNITUDE OF MASSES

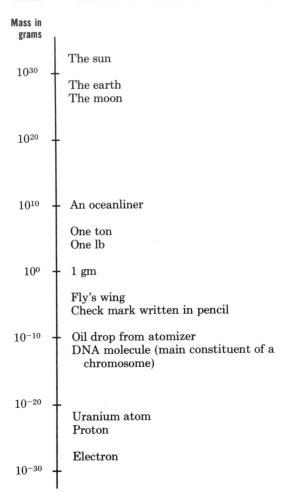

familiarity with mathematics. Fortunately, many of the fundamental principles and concepts in physics can be well understood by using only elementary algebra and geometry. This is a consequence of what seems to be a general property of the basic laws of nature; namely, the closer we get to the truth, the simpler the basic laws tend to become, a principle the philosophers call Ockam's razor. The more difficult mathematics usually enters when one tries to calculate something that is not basic, such as what is called the three-body problem (the motion of three mutually interacting bodies). The three-body problem is not basic because it is really the superposition of three interrelated two-body problems. Several hundred years ago Isaac Newton solved the really basic problem—the orbits of two bodies interacting under an inverse square force. The two-body problem of astronomy can be solved using elementary mathematics (see Chapter 4), but to do an accurate job of the three-body problem would require an enormous electronic computer. In one second an electronic computer can do the amount of calculating that would take one man several weeks to do by hand. Although the technology of electronic computers is still in its early stages, electronic computers are now widely used in physics, in other forms of basic research, and in industry.

A knowledge of trigonometry, calculus, or higher forms of mathematics is not needed for a complete mastery of this book. On the other hand, a good working knowledge of certain aspects of high school algebra is necessary. Many students have not worked with algebra for several years by the time they take physics in college. Also some of the algebra needed for this book is not stressed in many high schools. Hence, the mathematically weak reader should spend some time reviewing mathematics before proceeding with this book. Mathematic problems are provided at the end of this chapter for this purpose. The remainder of this section is a review of the mathematics pertinent to this book.

The following checklist is provided for readers who are unsure whether or not they need to review mathematics.

Check off each of the following mathematical statements that are untrue:

1. If $1/R = 1/R_1 + 1/R_2$, then $R = R_1 + R_2$

2. $(a + b)^2 = a^2 + b^2$

3. If the diameter of a circle is 10^{-8} cm, then its radius is 10^{-4} cm.

4. The radius of the above circle is 5^{-8} cm.

5. $\dfrac{A}{B} + \dfrac{X}{Y} = \dfrac{A + X}{B + Y}$

6. If the short side of a $60°$ right triangle is 1 cm, then the hypotenuse is $\sqrt{3}$ cm.

7. 4 divided by $\frac{1}{2}$ is 2.

8. $\sqrt{16\,ab} = 4\,ab$

9. $\sqrt{10^3} = 5^3$

10. $\dfrac{1}{a + b} = \dfrac{1}{a} + \dfrac{1}{b}$

11. Suppose a and b are negative numbers and b/a is greater than one, then $(a - b)$ is negative.

12. $\dfrac{10^{-10}}{10^{-5}} = 10^{-15}$

A person who misses more than one of the problems should consider himself as too rusty in mathematics to proceed without first reviewing and working the problems at the end of this chapter. The correct solution to the preceding checklist is that each item is untrue. You should now go back and calculate the correct answers. The correct answers are given at the end of this chapter.

Powers of ten

Most of the quantities encountered in physics are either much larger than one, or much smaller than one. For convenience, the standard practice is to write any quantity, no matter how large or small, as a number between one and ten times the appropriate power of ten. A physicist would write the United States national debt as 3.3×10^{11} dollars or 3.3×10^5 megabucks. The exponent of 10 tells the number of places to shift the decimal point to the right. If we must write the national debt without powers of ten, then the decimal point must be shifted 11 places to the right, giving

Q.1: (*This is the first of simple page by page questions which the reader should be able to answer before going on. Answers are always on the reverse side of the page.*)

What is the ratio of the largest to smallest distance in Table 1-2?

$330,000,000,000. If the exponent is negative, then the decimal point must be shifted the same number of places to the left. For example, the radius of the hydrogen atom is 5×10^{-9} cm. When we shift the decimal point 9 places to the left, this becomes 0.000000005 cm.

In multiplying and dividing numbers expressed in this way we make use of the relations $10^a \times 10^b = 10^{a+b}$, and

$$\frac{10^a}{10^b} = 10^{a-b}$$

Example 1

A nuclear reactor converts 1 gm of uranium per day into energy. What is its power output in watts?

Power is energy per unit time, and if we stick to MKS units the answer will be in watts. We can calculate the energy by using Einstein's famous mass-energy relation: $W = Mc^2$, where c is the speed of light that has the value 3×10^8 m/sec. The equivalence between mass and energy is studied in Chapters 5 and 11.

So   power $= \dfrac{Mc^2}{t}$

where $M = 10^{-3}$ kg and $t = 60 \times 60 \times 24$ sec $= 8.64 \times 10^4$ sec. When we put in the numbers,

$$\text{power} = \frac{10^{-3} \times (3 \times 10^8)^2}{8.64 \times 10^4} \text{ watts}$$

$$= \frac{10^{-3} \times (9 \times 10^{16})}{8.64 \times 10^4} \text{ watts}$$

The power of ten in the answer will be the sum of the exponents in the numerator minus the sum of the exponents in the denominator. The final exponent is $(-3 + 16) - 4$, or 9. Thus

$$\text{power} = \frac{9}{8.64} \times 10^9 \text{ watts}$$

$$\text{power} = 1.04 \times 10^9 \text{ watts}$$

or about one million kilowatts. This is the amount of power consumed by an average state in the United States.

Now that we have reviewed the rules for using powers of ten, let us see why $\frac{1}{2}$ of 10^{-8} is neither 10^{-4} nor 5^{-8}. The number 10^{-4} can be written

$$10^{-4} = 10^4 \times 10^{-8}$$

Thus the answer 10^{-4} is off by a factor of 20,000. The other

Ans. 1: Referring to Table 1-2, the distance to the farthest visible galaxy is about 10^{27} cm and the elementary particle diameter is about 10^{-13} cm. The ratio is $10^{27}/10^{-13} = 10^{27+13} = 10^{40}$.

common answer that college students give is 5^{-8}, which can be written

$$5^{-8} = \left(\frac{10}{2}\right)^{-8} = \frac{10^{-8}}{2^{-8}} = 2^8 \times 10^{-8} = 256 \times 10^{-8}$$

We see that this answer is too large by a factor of 512. The correct answer is

$$\tfrac{1}{2} \text{ of } 10^{-8} = \tfrac{1}{2} \times 10^{-8} = 0.5 \times 10^{-8}$$

This is also equal to 5×10^{-9}.

Simultaneous equations

We should be able to solve two or more simultaneous equations. As an example let us consider three equations which appear in Chapter 2 (at this point we need understand only the algebra, not the physics).

1. The formula for centripetal acceleration:

$$a_c = \frac{v^2}{R} \tag{1-1}$$

2. The relation between velocity v and the period T:

$$v - \frac{2\pi R}{T} \tag{1-2}$$

3. The relation between period T and frequency f:

$$f = \frac{1}{T} \tag{1-3}$$

Our sample problem is to find a formula for centripetal acceleration which contains only f and R. Starting with Eq. 1-1, we see that v on the right-hand side is unwanted. We can eliminate it by substituting the expression $\frac{2\pi R}{T}$ for v in Eq. 1-1:

$$a_c = \frac{\left(\frac{2\pi R}{T}\right)^2}{R}$$

$$= \frac{4\pi^2 R}{T^2} \tag{1-4}$$

But now T is unwanted. We can eliminate it by solving Eq. 1-3

Q.2: What is 10^{-2} divided by 10^{-5}? (Work these out in your head or on paper before looking at the answer. If your answer is wrong, do not go on without additional study.)

for T and substituting into Eq. 1-4. Equation 1-3 gives $T = 1/f$. Substituting this into Eq. 1-4 gives

$$a_c = \frac{4\pi^2 R}{(1/f)^2} = 4\pi^2 f^2 R$$

Fractional exponents

The general rule for multiplying N to the ath power by N to the bth power is

$$N^a \times N^b = N^{a+b}$$

If we put both a and b equal to $\frac{1}{2}$, we have

$$N^{1/2} \times N^{1/2} = N$$
Thus $\quad N^{1/2} = \sqrt{N}$

Likewise $N^{1/3}$ is the cube root of N. The quantity $N^{2/3}$ is $(N^{1/3})^2$ or the square of the cube root of N. It is also $(N^2)^{1/3}$ or the cube root of N squared.

Significant figures

Suppose in an experiment to measure velocity, an object is observed to move 10 cm in exactly 3 sec.

Then $\quad v = \dfrac{10 \text{ cm}}{3 \text{ sec}} = 3.33333 \text{ cm/sec}$

There is a problem of how many decimal places to use to express $\frac{10}{3}$ as a decimal. The convention is to use at most one more decimal place than the certainty of the result. Thus, if the 10 cm had been measured to 1% accuracy, the result could be expressed as $v = 3.33 \pm 0.03$ cm/sec. Since the true value of v lies somewhere between 3.30 and 3.36 cm/sec, the first two 3's are significant figures and the last decimal place is somewhat uncertain. It is poor practice to write the result as $v = 3$ cm/sec or $v = 3.333$ cm/sec. The form $v = 3.33$ cm/sec is preferred. To use more decimal places would not only be superfluous but misleading. We would be claiming that our result was better than it really was.

Suppose the velocity $v = 3.33$ cm/sec is to be added to another velocity $v' = 4.51 \times 10^2$ cm/sec and that v' is also known to 1% accuracy.

Ans. 2: $10^{-2} \div 10^{-5} = \dfrac{10^{-2}}{10^{-5}} = 10^{-2+5} = 10^3$

$$v = \quad 3.33 \text{ cm/sec}$$
$$\underline{v' = 451.00 \text{ cm/sec}}$$
$$v + v' = 454.33 \text{ cm/sec}$$

Note that if we quote the answer as 4.5433×10^2 cm/sec we are implying that the accuracy of our result is better than one part in 10^4. Since the accuracy of the sum can be no better than any of its parts, the result must be written as 454 ± 5 cm/sec, or as 4.54×10^2 cm/sec.

With multiplication or division the *percentage* accuracy of the result cannot exceed that of any of its parts. If one quantity has 1% accuracy and another has $\frac{1}{10}$ of 1% accuracy, then the product will have 1% accuracy.

Unless stated otherwise, quantities in this book will be written to 1% accuracy. The problems should be worked to this same accuracy, which can be obtained by using inexpensive pocket slide rules.

We make one exception to our rule of significant figures, and that is the speed of light. It has the value $c = 2.9979 \times 10^8$ m/sec. It will occur so often that we will write it simply as

$$3 \times 10^8 \text{ m/sec} \quad \text{or} \quad 3 \times 10^{10} \text{ cm/sec}$$

which is more accurate than one part in 10^3.

Conversion of units

Many calculations in physics require conversions of units. As an example let us convert a velocity of 60 mi/hr to meters per second.

$$60 \frac{\text{mi}}{\text{hr}} = 60 \times \frac{1 \text{ mi}}{1 \text{ hr}}$$

Now in place of the quantity (1 mi) substitute its equivalent 1.61 km.

$$60 \frac{\text{mi}}{\text{hr}} = 60 \times \frac{(1.61 \text{ km})}{1 \text{ hr}}$$
$$= 60 \times 1.61 \times \frac{1 \text{ km}}{1 \text{ hr}}$$

In place of (1 km) substitute 10^3 m and in place of (1 hr) substitute (60 min).

Q.3: If $A = 112$ and $B = 102$ are both known to 1% accuracy, what is the accuracy of the quantity $(A - B)$?

$$60 \frac{\text{mi}}{\text{hr}} = 60 \times 1.61 \times \frac{(10^3 \text{ m})}{(60 \text{ min})}$$

Now in place of (1 min) substitute (60 sec).

$$60 \frac{\text{mi}}{\text{hr}} = \frac{60 \times 1.61 \times 10^3}{60} \frac{\text{m}}{(60 \text{ sec})}$$

$$= 26.8 \frac{\text{m}}{\text{sec}}$$

We see that units may be converted by substitution of equivalent quantities. Because any unit has itself a numerical value, it is important to write out the units explicitly as in the above example. A numerical answer to a physics problem must never be given without explicitly writing down the units following the number.

Geometry

We should be able to calculate the areas and volumes of simple figures such as triangles, rectangles, circles, cubes, cylinders, and spheres. We should be familiar with what are called the "scaling laws." As an example consider different-sized orange juice cans where the can height is always equal to the diameter. In the language of geometry the different cans are similar figures. Suppose a small-sized can costs 10 cents and a large "double-sized" can (twice as tall) sells for 50 cents. Many, if not most, housewives would think five of the small cans hold more than one "double-sized" can. Such a housewife would lose 30 cents each time she bought eight small cans. The scaling law states that volumes of similar figures go as the cube of the linear dimension. The cube of 2 is 8. The "double-sized" can holds eight times as much as one small can. Since areas go as the square of the linear dimension, the "double-sized" can has four times the surface area of the small can. One saves money by buying the large "double-sized" pizza, provided it costs less than four small pizzas.

For this book one should also be familiar with the Pythagorean theorem and the two types of triangles shown in Fig. 1-4. These are the famous 30°–60° right triangle and the 45° right triangle. The Pythagorean theorem states that the square of the hypotenuse of a right triangle equals the

Fig. 1-4. Two common right triangles.

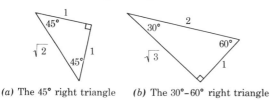

(a) The 45° right triangle   *(b)* The 30°–60° right triangle

Ans. 3: $(A - B) = (112 - 102) = 10 \pm 1$.
Both A and B have an uncertainty of about ± 1. The uncertainty of the result is about 1 part in 10 or 10% accuracy.

sum of the squares of the other two sides. Also one should know when two triangles are similar, and that in similar triangles the corresponding angles are equal and the corresponding sides are proportional. A useful theorem from geometry that will be used occasionally is: if two angles have their corresponding sides perpendicular, these two angles will be equal.

1-4 Graphical Representations

A picture is worth 10^4 words

We must know how to read and plot graphs. There will be occasions where the only way we can convey information is by plotting graphs. Quite often it is easier to visualize what is going on by looking at a graph rather than by looking at equations. Figure 1-5 is a graph for us to practice on. It represents the 20-min history of an automobile trip along a straight road where s is the distance from the driver's house and t is the time elapsed after leaving the house. Let us now see how much information we can squeeze out of this graph.

Any point on the curve tells us the value of s for that particular value of t. For example, point P is 3 mi "high" on the s-scale and 4 min "along" the t-scale. So we know from point P that the car reached the 3-mi point 4 min after leaving the house. Starting from the beginning ($t = 0$), we see that the car traveled about $\frac{1}{4}$ mi in the first minute and a total of 1 mi in the first 2 min. Since 1 mi/min is 60 mph, the average speed in the first 2 min is half this, or 30 mph. We conclude the driver was not trying for a particularly fast start (a fast car can accelerate to 60 mph in about 12 sec). After the second minute the driver kept up a pace of 1 mi/min which is a steady speed of 60 mph. But at $t = 5$ min, he made a quick stop in about one-fifth of a minute, or 12 sec. This would definitely require application of the brakes (the fastest possible stop from 60 mph takes about 4 sec). Then the driver stood still for about 3 min at a distance of about 4 mi from his house. At $t = 8$ min he starts back toward the house in less of a hurry (slower acceleration) than at the start. However, 2 min later he is slowly losing speed and continues to do so until he comes to a stop at $t = 16$ min.

Fig. 1-5.

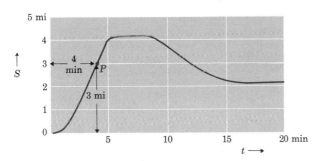

Q.4: In the above graph, where is the car at $t = 10$ min and what is its approximate speed?

He coasted about 1 min in this loss of speed. We conclude he lost engine power (probably ran out of gas).

You see, the graph tells us everything except why he applied his brakes when 4 mi from home, and why he decided to turn back. Even on this we can make a good guess. We suspect it was because he suddenly realized he was almost out of gas. Note that we have deduced the speed of the car from the steepness of the curve. The speed is proportional to the steepness or slope of the curve. When the curve is flat, the speed is zero.

1-5 Science and Society

The main goal

Almost every advance in our understanding of the physical world has ultimately some kind of practical application. However, physicists generally are not too interested in practical applications of their work. It is a surprise to many that the inevitable benefit of the physicist's work to mankind is not usually his main motivation. Physicists are more interested in discovering the secrets of nature, in getting closer to physical reality. Whether or not this leads to practical applications is of secondary importance.

For example, some scientists have criticized Project Apollo (the moon project) because it contributes little to basic science. To this author, sending a man to the moon is more pure exploration than pure science. However it is perhaps the greatest and most imaginative feat of exploration in the history of mankind. For centuries men have dreamed of it. Now that we have the technological capability and resources, we must achieve this pinnacle in the history of mankind. The United States is rich enough to pursue space exploration and at the same time to solve its social problems if it should so desire. If the expense looks too great, perhaps money could be saved by relaxing the pace of the program and by a less conservative approach to the degree of safeguards. Would we mind it if space exploration ever became as risky as climbing Mt. Everest?

To physicists our understanding of the world around us

Ans. 4: At $t = 10$ min, $s = 3.7$ mi from the driver's house, and the driver is returning at a rate of 1 mile in 3 min or 20 mph.

is a valuable goal in itself. Man is the only animal capable of such understanding. Those who graduate from college still illiterate in science are in this sense part man and part animal. Our scientific understanding is a central part of our modern culture and civilization. Those who are intellectually alive cannot help but strive to obtain this scientific understanding.

A secondary and more commonly expressed reason for the study of physics is the usefulness an understanding of science has to a person living in this modern, technological age—the age of automation, radiation, nuclear power, space travel, missiles, and nuclear bombs. Almost every edition of the daily newspaper contains articles that cannot be fully understood without a knowledge of physics. See the chapter opening for a random sample of such articles. Some of them are on subjects of vital importance. One wonders whether men who have little understanding of science are capable of making competent policy decisions on such vital topics. Yet, upon these decisions rests the survival of human civilization as we know it.

Answers to the quiz on page 9.

1. $R = \dfrac{R_1 R_2}{R_1 + R_2}$

2. $(a + b)^2 = a^2 + 2ab + b^2$

3 and **4.** The radius is 5×10^{-9} cm.

5. $\dfrac{A}{B} + \dfrac{X}{Y} = \dfrac{AY + BX}{BY}$

6. Hypotenuse is 2 cm.

7. $\dfrac{4}{\frac{1}{2}} = 8$

8. $\sqrt{16\,ab} = 4\sqrt{ab}$

9. $\sqrt{10^3} = 10\sqrt{10}$

10. $\dfrac{1}{a + b} = (a + b)^{-1}$

11. $(a - b)$ would be positive.

12. $\dfrac{10^{-10}}{10^{-5}} = 10^{-5}$

Problems

1. Find the cube root of 6^6.

2. Express the reciprocal of $\left(\dfrac{1}{A} + \dfrac{1}{B}\right)$ as a fraction which contains no fractions in the numerator or denominator.

3. What is 60 mi/hr in ft/sec?

4. The wavelength of light from a certain spectral line is 5.981×10^{-5} cm. How many wavelengths of this light are in 1 m?

5. If the diameter of a proton is 10^{-13} cm and its mass is 1.6×10^{-24} gm, what is its density in gm/cm^3?

6. Solve for x: $a = \dfrac{x-1}{x+1}$

7. Solve for β: $\dfrac{1}{1-\beta^2} = 1.25$

8. What is one-fifth of 10^{-10}? What is 10^{-10} divided by one-fifth?

9. What is $16^{1/4}$?

10. If there are $N_0 = 6.02 \times 10^{23}$ atoms in 4 gm of helium, what is the mass of the helium atom?

11. A certain brand of soap comes in two sizes, both of the same shape (they are similar figures). The large bathsize is 50% longer than the small size. How much more soap is in the large bar?

12. Solve for a in $v = \sqrt{2as}$.

13. The period of oscillation or time for one vibration of a crystal in a crystal oscillator is 2.5×10^{-6} sec. What is the oscillator frequency in vibrations per second?

14. Two grams of H_2 gas contains $N_0 = 6.02 \times 10^{23}$ molecules. What is the mass of the hydrogen atom (there are two atoms in each H_2 molecule)?

15. Using the following information, find W in terms of e and R only.

$$W = \tfrac{1}{2}mv^2 + U$$

$$U = -\frac{e^2}{R}$$

$$\frac{mv^2}{R} = \frac{e^2}{R^2}$$

16. Simplify the following expressions:

$$\frac{x^3 \times (4x)^2}{(2x)^4}, \quad \frac{(N^{2a})^3}{(N^6)^a}$$

17. Reduce the following fractions:

$$\frac{a^2 - b^2}{b - a}, \quad \frac{x^{2a} - x^a}{x^a + x^{2a}}$$

18. Solve for v:

$$M = \frac{M_0}{\sqrt{1 - \dfrac{v^2}{c^2}}}$$

19. Find x and y: $\begin{cases} x + y = -2 \\ x - y = 8 \end{cases}$

20. Subtract $(-U_0)$ from $(-W_0)$. Give algebraic form of the answer and evaluate for the values $W_0 = +2$ and $U_0 = +7$.

21. The altitude of an equilateral triangle is 5 cm. What is the side?

22. In a certain type of units the mass of the neutron is 939.506 and that of the proton is 938.213. In these same units, what is the neutron proton mass difference and how many significant figures does it contain?

Prob. 23

23. $D = 1$ m, $d = 0.1$ mm, $\lambda = 5 \times 10^{-5}$ cm. Find x in centimeters.

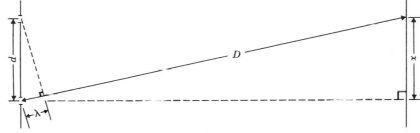

24. An object 100 m away subtends an angle of $1°$. How high is it? (*Hint:* $1°$ is 1/360 of a circle. Thus the object is 1/360 of the circumference of a 100 m circle.)

25. A certain highway rises 3 ft for every 100 ft. What is the angle of the highway from the horizontal?

26. Find an equation containing x and y but not t where

$$\begin{cases} x = v_x t \\ y = v_y t - \tfrac{1}{2} g t^2 \end{cases}$$

27. The speed of sound is 1100 ft/sec. What is the wavelength corresponding to a frequency of 60 cycles per sec (how far does the sound wave travel in 1/60 of a sec)?

28. In the following two simultaneous equations x' and t' are expressed in terms of x and t. We wish instead to express x and t in terms of x' and t'. So we must obtain an expression for x not containing t and an expression for t not containing x.

$$\begin{cases} x' = \dfrac{x + \beta c t}{\sqrt{1 - \beta^2}} \\[2ex] c t' = \dfrac{c t + \beta x}{\sqrt{1 - \beta^2}} \end{cases}$$

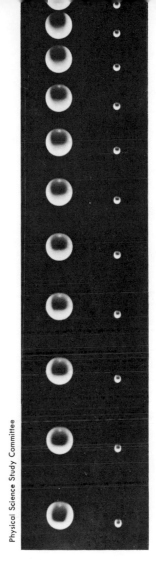

Kinematics

Chapter 2

Kinematics

2-1 Velocity

The speedometer reading

Kinematics is the study of motion. In kinematics we study position, velocity, and acceleration. We do not specify the nature of the particle or object whose motion is under study, nor do we study the forces which cause the acceleration. In spite of these temporary restrictions we will find ourselves almost immediately solving many kinds of practical problems.

In this age of automobiles, velocity is a concept obtained in childhood. The speedometer reads the instantaneous velocity in mph (miles per hour).

Constant velocity

If a car is moving with constant velocity, then the distance traveled is directly proportional to the time. For constant velocity v,

$$s = vt$$

Fig. 2-1. Plot of distance s versus time t of an object moving with constant velocity.

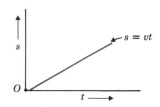

where s is the distance covered in time t. In Fig. 2-1 s is plotted against t for an object moving with uniform velocity. If we divide both sides of the above equation by t we obtain

$$v = \frac{s}{t} \qquad \text{if } v \text{ is constant} \qquad (2\text{-}1)$$

Instantaneous velocity

We shall now proceed to the more general case where the velocity is permitted to change with time. Until we come to Section 2-3 the discussion will be restricted to the situation where the magnitude, but not the direction, of the velocity may change. This type of motion in a fixed direction or along a straight line is called one-dimensional motion. We will, however, permit a reverse in direction. Then v will be negative which means s will be decreasing in value. Strictly speaking, we should use the word "speed" when we are only speaking of the magnitude of the velocity and not the direction. In our presentation of one-dimensional motion, we are keeping track of the direction by permitting the use of negative v.

2

Suppose a car is accelerating (speeding up or slowing down). Then Eq. 2-1 will give the wrong answer for the "speedometer reading" unless a very small value of s is used. We will use the symbol Δs to stand for a very small distance, and Δt for the time taken to travel the distance Δs.* The instantaneous velocity is defined to be

instantaneous velocity

$$v = \frac{\Delta s}{\Delta t} \qquad (2\text{-}2)$$

More correctly

$$v = \lim_{\Delta t \to 0} \left[\frac{\Delta s}{\Delta t} \right]$$

The above equation signifies that v is the limit of the ratio $\Delta s / \Delta t$ as Δt approaches zero. This is the mathematically rigorous definition of instantaneous velocity.†

Average velocity

Suppose a car in making a 60-mi trip travels at 20 mph for the first 30 mi and at 60 mph for the last 30 mi. We are tempted to say that the average velocity would then be $(20 + 60)/2$, or 40 mph. However, this would be unorthodox because of the convention that average velocity is defined with respect to time and not with respect to distance. If t_1 is the amount of time needed to make the first part of the trip (at velocity v_1) and t_2 is the amount of time for the second part, then, according to the definition of average, the average velocity with respect to time is

$$\bar{v} = \frac{v_1 t_1 + v_2 t_2}{t_1 + t_2} \qquad (2\text{-}3)$$

The preceding is an example of what is called a weighted average. In Eq. 2-3, t_1 and t_2 are the weighting factors. We can calculate the average velocity of the car by finding t_1

Fig. 2-2. Car which travels with velocity v_1 for first "half" of trip and velocity v_2 for second "half."

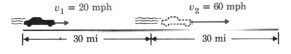

$v_1 = 20$ mph $v_2 = 60$ mph

|← 30 mi →|← 30 mi →|

---

*Here the symbol Δ is not an algebraic quantity which multiplies s. It is merely shorthand for a subtraction; namely, $\Delta s = (s_2 - s_1)$ where s_1 is the distance to the car at time t_1 and s_2 is the distance at time t_2. Similarly, $\Delta t = (t_2 - t_1)$.

† Those who know a little calculus will recognize our expression for instantaneous velocity as the definition of the derivative of s with respect to t

$$\left(v = \frac{ds}{dt} \right)$$

Q.1: Suppose on a hot summer day the temperature is 95°F for 16 hrs and 75°F for 8 hrs. What is the mean temperature for that day?

and t_2 from the relation $t = s/v$ that is obtained from Eq. 2-1. The result is that $t_1 = 30$ mi/20 mph $= 1.5$ hr, and $t_2 = 30$ mi/60 mph $= 0.5$ hr. If we substitute these values into Eq. 2-3, we obtain

$$\bar{v} = \frac{20 \times 1.5 + 60 \times 0.5}{1.5 + 0.5} \text{ mph} = 30 \text{ mph}$$

We then see that the average velocity for this trip is 30 mph rather than 40 mph.

We can use Eq. 2-3 to derive a simple formula for average velocity. We note that the quantity $v_1 t_1 = s_1$, the distance traveled when at velocity v_1. Similarly, $v_2 t_2 = s_2$. Now let us substitute s_1 for $(v_1 t_1)$ and s_2 for $(v_2 t_2)$ into Eq. 2-3:

$$\bar{v} = \frac{(v_1 t_1) + (v_2 t_2)}{t_1 + t_2} = \frac{(s_1) + (s_2)}{t_1 + t_2}$$

average velocity
$$\bar{v} = \frac{s}{t} \tag{2-4}$$

where s is the total distance, and t is the time required to cover the total distance s. This formula automatically includes all the weighting factors and hence gives the correct time average in all instances no matter how the velocity changes with time. We would, of course, get the same result for three or more intervals of time. Any changing velocity may be broken up into many small intervals where the velocity is essentially constant over each interval. We see then that Eq. 2-4 is completely general and covers all cases of one-dimensional motion.

Example 1

Suppose that a car traveling at 60 mph can stop in 4 sec after the brakes have been jammed on. We shall assume that in these 4 sec the velocity will decrease uniformly in time from 60 mph to 0 mph. Thus the average velocity will be 30 mph during the time the brakes are on. How much farther will the car travel from the time the brakes are jammed on until it stops?

If Eq. 2-4 is solved for s, we have

$$s = \bar{v}t$$

The average velocity $\bar{v}$ is 30 mph or 44 ft/sec. Thus

$$s = 44 \text{ ft/sec} \times 4 \text{ sec}$$

Ans. 1: The weighting factor for 95° is twice that for 75°; hence

$$\overline{T} = \frac{95 \times 2 + 75 \times 1}{2 + 1} = 88.3°\text{F}$$

or
$$s = 176 \text{ ft}$$

Should one keep about 170 ft spacing between cars when traveling at 60 mph? People would be better drivers if they would apply a knowledge of physics to their driving.

Example 2

A bicyclist is traveling in a region that is all hills. His uphill speed (magnitude of the velocity) is always 5 mph and his downhill speed is always 20 mph. What is his average speed if the uphill stretches are the same length as the downhill?

Let D be the length of the uphill stretches. Then $s = 2D$ is the total distance covered, and by Eq. 2-4 the average speed is

$$v = \frac{2D}{t} \qquad (2\text{-}5)$$

In order to solve this problem we must determine t, the time for the trip.

$$t = t_1 + t_2$$

where t_1 is the time spent traveling uphill and t_2 is the time spent traveling downhill. According to Eq. 2-1

$$t_1 = \frac{D}{v_1} \quad \text{and} \quad t_2 = \frac{D}{v_2}$$

Thus

$$t = \frac{D}{v_1} + \frac{D}{v_2}$$

Now substitute this expression for t into Eq. 2-5; then

$$\bar{v} = \frac{2D}{D\left(\dfrac{1}{v_1} + \dfrac{1}{v_2}\right)}$$

or

$$\bar{v} = \frac{2v_1 v_2}{v_1 + v_2}$$

Finally, we insert the numerical values $v_1 = 5$ mph and

$v_2 = 20$ mph:

$$\bar{v} = \frac{2 \times 5 \times 20}{5 + 20} \text{ mph} = 8 \text{ mph}$$

Note that in solving the above problem, algebraic symbols were used until the end. Generally it is a good policy not to substitute the numerical values until the very end. This general procedure of first solving a problem algebraically and then substituting numerical values will usually save a large amount of arithmetic and reduce the chance of making a mistake.

Q.2: A car is at position s_1 at a time t_1 and at s_2 at a time t_2. What is its average velocity when going from s_1 to s_2?

2-2 Acceleration

Speeding up or slowing down

By definition an object is moving with uniform acceleration if its velocity is uniformly increasing with time:

$$v - v_0 = at$$

uniform acceleration or

$$a = \frac{v - v_0}{t} \tag{2-6}$$

where $(v - v_0)$ is the increase in velocity during a time t. The initial velocity at the start of the time interval t is v_0. The constant a is defined as the acceleration and has the units of length divided by the square of time. In the MKS system, acceleration has units of m/sec/sec = m/sec^2. It can also be expressed in cm/sec^2 or ft/sec^2. If the acceleration is not uniform, then

instantaneous acceleration $$a = \frac{\Delta v}{\Delta t}$$

is the instantaneous acceleration in the limit of small Δt. However, in this chapter we shall limit the discussion to examples of uniform acceleration.

Example 3

An object starts from rest with a uniform acceleration of 9.80 m/sec^2. How long does it take to reach the speed of light which is 3.00×10^8 m/sec?

Solving Eq. 2-6 for t, we have

$$t = \frac{v}{a}$$

$$t = \frac{3.00 \times 10^8 \text{ m/sec}}{9.80 \text{ m/sec}^2} = 3.06 \times 10^7 \text{ sec}$$

which is almost one year.

Example 3 raises the question of a possible defect in Eq. 2-6. Most of us have already heard that no object can ever travel faster than the speed of light. Yet Eq. 2-6 puts no limit on the magnitude of v. The equations presented in the first five sections of this chapter were first discovered by Galileo in the early 1600's and are part of what is called classical mechanics. However, in 1905 Einstein proposed

Ans. 2: The car travels $s = (s_2 - s_1)$ in a time $(t_2 - t_1)$; hence,

$$\bar{v} = \frac{s_2 - s_1}{t_2 - t_1}.$$

modifications that become significant only at high velocities near the speed of light. His new theory is called the special theory of relativity and has been very thoroughly checked by many experiments. In the modern relativity theory the correct form for Eq. 2-6 for an object starting from rest is

$$v = \frac{at}{\sqrt{1 + \left(\dfrac{at}{c}\right)^2}}$$

where c is the speed of light and a is a uniform acceleration as measured by an observer on the moving object. We see that when the quantity (at) becomes much larger than c, the denominator approaches $(at/c)^2$ and then $v = at/(at/c) = c$. On the other hand, when the quantity (at) is much less than c, the square root in the denominator is very close to one, and then $v = at$ to great accuracy.

Since the modifications due to relativity are completely insignificant when dealing with "ordinary" velocities, we conclude it is reasonable to continue on with our study of classical mechanics using Eq. 2-6. The modifications due to relativity are discussed in Chapter 11.

Example 4

A driver traveling at 60 mph suddenly notices a stalled truck 100 ft ahead. He applies his brakes to give the maximum possible negative acceleration which happens to be about 16 ft/sec² (too much braking would cause the car to skid and then it would take even longer to stop). How long does it take the driver to stop, and does he hit the truck?

To find the time, we must solve Eq. 2-6 for t:

$$t = \frac{v - v_0}{a}$$

where the initial velocity is $v_0 = 88$ ft/sec, and the final velocity is $v = 0$. Note that the acceleration $a = -16$ ft/sec² is negative in sign because it opposes the motion. Inserting these values gives

$$t = \frac{0 - 88 \text{ ft/sec}}{-16 \text{ ft/sec}^2} = 5.5 \text{ sec}$$

During this time the average velocity is $v = \frac{1}{2}$ of 88 ft/sec and

$$s = vt$$

$$= (44 \text{ ft/sec}) \times 5.5 \text{ sec} = 242 \text{ ft}$$

Yes, the driver hits the truck.

Q.3: A car starts from rest and reaches a velocity v_1 in a time t_1. What is its average acceleration in terms of v_1 and t_1?

Fig. 2-3. Velocity v plotted against time t for an object moving with constant acceleration and starting with velocity v_0.

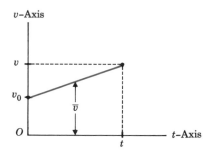

If we solve Eq. 2-6 for v, we obtain

$$v = v_0 + at \tag{2-7}$$

which is plotted in Fig. 2-3. In the time t, the average velocity will be the average height of the line between v_0 and v, which is $\frac{1}{2}(v_0 + v)$. Thus

$$\bar{v} = \frac{v_0 + v}{2} \tag{2-8}$$

But since the average velocity must also be s/t (Eq. 2-4), we have

$$\frac{s}{t} = \frac{v_0 + v}{2}$$

or

$$s = \frac{v_0 t + vt}{2}$$

Now we substitute the right-hand side of Eq. 2-7 for v and obtain

$$s = \frac{v_0 t + (v_0 + at)t}{2}$$

$$s = v_0 t + \tfrac{1}{2}at^2 \tag{2-9}$$

This is the distance traveled in time t for an object with uniform acceleration a. If the object starts from rest, then $s = \tfrac{1}{2}at^2$.

The derivation of Eq. 2-9 was first made by Galileo. Contrary to the accepted doctrine of that time, he observed that, neglecting air resistance, all bodies fall with the same acceleration toward the center of the earth (see Fig. 2-4). This particular value of acceleration is 9.8 m/sec² or 32 ft/sec², and is denoted by the special symbol g. Galileo conducted careful and fairly accurate experiments to check that the distance should be proportional to the square of the time. He measured elapsed time by collecting water flowing from a small pipe in the bottom of a large vessel. This device was similar to an hour glass, except that water was used in place of sand. He found that a ball rolling down an inclined plane resulted in a collection of twice as much water when it rolled four times as far; that is, in twice the time, it travels four times as far which agrees with Eq. 2-9.

Ans. 3: Just as average velocity is $\bar{v} = \dfrac{s_1}{t_1}$ where s_1 is the increase in distance during time t_1, so is average acceleration $\bar{a} = \dfrac{v_1}{t_1}$ where v_1 is the increase in velocity during t_1.

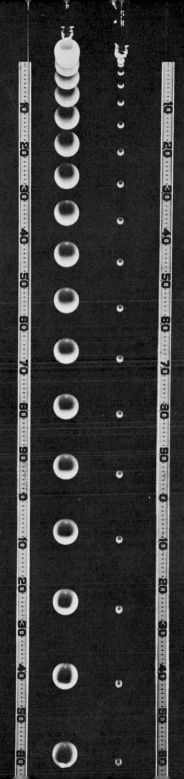

◀ **Fig. 2-4.** Stroboscopic photograph of two freely falling balls of unequal mass. Such a strobe photograph is taken by opening up the camera lens and flashing the light every thirtieth of a second. Note that the small mass hits the floor at the same time as the heavy mass. Both balls were released simultaneously. (Courtesy Physical Science Study Committee of Educational Services, Inc.)

Example 5

A bad tourist drops a stone from the rim of Grand Canyon.* Ignoring air resistance, how long does it take before the stone hits bottom? The canyon is 5000 ft deep.

In this problem we know s and a and we wish to find t. Equation 2-9 expressed a relationship between s, a, and t that corresponds to this situation of uniform acceleration in a straight line. Putting $v_0 = 0$ and $a = g$, we obtain

$$s = \tfrac{1}{2}gt^2$$

Solving for t:

$$t^2 = \frac{2s}{g}$$

$$t = \sqrt{\frac{2s}{g}} \tag{2-10}$$

Now that we have the algebraic solution for t in terms of known quantities, we insert the numerical values $s = 5000$ ft and $g = 32$ ft/sec² and obtain

$$t = \sqrt{\frac{10,000}{32}} \text{ sec} = 17.6 \text{ sec}$$

Example 6

A boy throws a ball straight up 64 ft and catches it when it comes down. For how long is the ball in the air?

This problem can be simplified by observing that the time the ball spends going up is the same as the time it spends falling. This is because the change in velocity or acceleration is the same in both cases, and in both cases the ball has zero velocity when at maximum height. Thus a motion picture of a ball falling from a height of 64 ft when run backwards will appear to have the same motion as a ball thrown upward to a height of 64 ft. The time it takes to fall 64 ft is immediately obtained from Eq. 2-10:

$$t = \sqrt{\frac{2s}{g}}$$

$$t = \sqrt{\frac{128}{32}} \text{ sec} = 2.0 \text{ sec}$$

The total time spent in the air is twice this amount, or 4 sec.

*He is a bad tourist because he threw a stone off a cliff. It might hit someone below.

In the above problem what is the acceleration of the ball when at its highest point? The immediate reaction of many is to say zero. It is true that at that instant of time the velocity is zero, but that does not mean the acceleration also has to be zero. In fact, the velocity is continuously decreasing. At the instant of zero velocity, the velocity is in the process of changing from a positive to a negative value and its rate of change is still $a = -32$ ft/sec^2.

Example 7

A certain type of rocket is capable of reaching escape velocity ($v = 11$ km/sec) after traveling 200 km. As we shall show in Chapter 5, this is the velocity necessary to escape from the influence of the earth's gravity. Assuming its acceleration is uniform, what would be the numerical value of its acceleration?

In this problem we are given v and s and we wish to find a. We would like a formula relating s, v, and a. Since we have no such formula, we can at least start by writing down the definition of a (Eq. 2-6).

$$a = \frac{v}{t}$$

But now the right-hand side contains another unknown, t. However, perhaps we can determine t by finding an equation containing t and the known quantities s and v. Equation 2-4 suits this purpose if we note that the average velocity of the rocket in time t is $v/2$ (see Eq. 2-8). Thus $\bar{v} = v/2$, which must also be s/t.

$$\frac{1}{2}v = \frac{s}{t}$$

or

$$t = \frac{2s}{v} = \frac{400}{11}\,\text{sec} = 36.4\,\text{sec}$$

If we substitute this value in the formula for a, we obtain

$$a = \frac{v}{t} = \frac{11 \times 10^3}{36.4}\,\text{m/sec}^2 = 302\,\text{m/sec}^2$$

This is thirty-one times the acceleration of a falling body. It is conventional to refer to this as an acceleration of 31g. This is too high an acceleration for astronauts to endure during blastoff.

Example 7 was more difficult than the previous ones in that it required solving for an auxiliary quantity t before the required quantity a could be evaluated. Mathematically speaking it was necessary to use two simultaneous equations that contained two unknowns, a and t. Many problems in physics are of such a nature that auxiliary quantities must be found or eliminated. The procedure of selecting the best equations and quantities to work with often involves trial and error. In this respect solving physics problems is analogous to working crossword puzzles.

To save time in problem solving, we will now derive an equation relating the quantities s, a, and v. For an object starting with initial velocity v_0 and undergoing uniform acceleration, the average velocity is

$$\frac{v_0 + v}{2} = \frac{s}{t}$$

Also

$$v - v_0 = at$$

We can eliminate the unwanted quantity t by multiplying these two equations together. We obtain

$$\left(\frac{v + v_0}{2}\right) \times (v - v_0) = \frac{s}{t} \times at$$

$$v^2 - v_0^2 = 2as \tag{2-11}$$

Example 8

In Example 4, with what velocity does the car hit the truck?

As before, $v_0 = 88$ ft/sec, $a = -16$ ft/sec^2, and the distance to the truck is $s = 100$ ft. We solve Eq. 2-11 for v and insert these values

$$v = \sqrt{v_0^2 + 2as} = \sqrt{(88)^2 + 2(-16)(100)} = 67.4 \text{ ft/sec}$$

$$= 46 \text{ mph}$$

We see that the first 100 ft of braking does not accomplish as much as the next 100 ft.

Example 9

Consider a sky-diver jumping out of an airplane. He will accelerate in free fall until air resistance causes him to reach a terminal velocity of about 140 mph or 200 ft/sec (at this velocity the force of air resistance on a human body is just strong enough to cancel out the force due to gravity). Suppose the parachute did not open, but the sky-diver was lucky to be over a soft snow bank which slows him down with constant acceleration. Assume he will survive if the acceleration is less than $30g$. How deep must the snow be?

In this problem $v_0 = 200$ ft/sec, the final velocity is $v = 0$, and the acceleration is $a = -30g = -960$ ft/sec^2. We solve Eq. 2-11 for

Q.4: At time t_1 a car at position s_1 has velocity v_1 and is slowing down with constant acceleration of magnitude a. What is its position s_2 at a time t_2?

s and insert these values:

$$s = \frac{v^2 - v_0^2}{2a} = \frac{0^2 - 200^2}{2(-960)} \text{ ft}$$

$$s = 21 \text{ ft}$$

Actually there have been cases of mountain climbers falling thousands of feet in free fall and landing on soft snow who have lived through it. Perhaps with a properly designed suit, sky divers would not need parachutes, but could arrest themselves by plunging 20 ft or more into water.

Fig. 2-5. Position of ball dropped from plane at successive time intervals. It falls the distance s_1 in the first time interval.

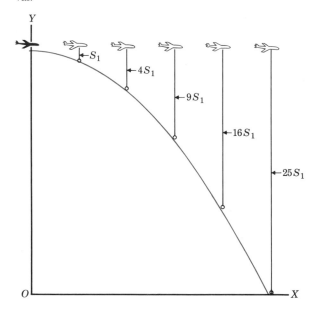

Ans. 4: Using Eq. 2-9, the distance covered $(s_2 - s_1)$ in a time $(t_2 - t_1)$ is $(s_2 - s_1) = v_1(t_2 - t_1) + \frac{1}{2}(-a)(t_2 - t_1)^2$. The acceleration is $(-a)$ because the car is slowing down.

2-3 Compound Motion

Vertical and horizontal, separate and together

If an airplane passenger drops an object, he sees it fall vertically with acceleration g until it hits the floor of the plane. Now suppose there was a hole in the floor. Then the airplane passenger would see the object continue falling in a vertical path directly beneath the plane (see Fig. 2-5). (Actually the object would eventually lag behind the plane due to air resistance.) What would an observer on the ground see? If y_0 is the height of the airplane, and y the height of the falling object, the ground observer would see the distance $(y_0 - y)$ increasing according to the square of the time where s in Eq. 2-9 corresponds to $(y_0 - y)$. Hence

$$(y_0 - y) = \tfrac{1}{2}gt^2$$

or

$$y = y_0 - \tfrac{1}{2}gt^2 \qquad (2\text{-}12)$$

Now let x be the horizontal distance of the object from the initial line joining the airplane and ground (x is distance from line OY in Fig. 2-5). If we ignore air resistance, the distance of the airplane, and thus of the object, from this vertical line will be

$$x = v_0 t \qquad (2\text{-}13)$$

where v_0 is the velocity of the plane or the initial velocity of the object as seen from the ground. The equation for the path of the falling object as viewed from the ground can be obtained by solving this equation for t and substituting into Eq. 2-12:

$$t = \frac{x}{v_0}$$

$$y = y_0 - \tfrac{1}{2}g\left(\frac{x}{v_0}\right)^2$$

$$y = y_0 - \frac{g}{2v_0{}^2}x^2$$

Those familiar with analytical geometry will recognize this as the equation of a parabola. The path of a falling ball given an initial horizontal velocity is shown in Fig. 2-6.

Galileo was the first to show that the trajectories of projectiles are parabolas (ignoring air resistance). He noted that the horizontal and vertical motions could be considered as independent. Because there is no source of acceleration in the horizontal direction, the horizontal distance is given by

$$x = (v_0)_x t$$

where $(v_0)_x$ is the initial velocity in the x-direction. The vertical displacement is always given by Eq. 2-12; or, if there is an initial vertical velocity $(v_0)_y$, the expression would be

$$y = y_0 + (v_0)_y t - \tfrac{1}{2}gt^2$$

Example 10

A bomber is flying at a height of 30,000 ft and velocity 600 mph. Ignoring air resistance, how many feet in advance of the target must the bombardier release the bomb?

One way to solve this problem would be first to find how long it takes the bomb to reach the target. We can find t using Eq. 2-10 where $s = 3 \times 10^4$ ft.

$$t = \sqrt{\frac{2s}{g}} = \sqrt{\frac{6 \times 10^4}{32}} \text{ sec} = 43.3 \text{ sec}$$

The horizontal distance traveled is

$$x = v_0 t = 880 \text{ ft/sec} \times 43.3 \text{ sec}$$

$$x = 38{,}100 \text{ ft} = 7.2 \text{ mi}$$

2-4 Vectors

Mathematics of arrows

Now that we are discussing motion in two dimensions rather than one dimension, it is important to recognize that velocity is what we call a vector. A vector has both magni-

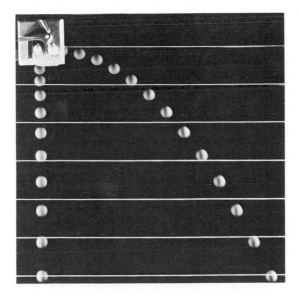

Fig. 2-6. Strobe photograph of two golf balls released simultaneously. At the time of release the right-hand ball was given a horizontal initial velocity by means of a spring. (Courtesy Physical Science Study Committee.)

Fig. 2-7. A displacement Δs and its x- and y-components Δx and Δy.

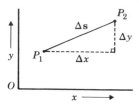

tude and direction. A velocity is not completely specified unless we know both its magnitude and direction. It is often useful to consider the x- and y-components of a velocity. The x- and y-components, v_x and v_y, are related to the total velocity v by the relation

$$v_x{}^2 + v_y{}^2 = v^2$$

This is seen in Fig. 2-7. Here an object moves from P_1 to P_2 in a time Δt. The distances Δx and Δy are the x- and y-components of the displacement Δs. By the Pythagorean theorem

$$(\Delta x)^2 + (\Delta y)^2 = (\Delta s)^2$$

By dividing both sides by $(\Delta t)^2$, we obtain:

$$\left(\frac{\Delta x}{\Delta t}\right)^2 + \left(\frac{\Delta y}{\Delta t}\right)^2 = \left(\frac{\Delta s}{\Delta t}\right)^2$$

or

$$v_x{}^2 + v_y{}^2 = v^2$$

Addition of vectors

Figure 2-7 also indicates a simple geometrical rule for finding the resultant velocity. We put the tail of v_y on the head of v_x. The resultant velocity is obtained by joining the tail of v_x to the head of v_y as shown in Fig. 2-8. Because any resultant displacement is the sum of the separate displacements, this geometrical rule can be generalized to add velocities pointing in arbitrary directions. This general rule for adding vectors is called the polygon rule and is illustrated in Fig. 2-9. In this case we wish to find the vector sum

$$\mathbf{v} = \mathbf{v}_1 + \mathbf{v}_2 + \mathbf{v}_3$$

The rule is to place the tail of each successive vector on the head of the preceding vector. The resultant is then given by joining the initial tail with the final head as shown in Fig. 2-9.

Subtraction of vectors

Suppose we wanted to find the vector $\mathbf{v}' = \mathbf{v}_2 - \mathbf{v}_1$ using the same $\mathbf{v}_2$ and $\mathbf{v}_1$ in Fig. 2-9. We note that subtraction is algebraically equivalent to the addition of a negative

Fig. 2-8. The x- and y-components of vector $\mathbf{v}$.

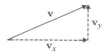

Fig. 2-9. How to obtain the sum of three vectors using the polygon rule of vector addition. $\mathbf{v} = \mathbf{v}_1 + \mathbf{v}_2 + \mathbf{v}_3$.

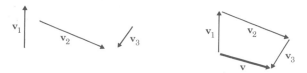

Fig. 2-10.

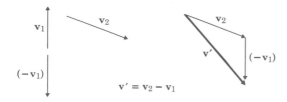

$$v' = v_2 - v_1$$

quantity; that is,

$$\mathbf{v}' = \mathbf{v}_2 + (-\mathbf{v}_1)$$

and the negative vector $(-\mathbf{v}_1)$ is just the same as $\mathbf{v}_1$, but pointing in the opposite direction (see Fig. 2-10).

Components

In addition to the polygon rule of addition of vectors, there is a second method which is called addition by components. To use this method we must find the x- and y-components of each vector. For example, the x- and y-components of the vector $\mathbf{v}_2$ in Fig. 2-9 are obtained by redrawing v_2 pointing out from the origin of our coordinate system. Now drop perpendiculars from the tip of the vector to the axes. Where these perpendiculars hit the axes measures off the x- and y-components of the vector. We see in Fig. 2-11 that $(v_2)_x = +5$ units and $(v_2)_y = -2$ units.

The sum $\mathbf{v} = \mathbf{v}_1 + \mathbf{v}_2 + \mathbf{v}_3$ is obtained by first summing the x-components together which then gives us the x-component of v:

$$v_x = (v_1)_x + (v_2)_x + (v_3)_x$$

Similarly,

$$v_y = (v_1)_y + (v_2)_y + (v_3)_y$$

The resultant vector $\mathbf{v}$ is formed by adding $\mathbf{v}_x$ and $\mathbf{v}_y$ as done in Fig. 2-8.

Many of the quantities used in physics are vectors and thus must be added in this way. Of the quantities studied so far, displacement, velocity, and acceleration are vectors.

Fig. 2-11. Finding the x and y components of vector $\mathbf{v}_2$.

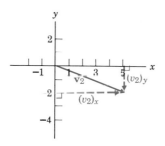

Example 11

As shown in Fig. 2-12, a ferryboat is sailing 30° west of north at 12 mph with respect to the water. The stream is flowing 6 mph due east. As viewed from the shore, what is the direction of travel and speed of the ferryboat?

We must form the vector addition

$$\mathbf{v}'_{\text{boat}} = \mathbf{v}_{\text{boat}} + \mathbf{v}_{\text{water}}$$

This addition is shown in Fig. 2-12b and results in a 30°-60° right triangle. The resultant velocity (heavy red vector) is pointing due north and its magnitude is given by the Pythagorean theorem:

Fig. 2-12. Ferryboat crossing stream.

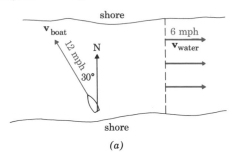

shore

$\mathbf{v}_{boat}$

12 mph

N

30°

6 mph

$\mathbf{v}_{water}$

shore

(a)

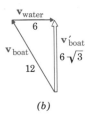

$\mathbf{v}_{water}$

6

$\mathbf{v}'_{boat}$

$\mathbf{v}_{boat}$

12

$6\sqrt{3}$

(b)

Fig. 2-13. Path of a projectile fired at an angle of 45°. Initial velocity is $\mathbf{v}_0$.

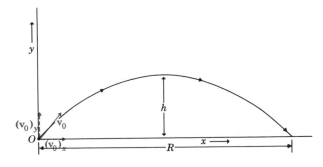

$$v'^2{}_{boat} = v^2{}_{boat} - v^2{}_{water}$$
$$v'{}_{boat} = \sqrt{12^2 - 6^2} \text{ mph}$$
$$= 6\sqrt{3} \text{ mph}$$
$$= 10.4 \text{ mph}$$

Hence we have the result that according to vector addition $(12 + 6)$ can equal 10.4.

2-5 Projectile Motion*

Military applications

We shall first consider the case of a cannon fired at an angle of 45° from the horizontal as shown in Fig. 2-13. We will assume the projectile velocity is $v_0 = 1$ km/sec as it leaves the cannon and we will ignore air resistance. We wish to determine how long the projectile is in flight, how high it goes, and how far it goes (what the range is of this cannon). We shall solve this problem by considering the horizontal and vertical motions separately. For a 45° angle, $(v_0)_x = (v_0)_y = v_0/\sqrt{2} = 707$ m/sec. First let us consider the vertical motion. According to Eq. 2-7

$$v_y = (v_0)_y - gt$$

where the acceleration is $a = -g$. The minus sign means that the acceleration slows down a positive velocity.

Now solve this equation for t, noting that when the projectile reaches its maximum height, $v_y = 0$:

$$t_1 = \frac{(v_0)_y - v_y}{g} = \frac{707 - 0}{9.8} \text{ sec}$$
$$t_1 = 72 \text{ sec}$$

where t_1 is the time required to reach maximum height. The total time of travel will be twice this or 144 sec. Let us call the maximum height h and the maximum horizontal distance R. Since the average vertical velocity is the distance h divided by 72 sec; that is, $\bar{v}_y = h/t_1$; and also $\bar{v}_y = \frac{1}{2}(v_0)_y$ (see Eq. 2-8), we have

$$\frac{h}{t_1} = \frac{1}{2}(v_0)_y$$

*This section may be omitted if desired.

$$h = \frac{(v_0)_y t_1}{2} = \frac{707 \times 72}{2} \text{ m}$$

$$h = 25.4 \text{ km}$$

The range R is immediately obtained from the equation

$$x = (v_0)_x t$$

$$R = (v_0)_x \times 144 \text{ sec}$$

$$R = 707 \times 144 \text{ m}$$

$$R = 102 \text{ km}$$

With a 45° firing it is interesting to note that the range R is exactly four times the maximum height h. If the cannon had been fired at an arbitrary angle θ, then $(v_0)_x$ would be $v_0 \cos \theta$ and $(v_0)_y$ would be $v_0 \sin \theta$. It can be shown by using simple trigonometry that the range is maximum when the firing angle is 45°. These early military applications of physics were first worked out by Galileo.

We shall now apply our knowledge of projectile motion to a more up-to-date problem—that of the intercontinental ballistic missile (ICBM). Let us consider a typical ICBM with maximum range of 5000 mi or 8000 km. Assume that the ICBM is fired at your city from a distance of 5000 mi and that it is first detected at its halfway point. How much warning time will you have? How fast will the missile be traveling when detected? With what velocity will it strike its target? What will be its maximum height? To solve this problem we must make some approximations. We shall assume that the earth is flat over this 5000 mi distance. We shall assume that g is the same at all heights and we shall ignore air resistance.

We know the ICBM entered the stratosphere at an angle of 45° because it was fired at its maximum range. Thus our initial information is that $(v_0)_x = (v_0)_y$, $R = 8000 \text{ km}$, and its vertical acceleration is $g = 9.8 \text{ m/sec}^2$. From this meager information we shall supply answers to all the preceding questions. Let us start with the equation for vertical motion:

$$v_y = (v_0)_y - gt$$

Let t_1 stand for the time to reach maximum height. At $t = t_1$, $v_y = 0$ and we have

$$0 = (v_0)_y - gt_1$$

Q.5: The best sports cars can accelerate up to 60 mph from rest in about 10 sec. What is this acceleration compared to g?

or

$$t_1 = \frac{(v_0)_y}{g}$$

In the above equation replace $(v_0)_y$ by $(v_0)_x$, which has the same value. Then

$$t_1 = \frac{(v_0)_x}{g}$$

We can find $(v_0)_x$ by noting that the horizontal velocity is R divided by $2t_1$, the total time of the trip. We now substitute the quantity $R/2t_1$ for $(v_0)_x$ in the preceding equation:

$$t_1 = \frac{(R/2t_1)}{g}$$

$$t_1{}^2 = \frac{R}{2g}$$

$$t_1 = \sqrt{\frac{8 \times 10^6}{2 \times 9.8}} \text{ sec}$$

$$t_1 = 639 \text{ sec}$$

Our result for the warning time is about $10\frac{1}{2}$ min. The velocity at maximum height is the total horizontal distance divided by the total time for the trip:

$$(v_0)_x = \frac{R}{2t_1} = \frac{8000}{2 \times 639} \text{ km/sec} = 6.26 \text{ km/sec}$$

The final velocity will have the same value as v_0, which is $\sqrt{2}$ times as much as $(v_0)_x$. Thus

$$v_0 = 1.41 \times 6.26 \text{ km/sec} = 8.85 \text{ km/sec} = 5.5 \text{ mi/sec}.$$

The maximum height h divided by t_1 is the average vertical velocity which is $\frac{1}{2}(v_0)_y$:

$$\frac{h}{t_1} = \tfrac{1}{2}(v_0)_y$$

$$h = \frac{(v_0)_y t_1}{2}$$

But $(v_0)_y$ has the same value as $(v_0)_x$, which is 6.26 km/sec. Thus

$$h = \frac{6.26 \times 639}{2} \text{ km} = 2000 \text{ km}$$

Ans. 5: $\bar{a} = \dfrac{v}{t} = \dfrac{88 \text{ ft/sec}}{10 \text{ sec}} = 8.8$ ft/sec^2 which is about $\frac{1}{4}$ of g. The ultimate limit no matter how powerful the engine is the point where the tires lose traction. This occurs when $a \approx \frac{1}{2}g$.

which is exactly one-fourth the range, as was also the case with the cannon fired at 45°. Despite the approximations made, the above calculation gives good agreement with the data we read about modern ICBM's. For a 5000-mi ICBM the total travel time is about 20 min and the required velocity $v_0 = 5.5$ mi/sec is comparable to the velocity of an earth satellite.

2-6 Centripetal Acceleration

Perpendicular acceleration

In projectile motion the acceleration **g** is perpendicular to the velocity at the time the projectile reaches its maximum height. As seen in the previous section, the projectile is then moving in a curved path. In this section we shall study the special case where the acceleration is *always* perpendicular to the velocity. First we shall show that if an object is moving uniformly in a circle, then its acceleration is always perpendicular to its velocity and thus points toward the center of the circle (the radius is perpendicular to the velocity because the velocity is tangent to the circle).

Figure 2-14a shows two successive positions of an object moving uniformly in a circle. Let Δt be the time required to move the distance Δs between these two positions. In this time the velocity direction has changed from $\mathbf{v}_1$ to $\mathbf{v}_2$.

To calculate the acceleration $\Delta \mathbf{v}/\Delta t$, we must first determine $\Delta \mathbf{v}$, the change in velocity during the time Δt. The vector $\Delta \mathbf{v}$ is the vector difference $\mathbf{v}_2 - \mathbf{v}_1$. This is obtained in Fig. 2-14b by adding the vector $(-\mathbf{v}_1)$ to $\mathbf{v}_2$. The vector $(-\mathbf{v}_1)$ is the same as $\mathbf{v}_1$, but pointing in the opposite direction.

$$\Delta \mathbf{v} = \mathbf{v}_2 + (-\mathbf{v}_1)$$
$$= \mathbf{v}_2 - \mathbf{v}_1$$

The calculation is completed by noting that the triangle in Fig. 2-14b with sides $\Delta \mathbf{v}$, $\mathbf{v}_1$, and $\mathbf{v}_2$ is similar to the triangle in Fig. 2-14a with corresponding sides Δs, R_1, and R_2. The triangles are similar because they are both isosceles and the sides $\mathbf{v}_1$ and $\mathbf{v}_2$ are mutually perpendicular to the sides R_1 and R_2; hence the angles labeled θ are equal. Since the cor-

Fig. 2-14. Two successive positions of an object moving uniformly in a circle are shown in (*a*). The vector $\Delta \mathbf{v} = \mathbf{v}_2 - \mathbf{v}_1$ is obtained in (*b*).

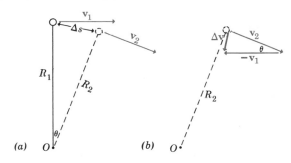

Q.6: Is it ever possible for an object to move in a curved path without accelerating?

responding sides of similar triangles are proportional, we have

$$\frac{\Delta v}{\Delta s} = \frac{v}{R}$$

$$\Delta v = \frac{v}{R} \Delta s$$

If we divide both sides by Δt, we obtain

$$\frac{\Delta v}{\Delta t} = \frac{v}{R} \times \left(\frac{\Delta s}{\Delta t}\right)$$

In the limit of small Δt, the two successive positions in Fig. 2-14 will be much closer and $\left(\frac{\Delta s}{\Delta t}\right)$ will equal v. Hence

$$\frac{\Delta v}{\Delta t} = \frac{v}{R} \times (v) = \frac{v^2}{R}$$

The quantity $\left(\frac{\Delta v}{\Delta t}\right)$ is the magnitude of the acceleration and is called the centripetal acceleration a_c. Centripetal acceleration occurs whenever an object is moving in a curved path. We see from the above equation that the formula for centripetal acceleration is

centripetal acceleration $$a_c = \frac{v^2}{R} \tag{2-14}$$

We note that centripetal acceleration gives rise only to a change in the direction of the velocity vector, whereas the magnitude of the velocity remains unchanged. Also note that the direction of a_c is pointing toward the center of the circle. (In Fig. 2-14b, Δv will point closer to the center of the circle as the two successive positions are made closer.)

For those familiar with the term "centrifugal force" we wish to point out that this term will never enter into our discussions. Centrifugal force is what is called a fictitious force which only occurs when the observer is in a rotating frame of reference. We will always observe rotating objects from the outside; we will never sit inside them.

Ans. 6: No, because the direction of velocity is changing and the vector $\mathbf{v}_1$ cannot possibly equal $\mathbf{v}_2$.

Example 12

What is the centripetal acceleration of an object on the earth's equator due to the earth's rotation? An object on the earth's equator travels the circumference of the earth (40,000 km) in one day (8.64×10^4 sec). Its velocity is then

$$v = \frac{4 \times 10^7 \text{ m}}{8.6 \times 10^4 \text{ sec}} = 463 \text{ m/sec}$$

According to Eq. 2-14 it must have an acceleration v^2/R pointing toward the center of the earth ($R = 6360$ km).

$$a_c = \frac{v^2}{R} = \frac{(463)^2}{6.36 \times 10^6} \text{ m/sec}^2$$

$$= 0.034 \text{ m/sec}^2$$

This is 0.34% of $g = 9.8$ m/sec². Because of this fact, as we shall see in the next chapter, people near the equator should weigh about 0.34% less than when near either pole. It is easier to do a 4 min mile on the equator than on the North Pole.

2-7 Earth Satellites

Fast "bullets"

Why doesn't an earth satellite "fall" toward the earth with an acceleration $g = 9.8$ m/sec² as do all other objects near the surface of the earth? Actually this is exactly what happens. A low-flying earth satellite is continuously falling toward the earth with the gravitational acceleration g. If it did not fall toward the earth; that is, if it were not under the influence of gravity, it would continue on unaccelerated in a straight-line tangent to the earth. To an observer on earth underneath the satellite, it would then appear to be moving up. According to Eq. 2-14 any object moving in a circle around the earth must have an acceleration $a_c = v^2/R$ pointing toward the center of the earth. An earth satellite orbiting near the earth's surface can be correctly considered as a freely falling body with constant acceleration $a_c = 9.8$ m/sec². We can now calculate the velocity and period of Sputnik I (see Fig. 2-15). The special name circular velocity, v_c, is given to this velocity needed to put a satellite into orbit. Since the centripetal acceleration has the value g,

$$\frac{v_c^2}{R_e} = g$$

where $R_e = 6500$ km is the distance to the center of the

Fig. 2-15. Full-scale model of Sputnik I on display in Moscow. (Courtesy Sovfoto.)

Fig. 2-16. Launching of Explorer I, first United States earth satellite, January 31, 1958. (Courtesy U. S. Army.)

Fig. 2-17. Proposal by Isaac Newton for an earth satellite.

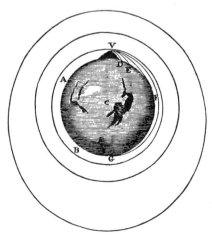

earth. Thus

$$v_c = \sqrt{gR_e}$$
$$= \sqrt{9.8 \times 6.5 \times 10^6}\,\text{m/sec}$$
$$= 8\,\text{km/sec} \quad \text{or} \quad 5\,\text{mi/sec}$$

This is the velocity achieved by Sputnik I and Explorer I (see Fig. 2-16). The time it takes a "low-flying" earth satellite to circle the earth is the circumference of the earth divided by v_c:

$$T = \frac{40{,}000\,\text{km}}{8\,\text{km/sec}} = 5000\,\text{sec} = 83\,\text{min}$$

Satellites in higher orbits will take longer to circle the earth.

Isaac Newton back around 1660 was the first to make the above calculations. Figure 2-17 shows a drawing of an earth satellite orbit made by Newton himself. He discussed the firing of a big cannon from a high mountain top. If a muzzle velocity of 5 mi/sec could ever be achieved, he predicted the cannon ball would circle the earth as shown. This drawing also shows the cannon ball path for smaller, more obtainable, muzzle velocities. We can see that the idea for an earth satellite is not a recent one. It has been a foremost dream in the minds of scientists since the time of Newton.

What would happen if we gave the projectile a bit too much velocity? This can be seen by solving Eq. 2-14 for R:

$$R = \frac{v^2}{a_c} \qquad (2\text{-}15)$$

Now a_c must still be g, since all objects, no matter how fast they are moving, experience the same gravitational acceleration as long as they are near the surface of the earth. So let us substitute $g = \dfrac{v_c{}^2}{R_e}$ for a_c in Eq. 2-15 with the result:

$$R = \frac{v^2}{g} = \frac{v^2}{\left(\dfrac{v_c{}^2}{R_e}\right)}$$

$$R = \left(\frac{v}{v_c}\right)^2 R_e$$

We see from this equation that if v is 10% too high, then $\left(\dfrac{v}{v_c}\right)^2 = 1.21$ and R will be 21% greater than the radius of the

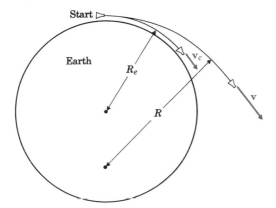

Fig. 2-18. The start of two possible satellite orbits. The satellite with velocity v_c circles about the center of the earth. The satellite with a greater velocity v starts out having an orbit of greater radius of curvature. R.

earth. Remember, R is the radius of curvature of the path of the projectile at the time when its velocity is v and acceleration is g. This situation is illustrated in Fig. 2-18. Such a projectile will start moving away from the earth. After traveling for a while, its velocity will have a component pointing away from the center of the earth. Hence there will be a component of the acceleration due to gravity pointing opposite to the direction of motion with the consequence that v will be reduced so that the projectile will eventually "fall back" toward the earth. As discussed in Chapter 4, the exact path will be an ellipse with one focus at the center of the earth.

2-8 Checking of Units

An ounce of prevention . . .

In working problems or in trying to check formulas, it is very useful to check units on both sides of the equation. If an error has been made, it will usually show up by giving a different set of units on the right-hand side from that on the left-hand side. For example, let us check the formula $v = \sqrt{2as}$ in the CGS system. The units on the left-hand side are cm/sec. If we correctly use acceleration in cm/sec², then the units on the right-hand side are

$$\text{units} = \sqrt{\left(\frac{\text{cm}}{\text{sec}^2}\right) \times (\text{cm})} = \sqrt{\frac{(\text{cm})^2}{(\text{sec})^2}} = \frac{\text{cm}}{\text{sec}}$$

Because the factor $\sqrt{2}$ on the right-hand side is a pure number and does not contain any units, we cannot check it by this method.

Example 13

A student remembers that the formula for the distance covered by a falling body is either $s = \frac{1}{2}at$ or $s = \frac{1}{2}at^2$. How can he make the correct choice by checking units?

For sake of variety let us assume he is working in English units. Then

$$\text{units of } (at) = \left(\frac{\text{ft}}{\text{sec}^2}\right) \times (\text{sec}) = \text{ft/sec}$$

and

$$\text{units of } (at^2) = \left(\frac{\text{ft}}{\text{sec}^2}\right) \times (\text{sec})^2 = \text{ft}$$

Q.7: An earth satellite cannot be in a stable orbit unless $v = \sqrt{gR_e}$. True or false?

We see that only the latter expression has the correct units for the distance s.

Example 14

Centripetal acceleration can be expressed in terms of the radius R and the period T (time to make one revolution). Check the units of the following expressions to determine which is the correct combination of T and R: (a) T^2/R, (b) T^2R, (c) R/T^2.

(a) Units of $\left(\dfrac{T^2}{R}\right) = \dfrac{\text{sec}^2}{\text{cm}}$

(b) Units of $(T^2R) = (\text{sec}^2) \times (\text{cm})$

(c) Units of $\left(\dfrac{R}{T^2}\right) = \dfrac{\text{cm}}{\text{sec}^2}$

We note that this last expression has the correct units for a_c. Actually the correct formula contains a numerical factor $4\pi^2$. The correct formula is therefore

$$a_c = \frac{4\pi^2}{T^2}\,R$$

The above formula can be derived directly by noting that the velocity v is the circumference of the circle $(2\pi R)$ divided by the period T. We substitute $v = 2\pi R/T$ into Eq. 2-14 and obtain

$$a_c = \frac{(2\pi R/T)^2}{R}$$

or

$$a_c = \frac{4\pi^2}{T^2}\,R \tag{2-16}$$

Problems

1. Equations 2-1 and 2-4 both have the same quantity s/t on the right-hand side; yet these are quite different equations. Explain.

2. An object starts from rest with uniform acceleration. In terms of s and t its instantaneous velocity will then be $v = 2s/t$. Derive this formula.

3. If in Problem 2, $v = 2s/t$, how is this compatible with Eq. 2-1 which says that $v = s/t$?

4. Suppose the object in Problem 2 starts out with initial velocity v_0. Derive a formula for v in terms of v_0, s, and t.

5. A car starts from $s = 0$ and reaches $s = 80$ ft in 4 sec as shown in the figure.
   (a) How far did it go in the first 3 sec?
   (b) What was the instantaneous velocity at $t = 1$ sec?
   (c) What was the average velocity for the first 4 sec?

Ans. 7: The satellite in Fig. 2-18 has a considerably higher velocity and is still in a stable orbit. At present there are about 1000 objects (including the moon) in stable orbit about the earth, and none of them would have v *exactly* equal to $\sqrt{gR_e}$.

Prob. 5

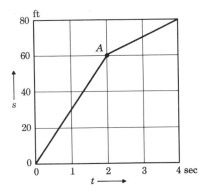

(d) What was the average velocity for the first 3 sec?

(e) At point A there is a sudden change in slope. Would this be possible for a real car?

6. At a time t_1, a body has a position x_1 and a velocity v_1. At a later time t_2 it has a position x_2 and a velocity v_2.

(a) What is its average velocity in terms of these quantities (do not assume uniform acceleration)?

(b) What is its average acceleration?

7. The engine of a certain rocket can exert a thrust or force exactly equal to the weight of the rocket. What will be the acceleration of the rocket when fired along a horizontal, frictionless track? What would be the acceleration when fired vertically from the surface of the earth?

8. At what point in its path does a projectile have its minimum velocity?

9. A car travels a distance s_1 at velocity v_1 and s_2 at velocity v_2. What is the "average" velocity with respect to *distance*? (Now s_1 and s_2 are the weighting factors.)

10. A boy throws a ball straight up in the air and catches it (at the same height from which he threw it) 2 sec later. How fast was the ball moving when it left his hand? How high did the ball rise above the point from which it was thrown?

11. On a trip along a straight road, an automobile travels 10 mi at the constant speed of 30 mph and another 10 mi at the constant speed of 60 mph. What is the average speed for the entire trip?

12. In order to get into orbit an astronaut has to accelerate from zero velocity during the time T to the orbital velocity of 8 km/sec. Let us assume that his acceleration during this take-off period is $4g$. How long does it take the rocket to reach orbital velocity and how far does it travel during this time?

13. A body moving with a velocity of 10 m/sec is uniformly decelerated, coming to rest in a distance of 20 m. What is its deceleration? How long a time was required for the body to come to rest?

14. The vector **C** equals (**A** + **B**; **A** − **B**; **B** − **A**; neither). The vector **Z** equals (**X** + **Y**; **X** − **Y**; **Y** − **X**; neither).

Prob. 14

15. A right fielder is 200 ft from home plate. Just at the time he throws the ball in to home plate, a runner leaves third base and takes 3.5 sec to reach home plate. If the maximum height reached by the ball was 64 ft, did the runner make it to home plate in time?

16. An earth satellite has an orbit 330 mi high. A super-powerful cannon is pointed vertically in an effort to shoot down the satellite. Assume the acceleration due to gravity is constant (32 ft/sec²) and ignore air resistance. What must be the muzzle velocity for the cannon shell to just barely reach the satellite? How long would it take the shell to reach the satellite?

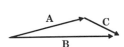
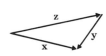

17. A steel ball is bouncing up and down on a steel plate with a period of oscillation of 1 sec. How high does it bounce?

18. A 60 mph car crashes into a solid wall. This is the same impact as if it had been dropped from what height?

19. Suppose the above car had a head-on collision with a heavy truck also traveling at 60 mph. Assume that during the collision the truck is not slowed down. Now what is the equivalent height from which the car is dropped?

20. Assume that for comfortable jetplane travel the jet should not have a horizontal acceleration greater than $2g$. If the distance from New York to Philadelphia is 100 mi, what is the fastest trip that could be made under these conditions for this "super-jet" to take off from New York and land in Philadelphia? What would be the maximum velocity during the trip?

21. A ball is dropped from a height of 64 ft to a flat surface and bounces to a height of 16 ft. What is the velocity of the ball just before it touches the surface? How much time elapses from the instant the ball is dropped to its arrival at the top of its bounce? What is the velocity of the ball just after it leaves the surface?

22. A certain United States Army rocket can obtain a velocity of 600 mph by the time it reaches 1000 ft. How many times g is its acceleration?

23. Assume an antimissle has 1 min warning time to intercept an incoming missile at a height of 200 km. If its rocket motors can supply a maximum acceleration of $10g$ and it is fired straight up, is the 1 min warning time adequate?

24. Consider an airplane whose cruising speed is 200 mph with respect to the air. It has a scheduled round trip flight between two points A and B, separated by 400 mi. Neglect the time to start, stop, and turn around.
   (a) How much time will the round trip flight take on a calm day?
   (b) How much time will it take on a day when a 50 mph wind is blowing from B to A?
   (c) How much time will it take if it is a 50 mph crosswind?

25. A boy drops a stone from a window 100 ft above the ground. Another boy drops a stone from a window 64 ft above the ground at the instant the first stone falls past him.
   (a) With what velocities will the stones hit the ground?
   (b) How much time will there be between the instants at which the two stones hit the ground?

26. An object starting from rest has a uniform acceleration of g for one year. According to relativistic mechanics what is its final velocity? If it accelerates for ten years, how close will it get to the speed of light?

27. If the distance from the earth to the moon is 240,000 mi, what is

Prob. 29

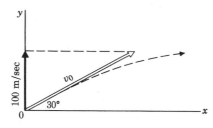

Prob. 30

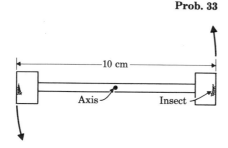

Prob. 33

the centripetal acceleration of the moon in its orbit around the earth?

28. An earth satellite which is sixty times closer to the center of the earth than the moon has a centripetal acceleration how many times greater than that of the moon? Would you say that the centripetal accelerations vary inversely as the square of the distance from the earth's center?

29. Consider a projectile fired at 30° from the horizontal. The vertical component of the initial velocity is 100 m/sec. Ignore air resistance.

    (a) What is the initial velocity?

    (b) Let T be the total time of flight. What is v_y at $t = \frac{1}{2}T$? What is the acceleration at this instant of time?

    (c) What is v_y just before $t = T$?

    (d) What is v_y at $t = \frac{1}{4}T$?

30. A particle starts from rest and undergoes accelerations as plotted in the figure for the first four sec.

    (a) Plot a graph of velocity versus time for this particle.

    (b) Plot a graph of distance versus time.

    (c) What will be its maximum velocity during the 4 sec?

    (d) How far will it go in the 4 sec?

31. A stream is flowing 5 mph due west. With respect to the water a ferryboat is sailing 30° east of north at 10 mph. What are the velocity and direction of the ferryboat as observed from the shore?

32. What is the average temperature with respect to time in a 24-hr period if from midnight to 10 A.M. the temperature increases uniformly from 30° to 60°, from 10 A.M. to 4 P.M. the temperature remains at 60°, and from 4 P.M. to midnight the temperature drops uniformly to 40°?

33. An apparatus is designed to study insects at an acceleration of $100g$. The apparatus consists of a 10 cm rod with insect containers at either end. The rod is rotated about its center.

    (a) What will be the insect velocity when at $100g$?

    (b) What will be the number of revolutions per second?

34. Assume the Project Apollo rocket is coasting in a circular orbit around the moon. If the radius of its orbit is $\frac{1}{3}$ the radius of the earth and the gravitational acceleration due to the moon is $g/12$ where $g = 980$ cm/sec², what is the velocity compared to that of a low-flying earth satellite? That is, what is the ratio of this lunar orbital velocity to that of an earth satellite circling close to the earth?

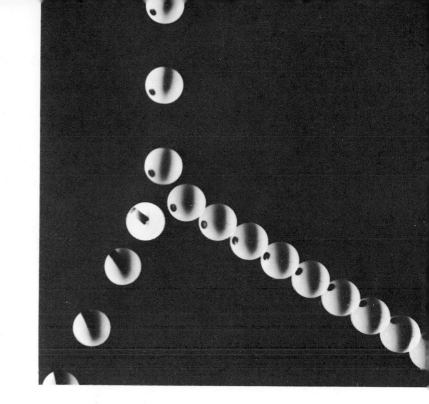

Dynamics

Chapter **3**

Dynamics

3-1 Newton's Laws of Motion

A general description

Dynamics is a study of certain general relations that describe the interactions of material bodies. We will find that a large class of phenomena can be described or explained in terms of Newton's laws of motion. One general property of a material body is its inertial mass. Another new concept useful in describing interactions of material bodies is force. These two concepts, inertial mass and force, were first defined in a quantitative manner by Isaac Newton. (In this book the terms mass and inertial mass have the same meaning and are used interchangeably.) The definitions of mass and force are contained in Newton's description of his three laws of motion. We shall first briefly state these laws and then discuss them along with the definitions of mass and force.

Newton's first law

A body remains in a state of rest or constant velocity (zero acceleration) when left to itself (the net force acting on it is zero). Mathematically this says

1st law $$a = 0 \quad \text{when} \quad \mathbf{F}_{\text{net}} = 0$$

Newton's second law

The net force on a body is the product of the mass of the body times its acceleration.

2nd law $$\mathbf{F}_{\text{net}} = M\mathbf{a} \tag{3-1}$$

Newton's third law

Whenever two bodies interact, the force on the first body due to the second is equal and opposite to the force on the second due to the first.

3rd law $$\mathbf{F}_{A \text{ due to } B} = -\mathbf{F}_{B \text{ due to } A} \tag{3-2}$$

First, let us discuss Newton's first law. We see that mathematically it is a special case of the more general second law. One reason for an explicit statement of this special case is that it seems at first to be contrary to common everyday experience. We observe that moving objects when not being pushed or pulled usually come to rest rather than continuing on with constant velocity. For example, an automobile will come to rest when the motor is turned off. If we are to believe Newton's first law, there must be a retarding force on the coasting automobile. Actually it is the external forces of air resistance and push of the road against the tires that act on the car and give it negative acceleration until it comes to rest. An automobile, or any object, coasting in free space would not come to rest. Until about the time of Newton the accepted scientific dogma was based on the old teachings of Aristotle. A basic principle in Aristotle's scheme of things was that all objects must come to rest in the absence of external forces. Another of his teachings was that objects fall at rates proportional to their weights. Galileo, as we have seen, was one of the first to criticize these two assertions of Aristotle.

One final point must be made concerning Newton's first law. Obviously, the first law appears to be violated if the observer himself is accelerating. Newton specified that the three laws of motion are valid only when the observer is in what he called an inertial system. Newton defined an inertial system as any system that is not accelerating with respect to the fixed stars. Actually it is not as simple as one might think to find an inertial frame of reference; it requires knowledge of the structure of the universe (cosmology). The problem of finding inertial frames of reference is discussed in detail in Chapter 11.

The careful reader might be disturbed by Newton's second law because it uses two new quantities, force and mass, neither of which have been rigorously defined. However, the combination of Newton's second and third laws uniquely defines both mass and force.

In this book we shall define inertial mass using a more modern equivalent of Newton's third law, the law of conservation of momentum. We now know that Newton's third law is not correct, gross violations of it having been observed. We shall see in the chapter on relativity that signals cannot travel faster than the speed of light. For this reason, Newton's third law cannot apply to force acting at a distance; otherwise forces could be used to transmit signals with infinite velocity. One simple example of a violation is that of a charged particle moving away from a wire carrying a current. As we shall see in Chapter 8, the wire exerts a magnetic force on the charged particle, while at the same time the net force on the wire due to the particle is exactly zero. This is contrary to Newton's third law which says the two forces should have the same magnitude. On the other hand, Newton's third law is still correct for objects at rest and for contact forces. As far as we know, the law of conservation of momentum is an exact law of nature. No violations of it have ever been found and it has been thoroughly checked by all kinds of experiments.

Actually Newton's laws can be derived from the laws of conservation of momentum and energy, or vice versa. Which one should be regarded as the basic law is a matter of taste. Furthermore, with the use of higher mathematics it can be shown that the laws of conservation of momentum and energy follow from the symmetry principles of homogeneity of space and time. By homogeneity of space, we mean the laws of physics are the same at all positions in space. Homogeneity in time means that the laws of physics do not change with time.

Before we can continue with the discussion of Newton's laws, we must first define inertial mass, and this we do by making use of the law of conservation of momentum.

3-2 The Conservation of Momentum

A sacred law of nature

The momentum **P** of an object is defined as the product of the mass of the object times its velocity.

definition of momentum

$$\mathbf{P} = M\mathbf{v} \tag{3-3}$$

The law of conservation of momentum states that in the absence of any external forces, the sum of the momenta of two particles remains constant. For example, if two particles M_A and M_B collide, then

$$\mathbf{P}_A + \mathbf{P}_B = \mathbf{P}_A{}' + \mathbf{P}_B{}'$$

conservation of momentum or

$$M_A\mathbf{v}_A + M_B\mathbf{v}_B = M_A\mathbf{v}_A{}' + M_B\mathbf{v}_B{}' \tag{3-4}$$

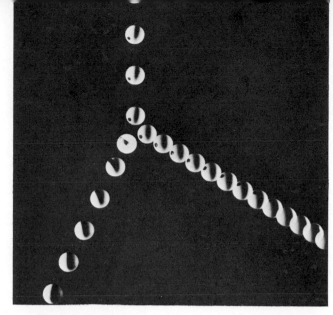

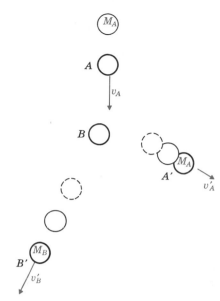

Fig. 3-1. Collision of two billiard balls of equal mass. Ball *B* was initially at rest. Ball *A* entered from the top and was deflected to the right. This strobe picture (flash rate 30 per second) was obtained through the courtesy of the Physical Science Study Committee.

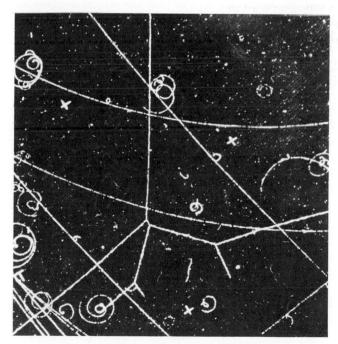

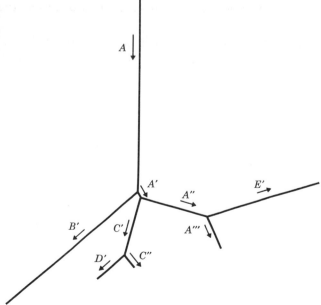

Fig. 3-2. Successive collisions of a proton against other protons initially at rest. Proton *A* enters the top and first collides with proton *B*. Protons *B*, *C*, *D*, and *E* were initially at rest as hydrogen nuclei in liquid hydrogen. This picture was taken by the liquid hydrogen bubble chamber group at the Lawrence Radiation Laboratory, University of California. In a bubble chamber moving protons will produce tracks of tiny bubbles.

where $\mathbf{v}_A$ and $\mathbf{v}_B$ are the velocities of masses M_A and M_B before the collision, and $\mathbf{v}_A{}'$ and $\mathbf{v}_B{}'$ are the velocities after the collision. A billiard ball example is shown in Fig. 3-1. The law is completely general—as shown in Fig. 3-2, it holds for elementary particles as well as billiard balls. Note that the velocities are vectors in Eq. 3-4; hence they need not be in the same direction.

Let us first consider the simplest possible application—where two objects are initially at rest as shown in Fig. 3-3; that is, $v_A = v_B = 0$. Then we let these two objects have an interaction by releasing a spring or making a small explosion between them. Ideally the mass of the spring should be very small compared to M_A or M_B. Since the initial momentum is zero, the left-hand side of Eq. 3-4 is zero and we have

$$0 = M_A\mathbf{v}_A' + M_B\mathbf{v}_B'$$
or
$$M_B\mathbf{v}_B' = -M_A\mathbf{v}_A'$$

where the minus sign tells us that the vectors are parallel, but pointing in opposite directions. In terms of the magnitudes of the vectors

$$M_B v_B' = M_A v_A'$$
or
$$M_B = M_A \frac{v_A'}{v_B'}$$

where v_A' and v_B' are the magnitudes of the vector velocities.

We now have an operational procedure for defining mass. Let M_A be a known standard mass. It could be the standard 1 kg platinum-alloy cylinder that is preserved in France. Then any unknown mass M_B can be determined by putting M_A and M_B together with a spring between them and observing the ratio of the final velocities. We see that by simple measurements of the ratio of final velocities, the inertial mass of any body can be determined.

It would seem that the law of conservation of momentum tells us that the ratio of final velocities is independent of the strength of the spring. Actually this is not quite true. If we

Fig. 3-3. Strobe picture of two unequal masses being pushed apart by spring. Ball A is twice the mass of ball B. (Courtesy Physical Science Study Committee.)

could find a super-strong spring that would give velocities comparable to the speed of light, we would find instead that the quantity

$$\frac{v_{A}'}{v_{B}'} \cdot \frac{\sqrt{1 - \left(\frac{v_{B}'}{c}\right)^2}}{\sqrt{1 - \left(\frac{v_{A}'}{c}\right)^2}}$$

would be independent of the strength of the spring. This experimental fact along with Eq. 3-4 forces us to conclude that the mass of an object is actually

$$M = \frac{M_0}{\sqrt{1 - \frac{v^2}{c^2}}} \qquad (3\text{-}6)$$

where M_0 is its mass when at rest and c is the velocity of light. Until Chapter 11 we shall make the very good approximation that mass is a constant independent of velocity. In these early chapters we will always be dealing with velocities much less than 1% the speed of light. Even at 1% of the speed of light Eq. 3-6 tells us that the mass increases by just one part in 20,000.

Another experimental consequence of the definition of mass contained in Eq. 3-5 is that masses are additive. By this we mean that if two masses M_B and M_C are joined together, then a similar determination of the combined mass M_D will numerically equal the sum of the separate determinations

Q.1: A billiard ball of velocity $\mathbf{v}_A$ in the x-direction collides with two others. Write an equation relating the y-components of the final velocities $\mathbf{v}_A'$, $\mathbf{v}_B'$, and $\mathbf{v}_C$.

M_B and M_C. Additivity of mass may seem obvious to the reader, but obvious or not, all speculations about nature must be checked by experiment. There are common physical quantities that are not additive. One example is the magnitudes of vectors; another is the addition of volumes. If 1 qt of alcohol is added to 1 qt of water, a combined volume noticeably less than 2 qt is obtained.

Before going on to the definition of force and discussion of Newton's second law, the widespread usefulness of the law of conservation of momentum is illustrated in the following examples.

Example 1

As shown in Fig. 3-4 a 3-kg rifle shoots 10-gm bullets with a muzzle velocity of 600 m/sec. If the gun is not held firmly against the shoulder, what is the recoil velocity of the rifle before it hits the shoulder? Initially the momentum of the gun plus the bullet is zero. Hence, according to the law of conservation of momentum the algebraic sum of the final gun momentum and bullet momentum must be zero.

$$M_g v_g' + M_b v_b' = 0$$

where $M_g v_g'$ is the gun momentum and $M_b v_b'$ the bullet momentum after firing.

$$v_g' = - \frac{M_b v'}{M_g} = - \frac{10(6 \times 10^4 \text{ cm/sec})}{3 \times 10^3}$$

$$v_g' = -200 \text{ cm/sec}$$

The minus sign indicates that the gun recoils in the opposite direction from that of the bullet.

Example 2

Eighty per cent of the mass of a 20-ton (20,000 kg) rocket is fuel. If this fuel is ejected as exhaust gases with an average velocity of 1 km/sec with respect to the ground, what will be the final velocity of the rocket? Ignore gravity and air resistance. This example is analogous to the preceding one. Here the recoiling rocket corresponds to the recoiling gun. Each molecule of exhaust gas can be thought of as a tiny bullet shot out of the rocket. The total momentum of the exhaust gas is the mass of the fuel M_f times the average gas velocity v_f' (see Fig. 3-5). As in the preceding example the total initial momentum is zero. Hence the sum of the two final momenta must also be zero.

$$P_f' + P_r' = 0$$

Fig. 3-4. Recoil velocity of rifle is such that the sum of rifle momentum plus bullet momentum is zero.

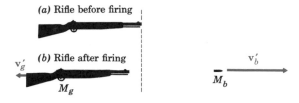

(a) Rifle before firing

(b) Rifle after firing

$\mathbf{v}_g'$

M_g

$\overline{M_b}$

$\mathbf{v}_b'$

Ans. 1: According to conservation of momentum the *y*-component of momentum will remain equal to zero. Hence

$$0 = (\mathbf{v}_A)_y + (\mathbf{v}_B')_y + (\mathbf{v}_C')_y$$

Fig. 3-5. Rocket after firing. Exhaust gases moving with average velocity v_f' to the left and empty rocket moving with velocity v_r' to the right.

or

$$M_f v_f' + M_r v_r' = 0$$

where M_r is the mass of the empty rocket, and v_r' is its final velocity. Solving for v_r', we have

$$v_r' = -\frac{M_f}{M_r} v_f'$$

Since the mass of fuel is four times that of the empty rocket and $v_f' = 1$ km/sec, we have

$$v_r' = -4(1 \text{ km/sec})$$

or

$$v_r' = -4 \text{ km/sec}$$

The minus sign means that the rocket moves in the direction opposite to its exhaust gases.

We see that the final velocity of a rocket can be made as large as desired by making the mass ratio of fuel to empty rocket correspondingly large. However, the fact that the rocket shell must be strong enough to hold the fuel sets an upper limit on the ratio M_f/M_r. We can get around this structural limitation by using multistage rockets. In the above problem suppose part of M_r contains a second stage. Assume that 80% of the mass of this second stage is also fuel. Then this second stage can acquire an additional v_r' of 4 km/sec. The final velocity of the empty second stage will then be 8 km/sec, which is the velocity necessary to go into orbit around the earth.

Example 3

As shown in Fig. 3-6 the third stage of a rocket consists of a 500 kg rocket motor and a 10 kg nosecone with a compressed spring between the two. When tested on the ground the spring gave the nosecone a velocity of 51 cm/sec relative to the rocket motor. If the nosecone is released when the third stage is in orbit with velocity $v_0 = 8$ km/sec, what will be the velocities of the nosecone and rocket motor?

Let $M_A = 500$ kg be the rocket motor and $M_B = 10$ kg be the nosecone. The initial momentum is $(M_A + M_B)v_0$. According to the conservation of momentum

$$M_A v_A' + M_B v_B' = (M_A + M_B)v_0$$

Also we know from the ground test that $v_B' - v_A' = 0.51$ m/sec, or

$$v_A' = v_B' - 0.51 \text{ m/sec}$$

Fig. 3-6. (*a*) Before nosecone is released. (*b*) Velocities after a spring pushes nosecone and rocket motor apart.

(*a*) 8000 m/sec

(*b*) 7999.99 m/sec   8000.5 m/sec

Q.2: If a rocket is fired from rest, could the exhaust gases have an average velocity of zero?

Substituting this in the momentum equation gives

$$M_A(v_B' - 0.51\ m/sec) + M_B v_B' = (M_A + M_B)v_0$$

$$v_B' = v_0 + \frac{M_A}{M_A + M_B} \times 0.51\ \text{m/sec}$$

$$v_B' = 8000.5\ \text{m/sec}$$

$$v_A' = (8000.5 - .51)\ \text{m/sec}$$

$$= 7999.99\ \text{m/sec}$$

We see that the spring imparts an additional velocity of 0.5 m/sec to the nosecone. The rocket motor is slowed down by 0.01 m/sec.

Example 4

Show that in an elastic billard-ball collision involving balls of equal mass, the cue ball and the target ball will have a 90° angle between them after the collision; that is, in Fig. 3-1 the vectors $\mathbf{v}_A'$ and $\mathbf{v}_B'$ must be at right angles. By elastic, we mean that the magnitude of the relative velocity after the collision is the same as it was before the collision.

We must show that the angle θ in Fig. 3-7a is 90°. According to Eq. 3-4 $\mathbf{v}_A = \mathbf{v}_A' + \mathbf{v}_B'$. We form this vector sum in Fig. 3-7b. The relative velocity is $(\mathbf{v}_A - \mathbf{v}_B)$ which was $\mathbf{v}_A$ before the collision. After the collision, it is $\mathbf{v}_C = \mathbf{v}_A' - \mathbf{v}_B'$ as shown in Fig. 3-7c. Note that the angle between sides v_A' and v_B' is the original angle θ. Since the collision is elastic, $v_C = v_A$, and the three sides of the triangle in Fig. 3-7c equal the three sides of the triangle in Fig. 3-7b, with the consequence that the two corresponding angles $(180° - \theta)$ and θ are equal:

$$\theta = (180° - \theta)$$

Hence $\quad 2\theta = 180°$

$$\theta = 90°$$

3-3 Force

The push and pull of physics

Now that we have a precise definition of mass and momentum, we can use Newton's second law to define force. Actually, Newton's original statement of this law was not $F = Ma$, but that the force on a mass M is the rate of change of momentum with respect to time of that mass:

$$F = \frac{\Delta P}{\Delta t} \qquad \text{or} \quad F = \frac{\Delta(Mv)}{\Delta t} \tag{3-7}$$

where ΔP is the change of momentum of mass M in a short

Fig. 3-7. Cue ball A strikes target ball B.

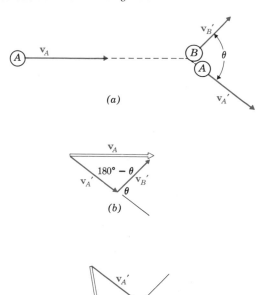

(a)

(b)

(c)

Ans. 2: No. If so, the momentum of the rocket would be equal to its initial value of zero.

time Δt. At ordinary velocities (much less than the speed of light), the mass M can be treated as a constant and factored out. Then

$$F = M \frac{\Delta v}{\Delta t}$$

or

$$F = Ma$$

since $\Delta v / \Delta t$ is the definition of acceleration. Of course, Newton did not know that mass increases with velocity according to Eq. 3-6; he thought that mass times acceleration must always be mathematically equivalent to rate of change of momentum with respect to time. According to modern relativistic mechanics the original definition of force (Eq. 3-7) is correct and the form $F = Ma$ is a very accurate approximation when dealing with ordinary velocities. We shall proceed to use the approximation $F = Ma$ as we continue on with our study of classical mechanics. The study of relativistic mechanics is deferred until Chapter 11. It is worth noting, however, that in this chapter on classical mechanics we have succeeded in giving rigorous definitions of mass and force (Eqs. 3-5 and 3-7) which are correct both in classical mechanics and relativistic mechanics.

It is important to remember that wherever the symbol F appears in this section, what is meant is the *net* force on a mass M. To test the significance of this, we will ask a few questions.

Question: It is possible for a body to remain at rest while being pushed by an external force?

The answer is yes. It is possible if at the same time there are other external forces such that the vector sum of all the external forces is zero; that is, $F_{net} = 0$.

Question: If the net force on a body is zero, then must the body be at rest?

The answer is no because the body may be moving with constant velocity.

The units of force are kg m/sec^2 in the MKS system and gm cm/sec^2 in the CGS system. These two units are given the special names newton and dyne, respectively. One newton is the amount of force required to give 1 kg an accelera-

Q.3: If a body is not at rest, the net force on it must not be zero. True or false?

tion of 1 m/sec². The conversion factor between newtons and dynes is obtained as follows:

$$1 \text{ newton} = \frac{1 \text{ kg} \times 1 \text{ m}}{(1 \text{ sec})^2}$$

Now we substitute 10^3 gm for 1 kg and 10^2 cm for 1 m. Then

$$1 \text{ newton} = \frac{(10^3 \text{ gm}) \times (10^2 \text{ cm})}{(1 \text{ sec})^2}$$

$$= 10^5 \frac{(1 \text{ gm}) \times (1 \text{ cm})}{(1 \text{ sec})^2}$$

$$= 10^5 \text{ dynes}$$

Contact force

When two objects are pushed into contact such as a block pushed against a wall or table, there are contact forces. Not only is there a force on the table due to the block, but there is a force on the block due to the table. The ultimate source of these two forces are the repulsive forces between atoms. When the electron clouds of two atoms begin to overlap, there is a repulsive force between them, and as the two atoms are pushed closer together the repulsive force increases. This repulsive force between atoms is electromagnetic in origin and can be very strong compared to gravitational forces. If we push a block harder against a table, it pushes the surface atoms of the block closer to those of the table until there is a net repulsive force equal and opposite to the applied force. We call such repulsive forces between surfaces, contact forces.

Consider the following "paradox." A wooden block of mass M is pushed against a solid wall with a force F. According to Newton's second law the acceleration is

$$a = \frac{F}{M}$$

This seems to say that the block should accelerate and start moving. However, we know from experience that the block will not move. What is wrong?

This paradox is resolved by noting that the F to be used in the equation $F = Ma$ must be the *net force*. If two forces F_1 and F_2 both act on the same mass M, then $\mathbf{F}_{\text{net}} = \mathbf{F}_1 + \mathbf{F}_2$.

Fig. 3-8. Block being pushed against immovable wall.

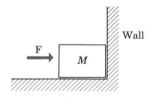

Ans. 3: False. The body could be moving at constant velocity and then $F_{\text{net}} = 0$.

In addition to the applied force $\mathbf{F}$ there is a second force $\mathbf{F'}$ of the wall pushing on the block in Fig. 3-8. The net force is then $\mathbf{F}_{net} = \mathbf{F} + \mathbf{F'}$. According to Newton's third law $\mathbf{F'}$ is equal and opposite to the force of the block pushing on the wall, so $\mathbf{F'} = -\mathbf{F}$. Thus $\mathbf{F}_{net} = \mathbf{F} + (-\mathbf{F}) = 0$; now Newton's second law gives the result

$$a = \frac{F_{net}}{M} = 0$$

As we proceed further we shall begin to appreciate the great simplicity and beauty of Newton's laws. However, the correct application of Newton's laws can often be subtle. The following paradox should serve as a warning to the careless thinker.

Consider two blocks, M_A and M_B, on a frictionless surface as shown in Fig. 3-9. A force F is applied to block A and transmitted through it to block B. By Newton's third law, block B must exert an equal and opposite force $(-F)$ on block A. Thus the net force on A is the sum of the applied force F plus the contact force $-F$ of block B on A. Thus $F_{net} = F + (-F) = 0$. Then

$$a = \frac{F_{not}}{M} = 0$$

The conclusion would be that block A can never move, no matter how large a force F is applied to it.

The preceding paradox contains a mistake, which is the assumption that the force F is transmitted through block A and is thus also applied to block B. There is nothing in Newton's laws saying this should be so. Instead we should assume an arbitrary value F' for the force of block A on block B. Then the net force on B is F' and the net force on A is $F - F'$. Newton's second law applied to blocks A and B respectively gives

$$F - F' = M_A a$$
$$F' = M_B a$$

Adding together the above equations we obtain

$$F = (M_A + M_B)a \qquad \text{or} \quad a = \frac{F}{M_A + M_B}$$

Fig. 3-9. Two blocks being pushed along frictionless surface.

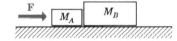

Q.4: In Fig. 3-8, is the force against the wall always equal to the force the wall exerts on the block?

Note that this result is merely the applied force F divided by the total mass of the two blocks.

The remaining three sections of this chapter are devoted to three applications of Newton's laws which should give some idea of how Newton's laws are used to solve problems in mechanics.

3-4 The Inclined Plane
Gravity reduced

Experimentally all freely falling bodies have the same acceleration. According to Newton's second law the gravitational force on any body of mass M near the earth's surface must then be $F_G = Mg$ where $g = 9.8$ m/sec^2.

Question: What is the acceleration of a mass M sitting on a table?

The answer is that since the velocity of the mass is not changing, by definition its acceleration must be zero.

Question: What is the net force on a mass M sitting on a table?

Since the acceleration is zero the net force is $F_{net} = Ma = M(0) = 0$. The contact force F' of the table pushing up against the mass is equal and opposite to the gravitational force pulling down on M. The net force is $F_{net} = F_G - F' = Mg - Mg = 0$. This is true of the vertical forces even if mass M is sliding along a frictionless table. In the absence of friction there will be no horizontal component of force.

Now if we tilt a frictionless table, there will still be zero component of force along the surface; that is, the only contact force the table can exert on M must be perpendicular to the surface as indicated by $\mathbf{F}'$ in Fig. 3-10. Suppose the table is tilted by an angle θ, what will be the acceleration of the mass as it slides down the inclined plane?

In this case the contact force $\mathbf{F}'$ of the incline on M must be of such a value so that $\mathbf{F}_{net}$ is pointed along the incline. Also $\mathbf{F}'$ must be perpendicular to the incline. The three vectors $\mathbf{F}_{net}$, $\mathbf{F}'$, and the gravitational force $\mathbf{F}_G$ are shown in Fig. 3-10. In Fig. 3-10b the vectors $\mathbf{F}'$ and $\mathbf{F}_G$ are added according to the polygon rule to give $\mathbf{F}_{net}$. The angle between $\mathbf{F}'$ and $\mathbf{F}_G$ must be θ since $\mathbf{F}'$ and $\mathbf{F}_G$ are mutually perpendicular to the incline and the horizontal. Since the sine of an

Fig. 3-10. Forces on block M sliding down inclined plane.

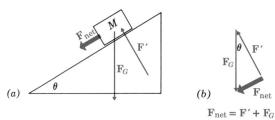

$$F_{net} = F' + F_G$$

Ans. 4: According to Newton's 3rd law the answer is yes.

angle is the opposite side divided by the hypotenuse, we have from Fig. 3-10 that

$$\sin \theta = \frac{F_{\text{net}}}{F_G}$$

Now substitute Ma for F_{net} and Mg for F_G:

$$\sin \theta = \frac{(Ma)}{(Mg)}$$

or

$$a = g \sin \theta$$

3-5 Atwood's Machine

Controlling gravity again

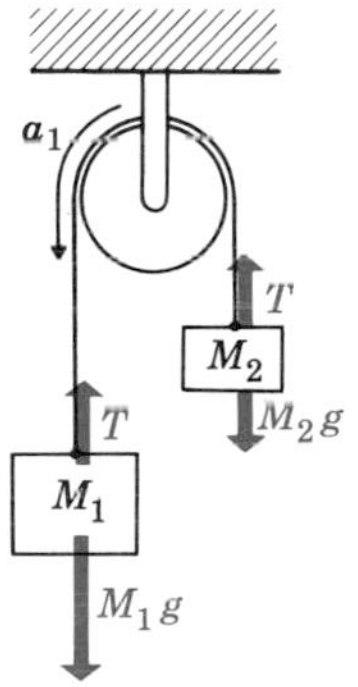

Atwood's machine consists of two masses, M_1 and M_2, connected by a light string over a pulley as shown in Fig. 3-11. If the mass of the string were actually zero, the net force on the string must also be zero; otherwise it would have infinite acceleration. In other words, if a string is pulled with a force T on one end, the other end must also be pulled with the same force T. In Fig. 3-11 let T be the force with which the string is pulling up on each of the two masses. The net force on M_1 is

$$M_1 g - T = M_1 a_1 \qquad (3\text{-}8)$$

The net force on M_2 is

$$M_2 g - T = M_2 a_2$$

Since the string is of fixed length, $a_1 = -a_2$, and the above equation becomes

$$M_2 g - T = -M_2 a_1$$

or

$$T = M_2 g + M_2 a_1$$

Now substitute this expression for T into Eq. 3-8:

$$M_1 g - (M_2 g + M_2 a_1) = M_1 a_1$$

Solving this for a_1 gives the result

$$a_1 = \frac{M_1 - M_2}{M_1 + M_2} g$$

Fig. 3-11. Atwood's machine.

Q.5: A tractor pulls a plow with a force F.

(a) Must the force the ground exerts on the plow also have the same magnitude F?

(b) Must the force of the plow pulling back on the tractor also have the same magnitude F?

We see that the acceleration can be made as small as desired by making the mass difference correspondingly small. Of course, if $M_1 - M_2$ is made too small, the mass of the string and pulley no longer can be considered as zero.

3-6 The Simple Pendulum and Simple Harmonic Motion

The usual type of motion

The simple pendulum is defined to be a small bob of mass M hanging by a string of length L. When one talks of the motion of a simple pendulum it is assumed that the displacement x is always much less than L. We shall now calculate the acceleration of M using Fig. 3-12. In Fig. 3-12b the force vectors $\mathbf{F}_G$ and $\mathbf{T}$ are added by the polygon rule to give $\mathbf{F}_{net}$. This vector triangle is similar to the triangle of Fig. 3-12a because its sides are all parallel to the corresponding sides of the triangle in Fig. 3-12a. Thus

$$\frac{F_{net}}{F_G} = \frac{x}{L} \text{ (in the limit of small } \theta) \tag{3-9}$$

Now substitute Ma for F_{net} and Mg for F_G. Then

$$\frac{Ma}{Mg} = \frac{x}{L}$$

or

$$a = -\left(\frac{g}{L}\right)x$$

We have inserted the minus sign to remind us that a and x are in opposite directions.

This is our first example of motion in which the acceleration changes in magnitude. Note that in this example the acceleration is related to the displacement x in a very simple way—it is directly proportional to x and always pointing in the direction opposite to x. This type of motion is defined as simple harmonic motion (SHM). The general condition for SHM is that the ratio of the acceleration to the diplacement is some fixed constant K:

$$\frac{a}{x} = -K$$

is the condition for simple harmonic motion. This condition

Fig. 3-12. A simple pendulum of length L.

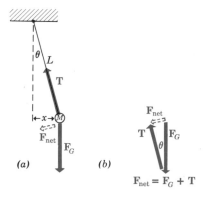

(a)   *(b)*

$$\mathbf{F}_{net} = \mathbf{F}_G + \mathbf{T}$$

is fulfilled by many common types of motion. For example, the force of a stretched spring is usually proportional to its displacement x. This experimental observation on the property of springs is called Hooke's law. Thus the motion of a mass attached to the end of a spring must also be simple harmonic motion. There are many other examples of SHM, such as the motion of any point on a plucked violin string, the motion of a small volume of air in an organ pipe, and the vibration of an atom in a solid. There is a good reason why much of the motion that occurs in nature is SHM: no matter in what complicated way a certain force varies with distance, as long as it is what we call a smooth function of the distance, the change in the force must be directly proportional to the displacement for very small displacements. Thus for a small displacement of almost any body from its equilibrium position, the condition for SHM is fulfilled and the body will oscillate back and forth in SHM.

Let the maximum displacement of the pendulum bob be x_0. At this position the restoring force is maximum and the mass is pulled back through $x = 0$ and overshoots to $x = -x_0$ (neglecting friction). Then the process is repeated. The time it takes for the bob to move from $x = x_0$ to $x = -x_0$ and back again to $x = x_0$ is called one complete period T. Figure 3-14 contains plots of x versus t, the force F versus t, and velocity v versus t for an object in SHM starting from the position $x = 0$.

The value of T is determined only by the ratio a/x. It will be shown in the following paragraph that

$$-\frac{a}{x} = \frac{4\pi^2}{T^2}$$

for all forms of simple harmonic motion. Since for a simple pendulum the quantity $\left(-\frac{a}{x}\right) = \frac{g}{L}$ we have

$$\frac{g}{L} = \frac{4\pi^2}{T^2}$$

or

$$T = 2\pi \sqrt{\frac{L}{g}}$$

◀ **Fig. 3-13.** Strobe photograph of swing of a simple pendulum.

Fig. 3-14. Curve (a) is a plot of the displacement x versus time in SHM. The time for one complete oscillation is T the period. Curve (b) shows the corresponding force, or acceleration versus t. Curve (c) shows the corresponding velocity v.

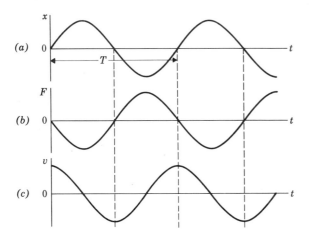

Fig. 3-15. The reference circle. Projection of ball position on vertical plane will have the same motion as a pendulum ball.

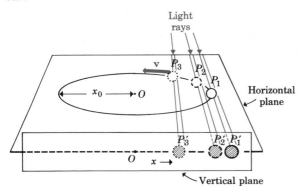

Note that the period of a simple pendulum is independent of the mass of the bob and also independent of how large a displacement x_0 is used. The suggestion of Galileo was followed in mechanical clocks by making use of the fact that in simple harmonic motion the period is independent of the size of the displacement.

The reference circle

Figure 3-15 shows the shadow of a ball traveling in uniform circular motion with velocity v and radius x_0. We shall show that the motion of the shadow on the screen in Fig. 3-15 is simple harmonic motion; that is, that the ratio a/x is a fixed constant for all values of x. Since the shadow is an edge-on view of the circle, the motion of the shadow is back and forth along a straight line (the x-axis), with a maximum displacement of x_0. The acceleration a of the shadow is the x-component of the centripetal acceleration a_c of the ball. We shall now determine a using Fig. 3-16, which is a top view of the circle in Fig. 3-15. By similar triangles

$$\frac{a}{x} = \frac{a_c}{x_0}$$

Since $a_c = -\dfrac{4\pi^2}{T^2} x_0$ (see Eq. 2-15), we have

$$\frac{a}{x} = -\frac{\dfrac{4\pi^2}{T^2} x_0}{x_0}$$

or

$$\frac{a}{x} = -\frac{4\pi^2}{T^2} \tag{3-10}$$

Thus we have shown that the acceleration of the shadow divided by its displacement is always constant and that it has the value $-(4\pi^2/T^2)$, where T is the period of oscillation. This is true for all forms of simple harmonic motion. The ratio a/x is always $-(4\pi^2/T^2)$.

Example 5
A hole is drilled through the center of the earth from the United States to Australia. Ignoring air resistance, how long will it take for a stone dropped in the hole at the United States end to reach Australia if the gravitational force on the stone is $F_G = Mg\dfrac{r}{R}$,

Fig. 3-16. Top view of the reference circle. The vector **a** is the projection in the x-axis of the ball's acceleration $\mathbf{a}_c$.

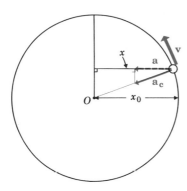

Fig. 3-17. Curve (c) is plot of typical force between two atoms in a molecule. Equilibrium position will be at $r = r_0$, where $F = 0$.

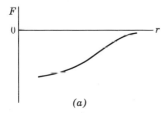

(a)

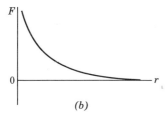

(b)

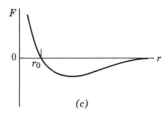

(c)

r is distance of proton from center of molecule

where r is the distance from the center of the earth and $R = 4000$ mi is the radius of the earth?

According to Newton's second law

$$Ma = Mg\frac{r}{R}$$

Thus

$$\frac{a}{r} = \frac{g}{R}$$

is the ratio of acceleration to displacement. We see that this ratio is independent of the position of the stone, thus satisfying the condition for simple harmonic motion. According to Eq. 3-10, this ratio of acceleration to displacement must then be $4\pi^2/T^2$; hence

$$\frac{4\pi^2}{T^2} = \frac{g}{R}$$

$$T = 2\pi\sqrt{\frac{R}{g}}$$

$$T = 5.1 \times 10^3 \text{ sec}$$

$$T = 85 \text{ min}$$

The time to reach Australia is one-half of a complete period, or 42.5 min.

Let us now consider in more detail the statement that an atom will oscillate in SHM if displaced from its equilibrium position. We will consider the hydrogen molecule which consists of two hydrogen atoms bound together due to the attractive forces between the electrons and the two protons. This net attractive force between two hydrogen atoms looks something like Fig. 3-17a when plotted versus r, the distance of one of the protons from the center of the molecule. But in addition there is a repulsive force between the two protons which looks something like Fig. 3-17b. The sum of the two curves gives a net force between the two hydrogen atoms looking something like Fig. 3-17c. The equilibrium position has the value $r_0 = 4 \times 10^{-9}$ cm, at which position the slope of the curve is $\Delta F/\Delta r = -1.1 \times 10^6$ dynes/cm.

Example 6

From the information given above, calculate the frequency of oscillation of the two hydrogen atoms in the hydrogen molecule.

Let $x = r - r_0$ be the displacement from the equilibrium position. Then from the slope of the curve at $r = r_0$ we have for small displacements $F = -(1.1 \times 10^6)x$ in dynes. Now divide both sides by the mass of the proton $(1.67 \times 10^{-24}$ gm) to obtain

$$a = -(0.66 \times 10^{30})x \quad \text{in cm/sec}^2$$

The ratio a/x which is also equal to $-4\pi^2/T^2$ has the value 0.66×10^{30} sec^{-2}. Solving for T gives 7.74×10^{-15} sec for each oscillation. In one second there will be $1/(7.74 \times 10^{-15})$ or 1.29×10^{14} of these oscillations. A charged particle oscillating at this frequency will emit electromagnetic radiation of this frequency. Such radiation appears as infrared radiation. If the frequency had been three and a half times greater, the radiation would appear as visible light. We conclude that heated hydrogen gas emits infrared radiation of frequency 1.29×10^{14} cycles/sec.

Problems

1. Can the two forces of Newton's third law ever act on the same body?

2. Suppose you were sitting in the middle of a completely frictionless surface such as an idealized pond of ice. Propose a method for getting out of such a predicament.

3. Consider as a system of mass M, a boy on a bike. What is the *external* force that can accelerate M? What is the external force that can bring M to rest? (The push against the pedals is not external to the system of boy plus bike.)

4. If $F = kx$; the force is proportional to the displacement. What are the units of k in gm, cm, and sec?

5. A tractor pulls a log attached by a chain with a constant speed of 5 mph and a force of 1000 newtons. The gravitational force on the log is $F_G = 2000$ newtons. According to Newton's first law what is the net force on the log?

6. What is wrong with the following reasoning? A tractor pulls a plow with a force F. By Newton's third law the contact force of the ground on the plow is $-F$. Since the sum of these two forces is zero, the plow cannot move.

7. A force F acts for a time t_0 on a mass M. What is the increase in momentum of M?

Prob. 8

8. Two blocks are tied together with a piece of string. Another string is tied to the top block. The blocks are pulled by the earth's gravity.

    (a) How much force F must be applied to the top string to suspend both blocks at rest?

    (b) How much force F must be applied to the top string to give both blocks an acceleration upward of 2 m/sec? What then is the tension in the string between the two blocks?

Prob. 10

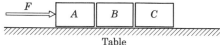

Table

Prob. 13

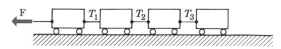

Prob. 14

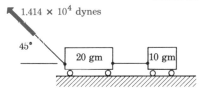

Prob. 16

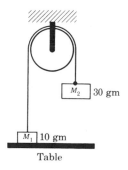

Table

9. A super ball is bouncing up and down without loss of energy to a height h.

    (a) Is this simple harmonic motion?

    (b) Derive a formula for the period of oscillation of the ball.

10. Three identical blocks of mass M each are pushed along a frictionless table.

    (a) What is the net vertical force on block A?

    (b) What is the net force on block A?

    (c) What is the acceleration of block C?

    (d) What is the force of block A on block B?

11. A block of wood having a mass of 2 kg is initially at rest on a smooth horizontal surface. A 5-gm bullet having a horizontal velocity of 500 cm/sec strikes the block and remains imbedded in it. With what final velocity do the block and bullet move after the collision?

12. Assume the hydrogen atom is an electron of mass 9×10^{-28} gm traveling in a circle of 10^{-8} cm diameter around a proton. The force of attraction is 10^{-2} dyne. What is the electron velocity? How many revolutions per sec does the electron make?

13. A four-car toy train is pulled by a child with a force F. The mass of each car is M. What are the tensions (or forces) T_1, T_2, and T_3 in the strings in terms of F and M? What is the acceleration in terms of F and M? Ignore friction.

14. A child pulls a frictionless pull-toy with a force of 1.414×10^4 dynes at an angle of 45° as shown. What is the acceleration of the 20-gm car? What is the tension in the string between cars? What is the tension in the string pulled by the child? With what force does the floor push the 20-gm car?

15. A pendulum of length 1 m swings back and forth 5 cm from the center position.

    (a) What is its velocity at the end of the swing (maximum displacement)?

    (b) What is the acceleration when at the end of the swing?

    (c) What is the acceleration and velocity when passing through the center?

16. M_1 and M_2 are attached by a string that runs over a frictionless pulley. M_1 is held down on the table. What force is required to hold M_1 down on the table? What is the tension in the string in dynes? What would be the tension in the string if M_1 were released?

17. A 500-gm mass is sitting on the end of a level, frictionless glass surface. If this end of the glass surface is raised 2.5 cm, what will be the velocity of the mass when it reaches the other end?

18. A boy on a rocking horse swings back and forth a distance of 2 ft from the equilibrium position. The maximum acceleration of the boy is 0.5 ft/sec². If the acceleration is proportional to the displacement, what is the period of oscillation?

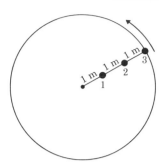

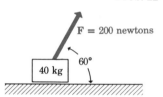

19. A rocketship of 10^6 kg mass and 2.94×10^7 newton thrust is fired vertically. What is its acceleration? (*Hint:* the net force on the rocket is not equal to the thrust.)

20. A boy whirls three balls, tied together with 1 m lengths of string, in a horizontal plane (ignore gravity). Each ball has a mass of 100 gm. If the outside ball is moving at a speed of 6 m/sec, what are the tensions in the three strings? If he continues to whirl them faster, which string will break first, assuming all strings are alike?

21. Assume the ball in the reference circle of Fig. 3-15 has a velocity v_0. Express the following quantities in terms of v_0 and x_0 only:
    (a) The maximum acceleration of the shadow.
    (b) The period of oscillation of the shadow.
    (c) The frequency or number of oscillations per second.

Prob. 22

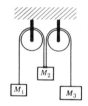

22. A block of 40 kg is on a frictionless surface under the influence of gravity and an external force F as shown in the figure. What are the magnitude and direction of the net force on the block? Suppose $F = 800$ newtons instead of 200. Now what is the magnitude of the net force?

Prob. 23

23. Derive a formula for the acceleration of M_1 in terms of M_1, M_2, M_3, and g. In this "double" Atwood's machine assume frictionless, massless pulleys.

Prob. 24

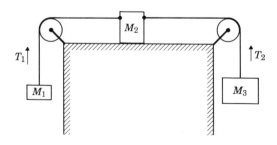

24. Consider the modified Atwood's machine shown in the figure. Assume frictionless, massless pulleys and that M_2 slides along the surface without friction. Let T_1 be the tension in the left string and T_2 the tension in the right string. Assume M_3 is heavier than M_1.
    (a) What is the net force on M_1? On M_2? Give answers in terms of M_1, M_3, g, T_1, and T_2.
    (b) What would be the net force on M_2 if it is held stationary by hand? What would then be T_1?
    (c) Now suppose that M_1 and only M_1 is held stationary by hand. What then would be T_1?
    (d) What is the acceleration of the three masses when they are free to move? Give answer in terms of the masses and g.

Phillippe Halsman

Gravitation

Chapter **4**

Gravitation

TABLE 4-1

Kind of force	Source of force	Relative strength (at small distances)	Range
1. Gravitational	Inertial mass	$\sim 10^{-38}$	Long
2. Weak	All elementary particles	$\sim 10^{-15}$	Short
3. Electromagnetic	Anything containing electric charge	$\sim 10^{-2}$	Long
4. Nuclear	Protons, neutrons, pions, the strange particles	1	Short

4-1 Newton's Universal Law of Gravitation

Source of the force

In Chapter 3 we studied general features of force without discussing just where forces come from. How are forces produced? According to our present knowledge there are only the four different kinds of forces listed in Table 4-1. Most everyday applications of physics deal with electromagnetic and gravitational forces. Only when we get to nuclear physics (Chapters 15 and 16) do we deal with the nuclear and the weak forces. We will start our formal study of the gravitational interaction in this chapter, and the electromagnetic interaction in Chapter 7.

According to Newton, the event that led him to his universal law of gravitation was an apple falling to the ground. This made Newton wonder whether the force that causes the apple to fall to the earth is the same force that causes the moon to "fall" toward the earth. Knowing the distance to the moon, and thus its velocity, Newton calculated v^2/R for the moon and found that it was about 1/3600 of the acceleration of the apple. He observed that this is the square of the ratio of the distances of apple and moon from the center of the earth. This led Newton to postulate that the gravitational force might vary inversely as the square of the distance.

Example 1

What is the acceleration due to the earth's gravity at the moon's position?

The moon is an earth satellite at a distance $R_m = 240,000$ mi $= 60\,R_0$ where R_0 is the radius of the earth. Any object, including the moon, at a distance of 240,000 mi from the earth will accelerate toward it with the same gravitational acceleration as the moon. According to Eq. 2-15, the acceleration of the moon toward the center of the earth is $a_m = \left(\dfrac{4\pi^2}{T_m^2}\right) R_m$, where T_m is the time for one complete revolution or 27.3 days. Now convert R_m to meters and T_m to seconds and we obtain the result $a_m = 2.73 \times 10^{-3}$ m/sec^2. Note that the acceleration of a low-flying earth satellite is 9.8 m/sec^2 which is about 3600 times larger. Newton was struck by the fact that 3600 is also the square of the ratio of the distances; that is, $g/a_m = R_m^2/R_0^2$. Hence, within the accuracy of our calculation, the acceleration due to gravity appears to decrease at a rate inversely proportional to the square of the distance from the center of the earth.

Since all masses fall toward the center of the earth with the same acceleration g, Newton also postulated that the gravitational force must be proportional to the mass. The final form of Newton's universal law of gravitation for the force between the two particles of mass M_1 and M_2 is

Newton's law of gravitation

$$F = G \frac{M_1 M_2}{r^2} \tag{4-1}$$

where r is the distance between the particles. The constant G has the value 6.67×10^{-11} newton m²/kg² in the MKS system and $G = 6.67 \times 10^{-8}$ dyne cm²/gm² in the CGS system. The value of G must be determined by experiment. Newton estimated G by guessing at the mass of the earth using its known volume and guessing its density. To show how G can be obtained in this way, we apply Eq. 4-1 to a mass M on the surface of the earth. Then

$$F_G = G \frac{M_e M}{R_e^2} \tag{4-2}$$

As mentioned in Chapter 2, we know that the gravitational force F_G on M will cause M to fall with an acceleration $a = g$. So in the equation

$$F_G = Ma$$

a has the value g. Making this substitution gives

$$F_G = M(g)$$

Now we substitute Eq. 4-2 for the gravitational force F_G obtaining

$$\left(G \frac{M_e M}{R_e^2} \right) = Mg$$

Now cancel out the mass M and solve for G:

$$G = \frac{gR_e^2}{M_e} \tag{4-3}$$

Q.1: If Newton's apple tree had been 240,000 mi high, with what acceleration would the apple leave the tree?

where M_e is the mass of the earth and R_e is the radius of the

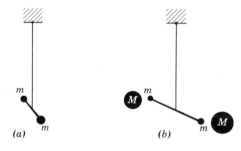

Fig. 4-1. Principle of the Cavendish experiment. Two small lead spheres of mass m are connected by a horizontal rod suspended by a thin quartz fiber in (a). In (b) two large lead spheres each of mass M are brought nearby. The gravitational attraction between m and M causes the small masses to move toward M and twist the quartz fiber. The force required to twist the quartz fiber this same amount is known from other previous measurements.

Ans. 1: The same as the moon's acceleration which is

$$a_m = \frac{g}{(60)^2}.$$

earth. We see that if M_e is known, then G is determined. Conversely, if G can be determined from a direct measurement of the force between two lead spheres, then M_e can be determined by solving Eq. 4-3 for M_e:

$$M_e = \frac{gR_e{}^2}{G}$$

The first direct determination of G was first made by Henry Cavendish in 1797 as illustrated in Fig. 4-1. We now understand why his experiment is called weighing the earth, even though the earth has nothing to do with the experiment itself. In fact, the Cavendish experiment would be easier to perform in the absence of the earth's gravity.

Example 1

What is the maximum gravitational force of attraction between two lead spheres, each of 45 kg mass and 20 cm diameter? Compare this force with the force of attraction to the earth.

According to Eq. 4-1 the force of attraction between the lead spheres is

$$F = G\left(\frac{M}{R}\right)^2$$

$$F = 6.67 \times 10^{-11}\left(\frac{45}{0.2}\right)^2 \text{ newton}$$

$$= 3.37 \times 10^{-6} \text{ newton}$$

The force is smaller than most frictional forces; hence it can only be detected under very special circumstances such as the Cavendish experiment. The attraction of the earth on one of the lead spheres is

$$F_G = Mg = 45 \times 9.8 \text{ newtons} = 440 \text{ newtons}$$

This is more than 100 million times larger than the attractive force between the two spheres.

We shall see in the next section that in addition to obtaining the mass of the earth from G, so could Cavendish obtain the mass of the sun, and the masses of all planets with observable satellites. All we need to know to get the mass of the sun is G, the distance from earth to sun, and the time it takes the earth to go once around the sun.

4-2 Kepler's Laws

Induction

Before Newton had postulated his universal law of gravitation, Johannes Kepler found that the "complicated" motions of the solar system could be described by three simple laws. Kepler's laws added strength to the hypothesis of Copernicus that the planets revolved around the sun rather than the earth.

Back in 1600 it was religious heresy to suggest that the planets revolve around the sun rather than the earth. In fact, in 1600 Giordano Bruno, an outspoken advocate of the Copernican heliocentric system and a religious heretic in general, was tried by the Inquisition and burned at the stake. Even the great Galileo himself was imprisoned, tried by the Inquisition, and made to renounce publicly his beliefs, in spite of the fact that the Pope was a good personal friend of his.

The dogma of the times holding sacred the teachings of Aristotle and Ptolemy taught that the orbits of the planets were described by complicated motions of circles within circles. A dozen or so circles of different sizes were needed to describe the orbit of Mars. Johannes Kepler's ambition was to "prove" that Mars and the earth must revolve around the sun. His approach was to find a simple geometrical orbit that would accurately fit all of the vast measurements of the position of Mars. After years of tedious work, he was able to discover three simple laws that very accurately agreed with the data on all the planets. Kepler's laws also apply to satellites revolving about a planet.

Kepler's first law

Each planet travels in an elliptical orbit with the sun at a focus of the ellipse.

An ellipse has several geometrical properties. One of these can be used to draw an ellipse using a string, pencil, and two pins as shown in Fig. 4-2. The two ends of the string are fastened at points A and B. Then the pencil at point P will trace out an ellipse. Points A and B are called the foci. An ellipse is often defined as a curve where the sum of the distances from any point on the curve to two fixed points (the

Fig. 4-2. Construction of an ellipse using two pins, string, and pencil. Points A and B are the foci of the ellipse.

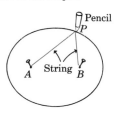

Q.2: What is G on the surface of the moon?

foci) is constant. Another geometrical property of an ellipse is that the lines AP and BP make equal angles with the tangent to the ellipse at point P. Thus any light or sound waves starting from A would be reflected to point B. This is the principle of the whispering gallery which may be found in some museums and science exhibitions. A whispering gallery has elliptically shaped walls. One person stands at point A and another, perhaps 50 ft away, at point B. If the two people are at the foci, they can whisper softly to each other without being heard by anyone else in the room.

Kepler's second law

The line joining the sun and a planet sweeps out equal areas in equal times.

In Fig. 4-3 the shaded areas are all equal. According to Kepler's second law, a planet must take the same amount of time to traverse distances s_1, s_2, and s_3. Thus, when the earth is closest to the sun (in early January) it is at its greatest velocity. According to Kepler's second law, winters in the Northern Hemisphere must be shorter than in the Southern Hemisphere where winters occur in July when the earth is farthest from the sun.

Kepler's third law

The cubes of the distances of any two planets from the sun are to each other as the squares of their periods:

$$\frac{R_1{}^3}{R_2{}^3} = \frac{T_1{}^2}{T_2{}^2} \tag{4-4}$$

where R_1 and T_1 are the distance and period of planet 1, and R_2 and T_2 are the distance and period of planet 2. Kepler specified that the distance R should be taken as one-half the major diameter of the ellipse.

The logical process by which Kepler obtained his three laws is called induction. We shall see in Section 4-3 how Newton obtained the same three laws by deduction. Newton was able to derive Kepler's laws from his universal law of gravitation. In this sense he explained why the planets move according to Kepler's laws.

Fig. 4-3. In the ellipse below, the shaded areas are equal. According to Kepler's second law, the distances s_1, s_2, and s_3 would all be covered by a planet in the same length of time.

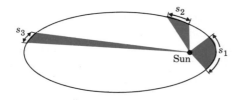

Ans. 2: The same as on the surface of the earth or anywhere else. It is a universal physical constant.

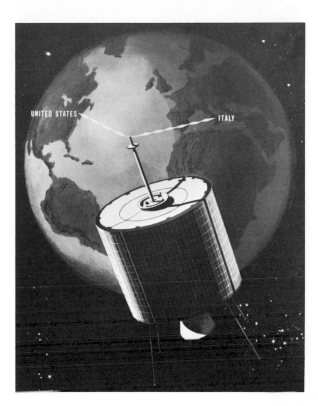

Fig. 4-4. Artist's conception of the "Early Bird" communications satellite in space. (Credit: Wide World Photos.)

Q.3: The net force acting on the moon is the sum of two forces, the centripetal force and the gravitational force. True or false?

Example 2

What must be the height of the Early Bird satellite? The Early Bird satellite hovers over the same point on the earth's equator indefinitely. This is because it has the same period as the period of rotation of the earth. We know that an earth satellite 4000 mi from the center of the earth, such as Sputnik I, has a period $T = 5000$ sec.

For this problem we will use Kepler's third law and let an earth satellite skimming the earth's surface be "planet" 1. Then $R_1 = 4000$ mi and $T_1 = 5 \times 10^3$ sec. We wish to find R_2 when $T_2 =$ one day, or 8.6×10^4 sec.

Solving Eq. 4-4 for R_2

$$R_2{}^3 = \frac{T_2{}^2}{T_1{}^2} R_1{}^3$$

$$R_2 = \sqrt[3]{\left(\frac{T_2}{T_1}\right)^2} R_1$$

$$R_2 = \sqrt[3]{\left(\frac{86}{5}\right)^2} \times 4000 \text{ mi}$$

$$R_2 = 26{,}600 \text{ mi}$$

We see that an earth satellite at a height of about 22,600 mi above the equator will hover over that same point forever if moving in the same direction as the earth's rotation.

Suppose the above problem had asked to find the height of a satellite having a period $T = 27.3$ days. The very same procedure would give the answer 240,000 mi. This happens to be the height of the world's "first" earth satellite. It is called the moon.

4-3 Derivation of Kepler's Laws

Deduction

About 300 years ago Newton was able to give complete derivations of Kepler's laws. Actually, the derivations given by Newton are above the mathematical level of this book. Let us try however to give simpler noncalculus derivations of Kepler's laws starting from Newton's universal law of gravitation. We will first consider a special case of Kepler's third law—that of planets moving in circles around the sun. Actually the orbits of all the planets except Pluto are very close to being circles with the sun at the common center. The orbits of the earth and Mars are accurately drawn to scale in Fig. 4-5. As can be best judged by the eye they appear to be circles, whereas the orbit of the first man-made planet appears more elliptical. In our derivation we will com-

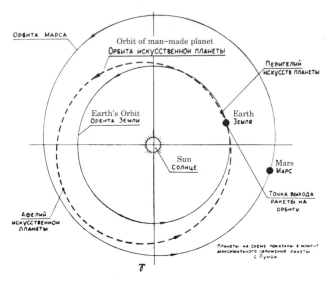

Fig. 4-5. Orbit of the first man-made planet compared to the orbits of Mars and Earth. All three orbits are ellipses although the orbits of Mars and Earth appear almost as circles. (Courtesy Sovfoto.)

Ans. 3: False. These two forces are one and the same.

pare the periods of planets 1 and 2 of Fig. 4-7. Our starting point, as is so often the case, is Newton's second law:

$$F_1 = M_1 a_1 \qquad (4\text{-}5)$$

Since planet 1 is moving uniformly in a circle, its only acceleration is centripetal acceleration. It will be to our advantage to use the formula (Eq. 2-16) for centripetal acceleration which contains the period T:

$$a_1 = \frac{4\pi^2}{T_1^2} R_1$$

Substituting this into Eq. 4-5 gives

$$F_1 = M_1 \left(\frac{4\pi^2}{T_1^2} R_1 \right) \qquad (4\text{-}6)$$

Now just where does this force F_1 come from? Gravity, of course—it is the gravitational interaction between the sun and planet 1, and this is given by Newton's law of gravitation, Eq. 4-1. So we can substitute

$$F_1 = G \frac{M_s M_1}{R_1^2}$$

into the left-hand side of Eq. 4-6, obtaining

$$G \frac{M_s M_1}{R_1^2} = M_1 \frac{4\pi^2}{T_1^2} R_1 \qquad (4\text{-}7)$$

Hence

$$\frac{4\pi^2}{T_1^2} R_1 = G \frac{M_s}{R_1^2}$$

or

$$\frac{R_1^3}{T_1^2} = \frac{GM_s}{4\pi^2} \qquad (4\text{-}8)$$

Note that this ratio is independent of the mass of planet 1. We can repeat this very same calculation for planet 2 at a distance R_2 from the sun. Likewise

$$\frac{R_2^3}{T_2^2} = \frac{GM_s}{4\pi^2} \qquad (4\text{-}9)$$

Equating the left-hand sides of Eq. 4-8 and Eq. 4-9 gives

$$\frac{R_1^3}{T_1^2} = \frac{R_2^3}{T_2^2} \qquad \text{or} \qquad \frac{R_1^3}{R_2^3} = \frac{T_1^2}{T_2^2}$$

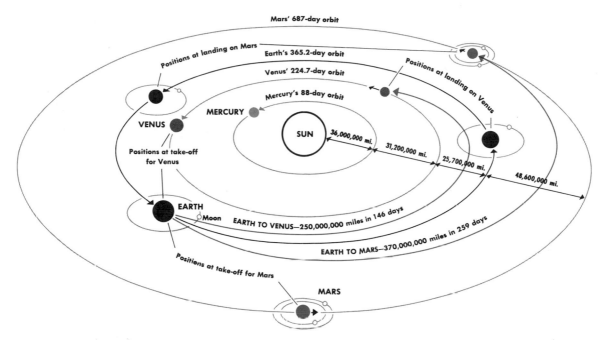

Fig. 4-6. Proposed orbits for space trips to Mars and Venus. Notice that the spaceship's path is always part of an ellipse with the sun at one focus. (After drawing by *Time*.)

which is Kepler's third law. The more general derivation for elliptical orbits is more complicated and will not be presented here.

Example 3

Mars is 52% farther from the sun than the earth. How long is a Martian year?

Since $\dfrac{R_{\text{Mars}}}{R_{\text{earth}}}$ is 1.52 we have

$$\frac{T^2_{\text{Mars}}}{T^2_{\text{earth}}} = (1.52)^3$$

$$T_{\text{Mars}} = \sqrt{3.5} \times T_{\text{earth}}$$

$$T_{\text{Mars}} = 1.87 \text{ years}$$

Before deriving Kepler's second law, we can squeeze more information out of Eq. 4-8. We can use it to obtain the mass of the sun. Solving Eq. 4-8 for M_s, we obtain

$$M_s = \frac{4\pi^2 R_1{}^3}{G T_1{}^2} \tag{4-10}$$

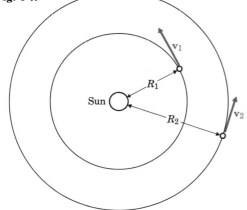

Fig. 4-7.

where R_1 and T_1 are the distance and period of any planet (the earth, for example). Now Cavendish and even prehistoric man knew what T_1 is for the earth—it is one year or 3.15×10^7 sec. We know from astronomical observations that R_1, the earth–sun distance, is 93 million miles or 1.50×10^{13} cm. Substituting these numbers into Eq. 4-10 gives $M_s = 2.0 \times 10^{33}$ gm.

The derivation of Eq. 4-10 would apply to any satellite revolving around any central object of mass M_s. Hence we can apply Eq. 4-10 to one of the moons of Jupiter with period T_1 and orbital radius R_1. Of course, we must replace M_s by M_j, the mass of Jupiter. Then we can write Eq. 4-10 as follows:

$$M_j = \frac{4\pi^2 R_1{}^3}{GT_1{}^2} \tag{4-11}$$

Let us take the moon of Jupiter named Io for our calculation. Galileo measured its period to be $T_1 = 1.77$ days or 1.53×10^5 sec. Its distance from Jupiter is measured to be $R_1 = 4.22 \times 10^{10}$ cm. If we substitute these values into Eq. 4-11, we obtain

$$M_j = 1.9 \times 10^{30} \text{ gm}$$

We begin to see the power of Newton's universal law of gravitation. So far we have calculated from it the masses of the earth, sun, and planets having satellites. Also we can obtain the distances of all the planets in terms of the earth–sun distance just by looking in the sky to see how long it takes a planet to go once around.

Example 4
What is the ratio of the velocities of the two planets in Fig. 4-7?
In this it is better to use the formula $a_1 = v_1{}^2/R_1$ for centripetal acceleration. Then Eq. 4-5 becomes

$$F_1 = M_1\left(\frac{v_1{}^2}{R_1}\right)$$

$$\left(G\frac{M_s M_1}{R_1{}^2}\right) = M_1\frac{v_1{}^2}{R_1}$$

or
$$v_1{}^2 R_1 = GM_s$$
Similarly $\quad v_2{}^2 R_2 = GM_s$

Equating the two left-hand sides gives

$$v_1^2 R_1 = v_2^2 R_2$$

or

$$\frac{v_1}{v_2} = \sqrt{\frac{R_2}{R_1}}$$

Next we shall present a simple derivation of Kepler's second law using plane geometry.

Kepler's second law, the law of equal areas, can be obtained with the aid of Fig. 4-8. Consider three close positions P_1, P_2, and P_3 on a planet's orbit. Let P_1, P_2, and P_3 each be 1 sec apart. Then the distance between P_1 and P_2 will be the planet's velocity v_1, and the distance P_2P_3 will be the velocity v_2 in the next second. According to Newton's first law of motion, the component of velocity perpendicular to the line SP_2 must be unchanged because the component of force in this direction is zero (the force is pointing along SP_2 toward the sun). Thus $(v_1)_\perp = (v_2)_\perp$. The area swept out by the planet in the first second is the triangle SP_1P_2. The area swept out in the next second is the triangle SP_2P_3. Since both these triangles have the same base SP_2 and equal altitudes $v_\perp$, they must have equal areas. Note that this derivation did not make use of the inverse square law of gravitation. All it depends on is that the interaction force is in the direction of the two interacting bodies. Thus Kepler's second law turns out to be more general than his other two laws. The law of equal areas must also hold for other kinds of forces that need not be inverse square forces.

Kepler's first law is the most difficult of the three to derive. The usual derivation is somewhat lengthy and involves higher mathematics (differential equations). A noncalculus derivation does exist, but is too lengthy and intricate to present here.

Before leaving this section let us discuss a trap that students tend to fall into. We know that the gravitational force on a planet is equal and opposite to the centrifugal force. Why then should not the net force be the vector sum of the two and hence be zero?

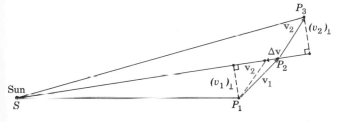

Fig. 4-8. Points P_1, P_2, and P_3 are successive positions of a planet 1 sec apart. The text shows that the area SP_1P_2 which is swept out in the first second must equal the area SP_2P_3 which is swept out in the second second.

Q.4: When an astronaut turns his rocket power off, the spacecraft stops accelerating. True or false?

Certainly, if the net force on a planet were zero, according to Newton's first law it would move in a straight line rather than a circle. As mentioned on page 40, the centrifugal force is a fictitious force which only exists in a rotating frame of reference. On the other hand, both the centripetal force and the gravitational force exist, but they are one and the same. According to Newton's second law $F_G = Ma_c$, where F_G is the force supplied by gravity and a_c is the acceleration (it happens to be centripetal in the case of a circle) caused by the force F_G.

4-4   Weight and Weightlessness

A person is as heavy as he feels

In physics-teaching circles, there is a controversy on how to define weight. In this book we will use the definition which makes sense physiologically. According to our definition, a man's weight would be what he would read if he would weigh himself on a spring balance such as bathroom scales. *We define the weight of an object as the force of the object against the floor.** The alternate definition used in some other books is that the weight of an object is the gravitational force on the object. But then an astronaut could never be weightless. In such cases the two definitions are incompatable.

Either way, weight is a force and is measured in newtons or dynes. In physics we do not express weight in grams because grams is mass and not force.† Let F_w be the weight of mass M. If mass M is resting on a floor attached to the surface of the earth as shown in Fig. 4-9, $\mathbf{F}_{\text{net}}$ must be zero (ignoring the small effect of the earth's rotation). $\mathbf{F}_{\text{net}}$ is the sum of $\mathbf{F}_G$ plus $\mathbf{F}'$, the reaction force of the earth.

---

*This definition is not completely general since it does not cover the case of a man in a swimming pool. In order to cover this special case, we must go completely physiological and say that the weight of a man is proportional to the force of the fluid on the nerve endings in the semicircular canals of the inner ear.

†The English system of units becomes particularly confusing at this point. There is an English unit of force called the pound defined such that 1 lb of force will give an acceleration of 32 ft/sec² to 1 lb of mass. The very same word, pound, is used for two quite different things! To avoid confusion, we shall never use an English unit of force in this book.

Ans. 4: False. With rocket power off the acceleration will be $a = \dfrac{F_G}{M}$, where F_G is the net gravitational force.

Fig. 4-9. Point of arrow indicates point of application of force. The solid red forces act on M. $\mathbf{F}_w$ pushes against the floor.

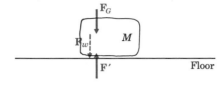

$$\mathbf{F}_{\text{net}} = \mathbf{F}_G + \mathbf{F}' \tag{4-12}$$

According to Newton's third law, the force of the floor on M is equal and opposite to the force of M on the floor (which we have defined as the weight F_w). That is,

$$\mathbf{F}' = -\mathbf{F}_w$$

Substituting into Eq. 4-12 we have

$$\mathbf{F}_{\text{net}} = \mathbf{F}_G - \mathbf{F}_w$$

or

$$\mathbf{F}_w = \mathbf{F}_G - \mathbf{F}_{\text{net}} \tag{4-13}$$

In our example of M at rest against the surface of the earth, F_{net} must be zero and we have

$$F_w = F_G$$

or

$$F_w = Mg$$

is the weight of mass M resting on the surface of the earth.

But suppose the floor is accelerating downward with an acceleration of magnitude a. This could happen if mass M were in an elevator. Then $F_{\text{net}} = Ma$ and the weight can be obtained by substituting into Eq. 4-13:

$$F_w = F_G - (Ma)$$

or

$$F_w = M(g - a) \quad \text{for downward acceleration}$$

We see that a person's weight is reduced whenever an elevator starts on a downward trip. Not only is this effect felt in the inner ear, but also in all other internal organs such as the stomach.

If the elevator is accelerating upward, we must reverse the sign on a. Then $F_w = M(g + a)$ for upward acceleration.

Suppose the elevator were freely falling; then $a = g$, and according to the equation $F_w = M(g - a)$, the weight would be zero. This is called the condition of weightlessness, and the passengers would look like the scientists in Fig. 4-10 or the lady in the chapter opening. All objects in the elevator would float freely in the air until the elevator hit ground. As we learned from Chapter 2, all earth satellites and projectiles are freely falling. Thus a rocket passenger must become

Fig. 4-11. Cartoonist version of elevator accelerating downward faster than g. (Reprinted by special permission from The Saturday Evening Post, Copyright 1959, by The Curtis Publishing Company.)

◀ **Fig. 4-10.** The airplane in this photograph has an acceleration $g = 32$ ft/sec² directed downward. The passengers are experiencing the "delights" of weightlessness. (Courtesy U.S. Air Force.)

weightless as soon as the rocket motors burn out or are turned off. Such a passenger would have difficulty drinking a glass of water because the water would float out of the glass, forming a blob or blobs of water floating in midair. Actually it is possible for ordinary airplanes to achieve the condition of weightlessness merely by following a parabolic path the same as a projectile having the same horizontal component of velocity. Some motor power must be used during such a maneuver in order to cancel the effects of air resistance. By this means passengers can be made weightless for more than 15 sec at a time. The experience of weightlessness is described in the following quotation by W. R. Young, a science editor of *Life*:

> Free floating at zero g seems to give simultaneous buoyancy to both body and spirit. I remember grinning ridiculously at others in the plane as I discovered the new world. It was hard to remember or to care which was "up" and which was "down." The third time the plane was put through the maneuver I was able to push off with my toes just vigorously enough to float a straight course the entire length of the compartment. When I had wafted to the other end, Major Brown snared me in midair and halted my forward flight, lest I sail on into the pilots' area. Although he had anchored himself to a piece of equipment with a firm handhold, he had to use some force to stop me, for weightlessness does not do away with the effects of momentum. Rebounding from his restraining hand, I caromed lightly off the floor and then off the ceiling.

If the downward acceleration of an elevator were greater than g, we would have the situation shown in Fig. 4-11. This figure is somewhat unrealistic; the passengers would be more comfortable standing on their feet (against the ceiling) than on their heads.

Example 5

Assume that for comfortable air travel passengers should never feel more than twice their normal weight ($\mathbf{F}_w = 2\,Mg$). What is the maximum horizontal acceleration permitted by this condition?

Let a be the maximum horizontal acceleration. The force of the seat on the passenger is always ($-\mathbf{F}_w$). Then

$$\mathbf{F}_{\text{net}} = \mathbf{F}_G - \mathbf{F}_w = M\mathbf{a}$$

or

$$\mathbf{F}_G = M\mathbf{a} + \mathbf{F}_w$$

where magnitudes of $\mathbf{F}_w$ and $\mathbf{F}_G$ are $2\,Mg$ and Mg, respectively. The direction of $\mathbf{F}_G$ is down and the direction of $M\mathbf{a}$ is horizontal.

Fig. 4-12. Airplane with a horizontal acceleration such that the passenger's weights are doubled.

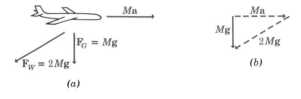

(a)

The two vectors $M\mathbf{a}$ and $\mathbf{F}_w$ are added together in Fig. 4-12 to give $\mathbf{F}_G$. By the Pythagorean theorem

$$Ma = \sqrt{(2\,Mg)^2 - (Mg)^2}$$
$$a = \sqrt{3}\,g$$

Example 6

Consider a roller-coaster shown in Fig. 4-13 which has a peak velocity $v = \sqrt{4gR}$, where R is the radius of curvature of the dip. What is the weight of the passengers at the bottom of the dip?

We can get the weight by calculating F_{net} and then substituting into Eq. 4-13. We can get F_{net} by using Newton's second law, $F_{net} = Ma$. Is the acceleration a the sum of g and the centripetal acceleration or is it just the centripetal acceleration alone? Even though gravity may be influencing the motion, acceleration is defined as the rate of change of velocity and we have shown in Fig. 2-14 that this is v^2/R. Substituting the value $\sqrt{4gR}$ for v into the equation $a = v^2/R$ gives

$$a = \frac{(\sqrt{4gR})^2}{R} = 4g$$

Hence $F_{net} = 4Mg$ pointing up.

The vector $-\mathbf{F}_{net}$ is $4Mg$ pointing down. To this must be added $\mathbf{F}_G$ which has the value Mg pointing down. According to Eq. 4-13 the weight is then $Mg + 4Mg$ or $5Mg$. The passengers would feel five times their normal weight.

4-5 Gravitational Mass

Who is accelerating?

Throughout this book we use the terms mass and inertial mass interchangeably. We saw in Chapter 3 that we could obtain the mass (or inertial mass) of an object by applying a force to it and measuring the corresponding acceleration. The ratio F_{net}/a will be the inertial mass of the object. Now there is no *a priori* reason why the gravitational force on an object must be proportional to the inertial mass. For all we know, it could just as well be proportional to the number of neutrons in the object. Actually the number of neutrons per atom is approximately proportional to the mass of the atom. However, hydrogen, whether in gaseous or liquid state, contains no neutrons, and then gravity would not act on hydrogen. Liquid hydrogen would be weightless. But when we do the experiment and measure the gravitational force on hydrogen (or anything else) we find that within the accuracy

Fig. 4-13. Roller coaster for Example 6.

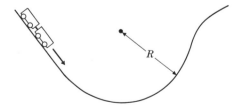

of our measurements, the gravitational force is proportional to the inertial mass of the object. This is not true for the electrostatic force, however. There $F = Q_1 Q_2 / r^2$ is the force between two objects 1 and 2, where Q_1 and Q_2 are called the electric charges of objects 1 and 2. We find that objects of large mass can have low electric charge and vice versa.

Now Newton, to be more general, could have written his universal law of gravitation as $F = G \left(\dfrac{Q_1 Q_2}{r^2} \right)$ where Q_1 and Q_2 are called the gravitational charges of objects 1 and 2 (not to be confused with electric charges). It then depends upon experimental observation to see whether Q is always proportional to M no matter what the object. The point we are trying to make is that there is no a priori reason why gravitational charge (also called gravitational mass) should be proportional to inertial mass. Furthermore, if gravitational mass and inertial mass turn out to be exactly the same, we have uncovered a new fundamental law of physics, just as fundamental as the conservation of momentum, or Newton's laws. The ratio of gravitational to inertial mass has recently been measured with great accuracy for various different elements and found to be the same for all substances within an experimental accuracy of about one part in 10^{10}. For these reasons it is assumed that gravitational mass and inertial mass are one and the same. Einstein's general theory of relativity uses this basic postulate as a starting point. This equivalence of gravitational and inertial mass is called the principle of equivalence and is discussed in more detail in Chapter 11.

Example 7

Show how an accurate measurement of the period of a simple pendulum depends on the ratio of gravitational to inertial mass.

Let Q_e be the gravitational mass of the earth and Q_A the gravitational mass of a pendulum bob made of substance A. Then the gravitational force on the bob is

$$F_G = G \frac{Q_e Q_A}{R_e{}^2}$$

The quantity $G \left(\dfrac{Q_e}{R_e{}^2} \right)$ will be a physical constant depending on the size of the earth. Let g_0 stand for the size of this constant:

$$g_0 = G \frac{Q_e}{R_e{}^2}$$

Q.5: A nonaccelerating person must be weightless. True or false?

Then
$$F_G = (g_0)Q_A$$

According to Eq. 3-9 in the derivation of the period of a simple pendulum, we had

$$\frac{F_{\text{net}}}{F_G} = \frac{x}{L}$$

or

$$\frac{(M_A a)}{(g_0 Q_A)} = \frac{x}{L}$$

then

$$\frac{a}{x} = \frac{Q_A}{M_A}\frac{g_0}{L}$$

In simple harmonic motion, this constant must be $4\pi^2/T^2$ (see p. 66). Then

$$\left(\frac{4\pi^2}{T^2}\right) = \frac{Q_A}{M_A}\frac{g_0}{L}$$

or

$$T = 2\pi \sqrt{\frac{L}{g_0}} \times \sqrt{\frac{M_A}{Q_A}}$$

If the ratio M_A/Q_A varies with different substances, then changing the material of the bob should change the period of oscillation.

According to the principle of equivalence, a laboratory undergoing acceleration is mathematically equivalent to a stationary laboratory under the influence of an equivalent gravitational force. One consequence of the principle of equivalence is that it is impossible to tell just what is the net gravitational force on the solar system. For all we know there may be a large, distant mass (so far away it cannot be seen) which exerts a gravitational attraction on our local region of the universe. According to the principle of equivalence, all local objects would "fall" in a condition of weightlessness toward the large, distant mass with the same acceleration and we would observe no local effect at all. We could not "feel" whether or not we were falling, and we could not observe any relative acceleration with respect to our neighbors. We would think our local system was an inertial frame of reference whereas, from the point of view of the distant mass, we would be accelerating. In this sense, the old definition of weight as the net gravitational force on an object would be meaningless. This is because in order to know the

Ans. 5: False. A person sitting on a stationary planet would be nonaccelerating, but would still have weight equal to the gravitational force of the planet on him.

exact net gravitational force on an object, we must take into account all the distance matter in the universe, even though it cannot be seen. Further implications of the principle of equivalence will be discussed in Chapter 11. The main point we want to make for now is that it looks like we cannot fully understand $F = Ma$ or gravitation without taking into account the influence of the distant matter in the universe.

Problems

1. Could the Cavendish experiment of weighing the earth be performed on Mars? Would it give the same result?

2. In Kepler's third law can we compare the period of the moon to the period of the earth? Can we compare the period of the moon to the period of a moon of Jupiter? Can we compare the periods of all of Jupiter's moons with each other?

3. What are the units of G in meters, kilograms, and seconds?

4. An elevator starts up from rest at the basement with an initial acceleration of 16 ft/sec².

(a) The passengers' apparent weight will be (increased, decreased, remain the same).

(b) The period of oscillation of a pendulum in such an elevator would (increase, decrease, remain the same).

(c) When the elevator reaches a velocity of 32 ft/sec, it maintains that constant upward velocity. Under this condition a passenger's apparent weight will be (increased, decreased, remain the same) as compared to his weight when at rest.

5. If the earth had half its present diameter, its mass would be one-eighth of its present mass. What would be the value of "g" on this "half-sized" earth?

6. If the moon had twice its present mass, but moved in its present orbit, what would be its period of revolution?

7. The mass of the moon is 0.012 that of the earth. Its diameter is one-fourth that of the earth. What is the value of "g" on the surface of the moon?

8. An elevator starts up with an acceleration of 16 ft/sec². What is the apparent weight of a 60-kg man during this acceleration? The elevator then attains a uniform upward velocity of 32 ft/sec. Now what is the apparent weight of the man? What would be his apparent weight if the elevator cable broke?

9. A man whose weight is 600 newtons gets into an elevator at the fiftieth floor of a 100-story building and steps onto some scales. When the elevator begins to move, he sees that the scales read 720 newtons for 5 sec, then 600 newtons for 20 sec, then 480 newtons for 5 sec, after which the elevator is at rest at one end of its track.

(a) Is he at the top or bottom of the building?

(b) How high is the building? (This same sort of method can be used to tell an astronaut how far his space capsule has traveled.)

10. What is the weight of a 60-kg man in the rocketship of Problem 19, Chapter 3?

11. A 50-kg man starts sliding down a rope. His downward acceleration is one-seventh of g.

(a) What is the apparent weight of this man?

(b) What is the tension in the rope above the man?

12. The newspapers claimed that the dog contained in Sputnik II was weightless. However, this satellite was observed to be tumbling. Assume the tumbling rate was one revolution every 10 sec and that the dog was 245 cm from the axis of rotation. What fraction of its normal weight would the dog then have?

13. A rocket containing a 0.1-kg mouse is fired vertically. When the rocket reaches a height of 50 km and velocity of 980 m/sec, the fuel runs out. Then what is the apparent weight of the mouse?

Prob. 14

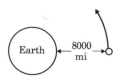

14. A new earth satellite is observed orbiting at a distance of 2 earth radii (8000 mi) above the surface of the earth.

(a) If the period of revolution of a low-flying earth satellite is 90 min, what is the period of this new satellite?

(b) What is the centripetal acceleration of this new satellite in terms of g?

(c) What is the centripetal acceleration of the moon in terms of g? (The moon is at a distance of 60 earth radii.)

15. The centers of two identical spheres are 1 m apart. What must be the mass of each sphere if the gravitational force of attraction between them is 1 newton?

16. At some point between the earth and the moon, the gravitational force on a spaceship due to the earth and moon together is zero. Where is this point? The earth–moon distance is 240,000 mi and the moon has 1.2% the mass of the earth. The passengers in Jules Verne's *From the Earth to the Moon* experienced weightlessness only when passing this point. Explain why this is incorrect.

Prob. 17

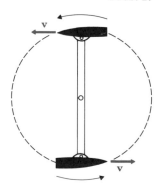

17. In the problem we shall design an amusement park ride capable of producing weightlessness for short periods of time. Two "rocket" cars are separated by a 20 m arm and rotated with velocity v in a vertical circle as shown. What must be the value for v for the passengers to be weightless when their car reaches the top? What would be their weight when their car reaches the bottom?

18. What is the value of "g" 100 mi above the earth's surface?

19. Knowing the value of G and that the distance to the sun is 93 million miles, calculate the mass of the sun.

20. If you know the distance from Jupiter to one of its moons, and the period of that moon, show how you could calculate the mass of Jupiter.

21. What is the velocity v of an earth satellite in a circular orbit of height h above the earth's surface? Express v in terms of R (the earth's radius), h, and g (the gravitational acceleration at the earth's surface). Does the velocity increase or decrease as the satellite is being "slowed down" by air resistance?

Prob. 22

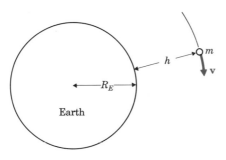

22. A satellite is in circular orbit at a height h above the earth's surface.

    (a) What is the gravitational force in terms of h and other appropriate quantities?

    (b) What is the centripetal force?

    (c) What is the net force?

    (d) What is the velocity v in terms of G, M_E, R_E, and h?

    (e) As h decreases, does v increase or decrease?

23. If the moon's orbital velocity were doubled in such a way to keep it in a circular orbit, what would be the radius of its new orbit? What would be its new period of revolution?

24. At the position of the moon, what is the gravitational force of the earth on a 1 kg mass? Also compute the gravitational force of the sun on a 1 kg mass at the position of the moon. The moon is 93 million miles from the sun and 240,000 miles from the earth. This problem can be easily worked without knowing G or the mass of the earth or the mass of the sun.

Angular momentum and energy

Chapter 5

Angular momentum and energy

There are several fundamental laws of nature that can be expressed mathematically as conservation laws. A conservation law states that in a closed system some physical quantity (such as the total momentum or energy) remains conserved (unchanged) forever. By a closed system we mean a system of particles under no outside influences whatsoever. There must be no external forces. This does not imply any restrictions on internal forces, however. The particles are permitted to interact with each other as they please. The conservation laws studied in this chapter are all believed to be exact, no violations having ever been observed. One of these conservation laws, the conservation of momentum, has already been presented in Chapter 3. We shall next study a closely related law, the conservation of angular momentum.

5-1 Conservation of Angular Momentum

Persistent rotation

At the end of Chapter 4 it was noted that the derivation of the law of equal areas did not depend on Newton's universal law of gravitation at all. Actually the law of equal areas is a special case of the more general law of conservation of angular momentum. In Fig. 5-1 consider any two triangles swept out by a planet in 1 sec. The areas of these two triangles are $\frac{1}{2}R_1(v_1)_\perp$ and $\frac{1}{2}R_2(v_2)_\perp$, where $v_\perp$ is the component of v perpendicular to R. Thus

$$R_1 \times M(v_1)_\perp = R_2 \times M(v_2)_\perp$$

The quantity $M(v_1)_\perp$ is $(P_1)_\perp$, the component of momentum perpendicular to the line R_1. Thus

$$R_1 \times (P_1)_\perp = R_2 \times (P_2)_\perp \tag{5-1}$$

The quantity $R \times P_\perp$ is defined as the angular momentum. Equation 5-1 says that the quantity $R \times P_\perp$ stays constant for every position of a planet on its orbit. In other words, the total angular momentum of the solar system is conserved (it never decreases or increases). This is true whether the distance R is measured to the sun or to any other non-

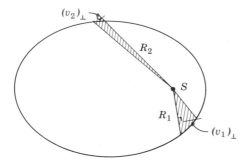

Fig. 5-1. Shaded "triangles" are areas swept out by planet in 1 sec at two different positions of the planet.

Fig. 5-2. (*a*) The "three dumbbells" demonstration. (*b*) Student is handed a spinning wheel and then stops it. (*c*) Student starts up a wheel and then turns it over.

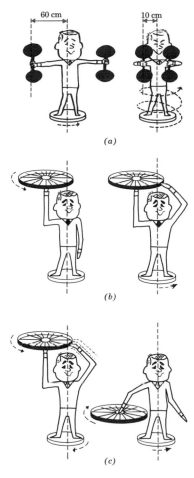

60 cm   10 cm

(*a*)

(*b*)

(*c*)

accelerating point in space. The law of conservation of angular momentum states that the total angular momentum of any closed system must remain constant forever. This law holds no matter what the nature of the interactions between the particles. It has been well tested and no violations have ever been found.

Example 1

A lightweight student holds two heavy dumbbells with outstretched arms while standing on a turntable. He is given a push until he is rotating at a rate of one revolution per second. Then the student pulls the dumbbells in toward his chest. What happens? Assume the dumbbells are originally 60 cm from his axis of rotation and are pulled in to 10 cm from the axis of rotation. Ignore the angular momentum of the student compared to that of the dumbbells.

The initial velocity of the dumbbells is $v_1 = 2\pi R_1/T_1$, where $R_1 = 60$ cm and $T_1 = 1$ sec. The initial total angular momentum is

$$\text{ang. mom.} = 2R_1Mv_1 = 4\pi M \frac{R_1^{2}}{T_1}$$

where M is the mass of each dumbbell. After the dumbbells are brought to a distance $R_2 = 10$ cm, the total angular momentum is

$$\text{ang. mom.} = 2R_2Mv_2 = 4\pi M \frac{R_2^{2}}{T_2}$$

According to the law of conservation of angular momentum, these two expressions must be equal; so

$$\frac{R_1^{2}}{T_1} = \frac{R_2^{2}}{T_2}$$

$$T_2 = \frac{R_2^{2}}{R_1^{2}} T_1$$

$$T_2 = \frac{100}{3600} \times 1 \text{ sec}$$

$$T_2 = \frac{1}{36} \text{ sec}$$

We see that the student should end up spinning at a rate of 36 revolutions per second. Some physics teachers call this the three dumbbells demonstration.

Another classic demonstration involving a student on a turntable is the spinning bicycle wheel (the tire is loaded with lead). The nonrotating student is handed a spinning

Q.1: Does $R_1 \times (P_1)_\perp$ of planet 1 equal $R_2 \times (P_2)_\perp$ of planet 2?

bicycle wheel and asked to hold it over his head and then to grasp the rim and stop the wheel as shown in Fig. 5-2b. Since there is no way to transfer the angular momentum to the earth, the student will then start spinning with the same angular momentum as that originally held by the wheel.

Another form of this demonstration has a stationary student starting to spin the wheel while holding it over his head. He, of course, will start spinning in the opposite direction with equal and opposite angular momentum. Now the student is asked to turn the wheel over. Upon doing so the student will abruptly stop rotating and start rotating in the direction opposite to his original rotation so that his angular momentum is always equal and opposite to that of the wheel (the net angular momentum must remain equal to its original value which was zero). Now if the student grasps the rim and stops the wheel, he will return to his original stationary situation.

5-2 Center-of-Mass

A weighted average

Figure 5-3 shows a closed system in motion. The closed system is a freely moving wrench that has a net external force of zero acting on it. Note that the angular momentum of the wrench is constant; the wrench rotates at a uniform rate about a point marked with a black "x." We shall now show that this special point which has no rotational motion in the absence of external forces is the center-of-mass. The x-component of the center-of-mass of a system of N particles is defined to be

$$x_c = \frac{M_1 x_1 + M_2 x_2 + \cdots + M_N x_N}{M} \tag{5-2}$$

where M is the total mass of all the particles, and x_1 is the x-component of $\mathbf{R}_1$, the distance to the first particle. The x-component of the velocity of the center-of-mass is obtained by dividing both sides of the above equation by t. Then

$$\frac{x_c}{t} = \frac{M_1 \dfrac{x_1}{t} + M_2 \dfrac{x_2}{t} + \cdots + M_N \dfrac{x_N}{t}}{M}$$

Ans. 1: No. The angular momentum of the *same* planet stays the same. Two different planets will have different angular momenta.

Fig. 5-3. A freely moving wrench. The net external force on this wrench is zero. Note that it rotates uniformly about its center-of-mass marked with black tape. (Courtesy Physical Science Study Committee.)

$$(v_c)_x = \frac{M_1(v_1)_x + M_2(v_2)_x + \cdots + M_N(v_N)_x}{M}$$

$$= \frac{(P_1)_x + (P_2)_x + \cdots + (P_N)_x}{M}$$

$$= \frac{P_x}{M}$$

where P_x is the x-component of the total linear momentum of the system. According to the conservation of momentum, P_x, P_y, and P_z must remain constant if there are no external forces. Thus, whether or not the system is rotating, all three components of the velocity of the center-of-mass must be constant, and thus the center-of-mass continues to move in a straight line in the absence of external forces. If a rigid body, such as a wrench, is freely moving and rotating in the absence of external forces, we see that the definition of center-of-mass requires that the center-of-mass have no acceleration or rotational motion. This is why rigid bodies and systems of particles always rotate about their centers-of-mass. The law of conservation of angular momentum requires that the earth rotate about its center-of-mass forever at the same rate, ignoring the tidal forces on the earth which are external forces coming from the moon and the sun.

Example 2

The mass of the moon is 1.2% of the earth's mass. If the distance from the earth to the moon is 240,000 mi, where is the center-of-mass of the earth–moon system?

Let us measure x from the center of the earth out toward the moon. Equation 5.2 then gives

Q.2: What happens to a stationary student on a turntable who is handed a rotating wheel, and then turns the wheel upside down?

$$x_c = \frac{M_e x_e + M_m x_m}{M_e + M_m}$$

where $x_e = 0$ is the position of the earth and $x_m = 240{,}000$ mi is the position of the moon.

$$x_c = \frac{M_m}{M_e + M_m} x_m$$

$$x_c = \frac{0.012}{1.012} \times 240{,}000 \text{ mi}$$

$$x_c = 2850 \text{ mi from the center of the earth}$$

Since both the earth and moon must revolve about this point once each month, we see that the earth actually revolves about a point within itself once each month.

5-3 Statics*

How to keep from rotating

We have seen that the total momentum and angular momentum of a system will remain unchanged in the absence of any external force. Now we ask the question: Is it ever possible to apply external forces and still have momentum and angular momentum conserved? The answer is yes. Recall the example on page 60 of the wooden block being pushed against the wall. Then there were two forces on the block and their sum was zero. It is clear from Newton's laws that if the vector sum of all the external forces is zero, then the momentum of the system will remain unchanged. If the initial momentum of a rigid body is zero, it will remain zero if the vector sum of all the external forces is zero.

But what about rotational motion? We know we can spin a coin by applying two equal and opposite forces $\mathbf{F}_1$ and $\mathbf{F}_2$, as shown in Fig. 5-4. Even though the vector sum of these external forces is zero, the angular momentum of the coin is not conserved. Angular momentum is distance times perpendicular component of momentum. Since rate of change of momentum is force, it can be shown that the rate of change of angular momentum is distance times the perpendicular component of the rate of change of momentum or force. The quantity $R \times F_\perp$ (distance times perpendicular component

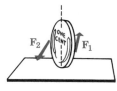

Fig. 5-4. The coin starts moving (spinning) even though the vector sum of $\mathbf{F}_1$ and $\mathbf{F}_2$ is zero.

Ans. 2: The student will spin on the turntable with *twice* the angular momentum of the wheel in the original direction.

*This section may be omitted if desired. None of the subsequent material depends upon it.

of force) is given the name torque. The rate of change of angular momentum equals the torque. Torques that would tend to give counterclockwise rotation are defined to be positive in sign, whereas torques that would tend to give clockwise rotation are negative. The condition required for the angular momentum of a rigid body to remain fixed is that the sum of the torques must be zero. A rigid body is said to be in static equilibrium if the forces on it are such that it will remain at rest. We can see that there are two conditions to be satisfied for static equilibrium.[*]

Condition I: The sum of the torques must be zero.
Condition II: The vector sum of the forces must be zero.

In the case of three external forces applied to positions R_1, R_2, and R_3 on a rigid body, these conditions give the equations

$$R_1 \times (F_1)_\perp + R_2 \times (F_2)_\perp + R_3 \times (F_3)_\perp = 0$$
$$\mathbf{F}_1 + \mathbf{F}_2 + \mathbf{F}_3 = 0$$

The distance R_1, R_2, and R_3 can be measured from any arbitrary point.

Example 3

As shown in Fig. 5-5 a 70-kg father and a 50-kg mother are at the ends of a 6 m seesaw. Where should their 25-kg child sit so that the seesaw can be balanced at the center?

Let F_1, F_2, and F_3 be the weights of the father, mother, and child, respectively. Let us measure the torques about the pivot. The father exerts a counterclockwise torque of R_1F_1, where $R_1 = 3$ m. The mother and child exert clockwise or negative torques of $-R_2F_2$ and $-R_3F_3$, where $R_2 = 3$ m and R_3 is to be determined. Condition I gives

$$R_1F_1 - R_2F_2 - R_3F_3 = 0$$

Solving for R_3

$$R_3 = \frac{R_1M_1g - R_2M_2g}{M_3g}$$

$$R_3 = \frac{3 \times 70 - 3 \times 50}{25} \text{ m} = 2.4 \text{ m}$$

Fig. 5-5. Family on a seesaw.

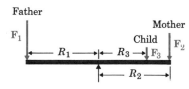

Q.3: If all forces are in the same plane, how many useful simultaneous equations are given by Conditions I and II?

[*]As outlined above, Condition I is a direct consequence of Newton's laws. It is not an independent law of nature as implied in some introductory physics textbooks.

Fig. 5-6. Forces of man, wall, and ground on a ladder.

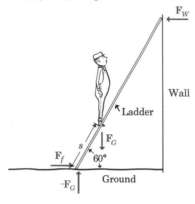

Example 4

A ladder of length $L = 4$ m is leaning against a frictionless wall as shown in Fig. 5-6. The maximum frictional force the ground can supply before the ladder slips is $F_f = 200$ newtons. How far along the ladder can a man weighing 600 newtons climb before the ladder slips? Ignore the weight of the ladder compared to the man.

Let the distance s denote the position of the man on the ladder. Let us measure torques from the point where the ladder rests on the ground. Then there are two torques: a positive torque due to the force F_w of the wall on the ladder, and a negative torque due to the weight F_G of the man. The components of F_w and F_G perpendicular to the ladder are $\frac{\sqrt{3}}{2} F_w$ and $\frac{F_G}{2}$, respectively. According to Condition I the sum of these two torques is

$$L \times \left(\frac{\sqrt{3}}{2} F_w \right) - s \times \frac{F_G}{2} = 0$$

$$s = \frac{\sqrt{3} F_w}{F_G} L$$

According to Condition II, the sum of all the horizontal forces on the ladder must be zero, which tells us that $F_f = F_w$. Then

$$s = \frac{\sqrt{3} F_f}{F_G} L = \sqrt{3} \times \frac{200}{600} \times 4 \text{ m} = 2.30 \text{ m}$$

In problems on statics there may be two or three unknown forces or distances to be determined. Conditions I and II can yield three simultaneous equations which are sufficient for solving up to three unknown quantities. The solution to such a problem is usually simplified by measuring the torques from a point at which one or more of the forces are applied.

5-4   Energy

Joules and ergs

In the MKS system energy is defined quantitatively as follows: a body is given one unit of energy if a net force of 1 newton is exerted on it for a distance of 1 m in the direction of the force. The units of energy would then be kg $\times$ m²/sec². However, as with force, the unit of energy is given a special name, the joule. In the CGS system the unit of energy is called the erg; a net force of 1 dyne exerted on a body over a

Ans. 3: Three. Condition II gives two equations—one for each component of force.

distance of 1 cm is 1 erg. The conversion is

$$1 \text{ joule} = 1 \text{ newton} \times 1 \text{ m}$$
$$= 10^5 \text{ dynes} \times 10^2 \text{ cm}$$
$$= 10^7 \text{ ergs}$$

Kinetic energy

Consider the ideal situation of a mass M floating freely in outer space. Then an external force F_{ext} is suddenly applied to M. According to Newton's second law the mass will suddenly start accelerating with acceleration $a = F_{\text{ext}}/M$. According to Eq. 2-11 it will achieve a velocity v after moving a distance s where

$$v^2 = 2as$$

or

$$v^2 = 2\left(\frac{F_{\text{ext}}}{M}\right)s$$

or

$$\tfrac{1}{2}Mv^2 = F_{\text{ext}} \cdot s$$

We see that the quantity $\tfrac{1}{2}mv^2$ is numerically equal to the energy given to M. *One-half the mass of a body times its velocity squared is defined as the kinetic energy (KE) of a body.*

definition of kinetic energy $\qquad$ $\text{KE} = \tfrac{1}{2}Mv^2$

If the mass had initial velocity v_0, then according to Eq. 2-11

$$v^2 - v_0{}^2 = 2as = 2\left(\frac{F_{\text{ext}}}{M}\right)s$$

or

$$\tfrac{1}{2}Mv^2 - \tfrac{1}{2}Mv_0{}^2 = F_{\text{ext}} \cdot s$$

In this case we see that whatever energy $F_{\text{ext}} \cdot s$ is given to M, it must show up as an increase in the kinetic energy of M. The energy given to a mass M by means of an external force is defined as the work done on M. "Work" is just another word for the energy given to an object by an external agent or force. We have just seen in the ideal case of a free particle that the work done on the particle shows up as kinetic energy of the particle.

Q.4: Can kinetic energy be negative?

5-5 Potential Energy

Energy potentially available

We shall see that potential energy means just what it says. First, let us consider the case of a mass M on the surface of the earth. Now an external force $F_{ext} = -F_G$ is used to lift slowly the mass to a height h. The work done will be

$$W = F_{ext} \cdot h$$

or

$$W = Mgh$$

In this instance the increase in kinetic energy will be zero. Where did the energy Mgh go? The answer is that an amount of energy Mgh is potentially available to be extracted in the form of kinetic energy. All we need do to extract it is to let go of the mass. Then M will drop a distance h, reaching a velocity given by the formula $v^2 = 2gh$. Let us calculate how much kinetic energy this will be.

$$KE = \tfrac{1}{2}mv^2$$
$$= \tfrac{1}{2}m(2gh)$$
$$= Mgh$$

We see that the entire work done ($W = Mgh$) can be converted back into kinetic energy at will.

Definition: *The energy potentially obtainable by virtue of the position of M is defined as the potential energy U.* An alternate form of this is that potential energy is the work that must be done on M to move it against a conservative force; by conservative force we mean a force that depends on the position of M. In the above example $U = Mgh$. Whenever M is at height h, the potential energy U can be converted into KE by letting M drop back down the distance h. Potential energy means just what it says—it is energy that is potentially available.

Note that the force required to move slowly M horizontally is zero (neglecting friction) and hence no work is done in moving M horizontally. If the mass is lifted to a height h by a roundabout path, the work done by the vertical component of force is Mgh and that done by the horizontal component is zero. Hence the work done in moving M a height h is Mgh independent of the path taken.

Ans. 4: No. Mass is always positive, and v^2 must be positive.

Potential energy occurs whenever a mass M is under the influence of a force F_c that depends only on the position of the mass. Such a force is called a conservative force. The force of air resistance which depends on velocity would be a nonconservative force. The most general definition of potential energy is given by the equation

definition of potential energy

$$\Delta U = -F_c \cdot \Delta s \cdot \cos \theta \tag{5-3}$$

where θ is the angle between $\mathbf{F}_c$ and $\mathbf{\Delta s}$, the displacement. If the object is free to move starting from rest, the force $\mathbf{F}_c$ will move it a distance $\mathbf{\Delta s}$ in the same direction as $\mathbf{F}_c$ and according to the minus sign in Eq. 5-3 its potential energy will decrease. As the force speeds it up, its kinetic energy will correspondingly increase. In the next section we shall show that the increase in kinetic energy must be exactly equal to the decrease in potential energy.

5-6 Conservation of Energy

Decrease equals the increase

Consider a closed system of two particles of mass m and M. In general, there will be a force F_c between them. Actually, all objects (even light) are made up of elementary particles. Any two elementary particles have either an electric or a gravitational force between them. (If they are closer than 10^{-12} cm they may also have an additional short-range force because of the strong and/or the weak interaction.) Let us now calculate the increase in kinetic energy of our two freely moving objects. First consider mass m and for simplicity assume it moving in the same direction as the force from M acting on it. The acceleration of mass m as it freely moves a distance Δs is

$$a = \frac{F_c}{m}$$

Using Equation 2-11 which says that $a = \dfrac{v^2 - v_0{}^2}{2\,\Delta s}$,

we obtain

$$\left(\frac{v^2 - v_0{}^2}{2\,\Delta s} \right) = \frac{F_c}{m}$$

Q.5: If two masses are pushed closer together, will their gravitational potential energy increase?

Conservation of mechanical energy

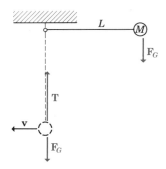

Fig. 5-7. A pendulum bob released from the horizontal position.

Ans. 5: No. In this case $F_c = F_G$, the gravitational force, and Δs is in the direction of F_c. Hence the product $F_c \cdot \Delta s \cdot \cos \theta$ is positive in Eq. 5-3.

$$\tfrac{1}{2}mv^2 - \tfrac{1}{2}mv_0{}^2 = F_c \cdot \Delta s$$

But according to Eq. 5-3 the right-hand side is $-\Delta U$; hence

$$\tfrac{1}{2}mv^2 - \tfrac{1}{2}mv_0{}^2 = -(U - U_0)$$

or

$$\tfrac{1}{2}mv^2 + U = \tfrac{1}{2}mv_0{}^2 + U_0 \qquad (5\text{-}4)$$

We see that the increase in kinetic energy is exactly equal to the decrease in potential energy. Another way of stating this is that the sum of kinetic energy plus potential energy stays the same, no matter where the particle goes. The same argument also holds for the second particle M. If there are three or more particles, a similar argument shows that the sum of the total kinetic energy plus the total potential energy of the system remains constant as long as there are no external forces.

This in words is what is called the law of conservation of energy. It is an extremely useful relation. No matter how complicated the motions of the particles, if at any given time we observe the positions of all the particles, we then can deduce their total kinetic energy because the potential energy depends only on the positions. The following is an example of how to determine the kinetic energy or velocity of an object without knowing its equation of motion.

Example 5

A pendulum bob of mass m is held out in a horizontal position and then released. (This is not simple harmonic motion because the approximation of small displacement does not apply in this case.) If the string is of length L, what is the bob velocity and what is the force on the string when the bob reaches its lowest position? See Fig. 5-7.

The initial conditions are $v_0 = 0$ and $U_0 = mgL$. When at its lowest position the bob has dropped a vertical distance L and now $U = 0$. If v is the velocity when at the lowest position, Eq. 5-4 says

$$\tfrac{1}{2}mv^2 + 0 = 0 + mgL$$

or

$$v^2 = 2gL$$

$$v = \sqrt{2gL} \qquad (5\text{-}5)$$

Let T be the force of the string on m. Then the net force on m is

$$F_{\text{net}} = T - F_G = ma_c$$

where $a_c = v^2/L$ is the centripetal acceleration of m. Since $F_G = mg$,

$$T - mg = m\left(\frac{v^2}{L}\right)$$

Now substitute the right-hand side of Eq. 5-5 for v^2:

$$T - mg = m\frac{(2gL)}{L}$$

$$T = 3mg$$

We now define total mechanical energy as the sum of the total kinetic energy plus the total potential energy. We will use the symbol W for total energy rather than E because we will use the letter E later for electric field.

$$W = \text{KE}_{\text{total}} + U_{\text{total}}$$

The law of conservation of energy is then

$$W_1 = W_2$$

for any two times t_1 and t_2 as long as no external forces have been acting.

So far our discussion has been restricted to conservative forces only. But if we maintain the microscopic or elementary particle point of view, all forces are conservative because there are only four different kinds of forces (electric, gravitational, strong, and weak) and they are all conservative. Air resistance becomes conservative if we explicitly take into account the kinetic and potential energies of each air molecule. So from the microscopic point of view, the law of conservation of energy is completely general and applies to all phenomena.

But sometimes it is convenient not to have to consider the kinetic and potential energies of each air molecule. We can then lump together all the kinetic and potential energies of the air molecules and for bookkeeping purposes call the energy of such random particle motion, heat energy. This macroscopic approach to energy conservation is discussed in Section 5-10.

5-7 Potential Energy Diagrams

Worth ten thousand words

Let us now calculate and plot the potential energy of a mass m attached to the end of a spring. Here $F_c = -kx$

Fig. 5-8. An example of the conservation of energy. The quantity $(\frac{1}{2}Mv^2 + Mgh)$ remains constant. (Drawing by O. Soglow. © 1959 *The New Yorker,* Inc.)

Fig. 5-9. The force required to pull a spring a distance x plotted as a function of x.

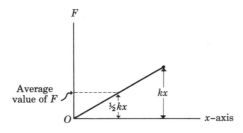

Fig. 5-10. Potential diagram for a spring stretched a distance x_0.

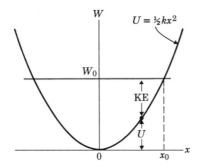

according to Hooke's law. The constant k is called the spring constant. According to the definition of potential energy, U is the work done in stretching the spring a distance x. As the spring is being stretched, the force is increasing uniformly with distance from zero to a value kx. The work done will be the value of F averaged over the distance x times x. F as a function of x is shown in Fig. 5-9. We see that the average value of F over the distance x is $\frac{1}{2}kx$. The work done is this value $\frac{1}{2}kx$ times the distance x or

$$U = \tfrac{1}{2}kx^2$$

Suppose the mass m on the end of the spring is pulled (or pushed) until the spring is stretched (or compressed) a distance x_0. Then the work $W = \frac{1}{2}kx_0^2$ will have been done on m. This will be stored in the form of potential energy $U = \frac{1}{2}kx_0^2$. When we let go of the spring, in the absence of friction, m will now oscillate in simple harmonic motion back and forth forever. As x starts decreasing from its maximum value x_0, the potential energy will decrease and the kinetic energy will increase.

Since $\text{KE} + U$ is always a fixed value of energy for a closed system, the value of KE can be obtained conveniently from what is called a potential diagram. The parabola in Fig. 5-10 is a plot of U versus x for a stretched spring. The horizontal line corresponds to the energy value $W_0 = \frac{1}{2}kx_0^2$, which is the energy of a spring initially stretched to a distance x_0. Because of the conservation of energy

$$\text{KE} + U = W_0$$

or

$$\text{KE} = W_0 - U$$
$$= \begin{pmatrix} \text{height of} \\ \text{red line} \end{pmatrix} - \begin{pmatrix} \text{height of} \\ \text{parabola} \end{pmatrix}$$

which says that the vertical distance from the curve (parabola) to the horizontal red line must be the kinetic energy at that value of x.

Note that this distance plus U (the distance from the curve to the x-axis) is W_0, the total energy of the system, which must be conserved. Thus the value of KE for any position x can be immediately obtained from the potential diagram.

We shall find potential diagrams very useful in later chapters where we treat more complicated forces than those of a spring.

Actually, all that ever enters in physics are potential energy differences. Hence the position from which zero energy is measured is arbitrary. We shall show this by changing the zero on the energy scale of Fig. 5-10, keeping everything else the same (see Fig. 5-11). Now we wish to determine the kinetic energy when $x = x_1$. The rule is still the same; merely measure the vertical distance from the curve to the red line. In Fig. 5-11 this is labeled a. Note that this distance on the vertical axis is independent of where we choose our zero. Here we have chosen our zero in such a way as to make the potential energy and even the total energy negative. If a and b are taken as positive numbers, the potential energy at $x = x_1$ is

$$U = -(a + b)$$

and the total energy is

$$W_0 = -b$$

As a check, we take the difference

$$W_0 - U = (-b) - [-(a + b)]$$
$$= a$$

which is the kinetic energy as it should be.

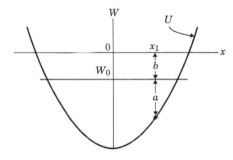

Fig. 5-11. Same as Fig. 5-10 except that the zero position on the energy scale has been changed.

5-8 Gravitational Potential Energy

The work involved in getting off the earth

So far our examples of potential energy have included the stretching of a spring ($U = \frac{1}{2}kx^2$) and the lifting of a mass m at the surface of the earth ($U = mgh$). Let us now consider the more general situation of lifting a mass to a distance far from the surface of the earth. Here the gravitational force F_G is not constant but decreases inversely as the square of the distance from the center of the earth: $F_G = GM_em/r^2$, where M_e is the mass of the earth. We wish to calculate the work done in moving a mass m from a distance R (radius of the earth) to a distance r, both measured from the center of

Q.6: If the energy scale of Fig. 5-10 had been chosen such that the red line became the new x-axis, what would be the value of the total energy?

Fig. 5-12. In order to calculate the work done in lifting a mass away from the earth, the distance $(r - R)$ is divided into small intervals.

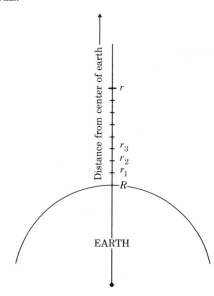

Distance from center of earth

r

r_3
r_2
r_1
R

EARTH

Ans. 6: Zero.

the earth. This calculation is not as simple as before because the force is neither constant nor changing uniformly with distance as with the spring.

In this calculation we are forced to break up the distance from R to r into many small intervals so that over each interval the force F_G will be almost constant. Then we are able to calculate the work done in each interval. The work done over the entire distance is obtained by adding up that done in each of the intervals. At the beginning of the first interval the force F_G is GM_em/R^2, and at the end of the first interval it is GM_em/r_1^2. See Fig. 5-12. These two values are almost exactly the same. For convenience we will use an average value $F_G = GM_em/Rr_1$. The work done in the first interval is

$$W_1 = F_G(r_1 - R) = \frac{GM_em}{Rr_1}(r_1 - R)$$

or

$$W_1 = GM_em\left(\frac{1}{R} - \frac{1}{r_1}\right)$$

In the second interval the average force is

$$F_G = \frac{GM_em}{r_1r_2}$$

and

$$W_2 = \frac{GM_em}{r_1r_2}(r_2 - r_1) = GM_em\left(\frac{1}{r_1} - \frac{1}{r_2}\right)$$

Similarly, the work done in covering the third interval is

$$W_3 = GM_em\left(\frac{1}{r_2} - \frac{1}{r_3}\right)$$

We now add together the work done in the first three intervals:

$$W_1 + W_2 + W_3$$
$$= GM_em\left[\left(\frac{1}{R} - \frac{1}{r_1}\right) + \left(\frac{1}{r_1} - \frac{1}{r_2}\right) + \left(\frac{1}{r_2} - \frac{1}{r_3}\right)\right]$$
$$= GM_em\left(\frac{1}{R} - \frac{1}{r_3}\right)$$

Note that the intermediate values r_1 and r_2 cancel out and only the values of r at the two extreme ends (R and r_3) remain. This must also be the case if we add in all the remaining intervals from r_3 to r. Then all that remains is

$$W = GM_e m \left(\frac{1}{R} - \frac{1}{r} \right)$$

The preceding is the work done in moving a mass m against the earth's gravitational field out to a distance r. By definition this is the gravitational potential energy of m in the field of the earth:

gravitational potential energy

$$U = GM_e m \left(\frac{1}{R} - \frac{1}{r} \right) \tag{5-6}$$

is the gravitational potential energy. This equation is plotted in Fig. 5-13.

Example 6

Show that for $(r - R)$, small compared to R, U approaches mgh where $h = (r - R)$.

According to Eq. 4-3,

$$GM_e = gR^2$$

Substituting gR^2 for GM_e in Eq. 5-6 gives

$$U = (gR^2)m \left(\frac{1}{R} - \frac{1}{r} \right) \tag{5-7}$$

$$U = \frac{mg\bar{R}(r - R)}{r}$$

When r is almost equal to R, the denominator r can be replaced by R. Then

$$U = \frac{mgR(r - R)}{R} = mg(r - R) = mgh$$

5-9 Velocity of Escape

How to get away from it all

After a rocket's motors are turned off, the sum of its KE and potential energy must remain constant. Initially the rocket's potential energy is zero and its KE is $\frac{1}{2}mv^2$, where v is the velocity given to the last stage by the rocket motors. If the rocket is aimed straight up, it will travel away from the earth, slowing down all the while until it reaches a maximum distance $r_{\max}$ and then falls back. At this maximum distance all the rocket's kinetic energy has gone into

Fig. 5-13. Plot of gravitational potential energy as measured from the surface of the earth ($U = 0$ when $r = R$). Dashed line is approximation $U = mgh$ which holds near the surface of the earth.

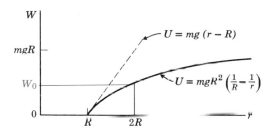

Q.7: In Fig. 5-13 what is the value of W_0 in terms of m, g, and R?

gravitational potential energy. According to Eq. 5-7 the initial kinetic energy is equal to

$$\tfrac{1}{2}mv^2 = mgR^2\left(\frac{1}{R} - \frac{1}{r_{\max}}\right) \tag{5-8}$$

If the initial velocity v is known, this equation can be used to determine the maximum possible height of the rocket.

Example 7

A rocket given the circular velocity $v_c = 5$ mi/sec is fired straight up rather than parallel to the earth's surface. How far from the earth will it go?

The condition for circular velocity is that the centripetal acceleration $v_c^2/R = g$ (see page 41). The initial kinetic energy is obtained by substituting gR for v^2:

$$\tfrac{1}{2}mv_c^2 = \tfrac{1}{2}m(gR)$$

Since the initial potential energy is zero, the total energy is the initial kinetic energy or

$$W_0 = \tfrac{1}{2}mgR$$

This is indicated by the red line in Fig. 5-13. When $r = r_{\max}$, KE $= 0$, and using Eq. 5-8 we have

$$W_0 = 0 + mgR^2\left(\frac{1}{R} - \frac{1}{r_{\max}}\right)$$

$$\tfrac{1}{2}mgR = mgR^2\left(\frac{1}{R} - \frac{1}{r_{\max}}\right)$$

$$\frac{1}{2} = \frac{R}{R} - \frac{R}{r_{\max}}$$

$$r_{\max} = 2R$$

Thus the maximum height attained would be 4000 mi.

An interesting phenomenon occurs when the initial kinetic energy is equal to or greater than mgR. If the value mgR is used for the left-hand side of Eq. 5-8, the quantity $1/r_{\max} = 0$ or $r_{\max} = \infty$. Physically this means the rocket would never come back—its velocity would never be slowed down to zero. This special value of velocity that makes the kinetic energy equal to mgR is called the velocity of escape v_R. The kinetic energy

$$\tfrac{1}{2}mv_R^2 = mgR$$

$$v_R = \sqrt{2gR}$$

Ans. 7: At $r = 2R$, $W_0 = U$; hence $W_0 = mgR^2\left(\dfrac{1}{R} - \dfrac{1}{2R}\right) = \tfrac{1}{2}mgR^2$.

velocity of escape

Note that the velocity of escape is exactly $\sqrt{2}$ times the circular velocity. Thus the velocity of escape is 1.414 times 5 mi/sec, or about 7 mi/sec.

Example 8

What fraction of the escape velocity is needed to send a rocket as far out as the moon?

The distance to the moon is 60 earth radii. Substituting $r_{max} = 60R$ into Eq. 5-8 gives

$$\tfrac{1}{2}mv^2 = mgR^2 \left(\frac{1}{R} - \frac{1}{60R} \right)$$

$$v^2 = 2gR \left(1 - \frac{1}{60} \right)$$

$$v = \sqrt{\frac{59}{60}}\, v_R$$

$$v = 0.99 v_R$$

The velocity needed to reach the moon is 99% of the escape velocity.

5-10   Friction and Heat

Energy of the microscopic world

There are many common forces that do not depend on position alone. Air resistance is one example. The forces of air resistance and other forms of friction are always opposed to the direction of motion. These frictional forces depend on the direction of the velocity and on the magnitude of the velocity. Suppose a wooden block is slowly pushed a distance x along a rough table. Let the frictional force of the table on the block be F_f. Then the work done to overcome this force and move the block a distance x is $W = F_f \cdot x$. After this much energy is expended, the kinetic energy of the block has not increased and its potential energy also has not increased. Where did the energy go? We observe that some heat is generated whenever the block is pushed. When physicists first learned how to measure heat quantitatively, they discovered that the amount of heat produced is always proportional to the amount of work done against frictional forces. This proportionality factor is known as the mechanical equivalent of heat. As we shall learn in the next chapter, heat energy is in reality the same mechanical kinetic and potential

Q.8: If the kinetic energy of a low-flying earth satellite is doubled, will it escape?

energy that we have just studied; but it is the kinetic and potential energy of the individual molecules that make up the physical object. Thus the heat energy is not as readily apparent as the kinetic and potential energy of a macroscopic physical object.

Heat energy should be thought of as microscopic kinetic and potential energy, whereas ordinarily the terms kinetic and potential energy are macroscopic in usage. Because the conservation of energy must hold for small particles as well as for large objects, the law of conservation of energy must also include heat energy. So the sum of the kinetic plus potential plus heat energy of any closed system must remain conserved. If an external force F_{ext} is applied over a distance Δx, then the law of conservation of energy takes the form

conservation of energy
$$F_{ext} \cdot \Delta x = (\tfrac{1}{2}Mv^2 - \tfrac{1}{2}Mv_0{}^2) + (U - U_0) + F_f \cdot \Delta x \qquad (5\text{-}9)$$

or

(work done) = (increase in KE)
$$\qquad\qquad\quad + \text{(increase in } U\text{)} + \text{(increase in heat energy)}$$

Note that mathematically a decrease is a negative increase. Equation 5-9 can be obtained by applying Newton's second law to the small displacement Δx of a mass M. The net force is the applied force F_{app} plus any force F_c that gives rise to a potential energy minus the force of friction F_f:

$$F_{net} = F_{app} + F_c - F_f$$
$$(F_{ext} + F_c - F_f) \cdot \Delta x = Ma \cdot \Delta x$$

According to Eq. 2-11 we can substitute $\tfrac{1}{2}(v^2 - v_0{}^2)$ for the quantity $(a \cdot \Delta x)$. Then

$$F_{ext} \cdot \Delta x + F_c \cdot \Delta x - F_f \cdot \Delta x = M \cdot \tfrac{1}{2}(v^2 - v_0{}^2)$$
$$F_{ext} \cdot \Delta x = \tfrac{1}{2}Mv^2 - \tfrac{1}{2}Mv_0{}^2 - F_c \cdot \Delta x + F_f \cdot \Delta x$$

According to Eq. 5-3, the quantity $-F_c \cdot \Delta x$ is the increase in potential energy. Replacing it with $(U - U_0)$ gives

$$F_{ext} \cdot \Delta x = (\tfrac{1}{2}Mv^2 - \tfrac{1}{2}Mv_0{}^2) + (U - U_0) + F_f \cdot \Delta x$$

In our discussion of energy we have so far only used forces that are parallel to the displacement. However, the above derivation can be easily adapted to the more general case

Ans. 8: Yes, but just barely. By doubling the kinetic energy we double v^2, or increase v by $\sqrt{2}$ which is exactly the velocity of escape.

where F and Δx are not parallel. The adaptation is merely to replace each force by the component of that force along Δx. Thus in the general case, work is the product of displacement times the component of force in the direction of the displacement; that is,

$$(\text{work}) = F_{ext} \cdot \Delta x \cdot \cos \theta$$

Note that the displacement of an object is in the direction of its velocity. If the force is always perpendicular to the velocity, as in the case of a centripetal force, $\cos \theta = 0$ and no work is done.

Example 9

The mass of a boy plus his sled is 20 kg. The snow exerts a frictional force of 50 newtons.

(a) How much work is required to pull the boy and sled for 100 m along a 30° slope?

(b) Now that he has reached the top, the boy slides back down on the sled. What will be his velocity and kinetic energy when he reaches the bottom?

The work done in pulling the sled up the hill will be the increase in KE (which is zero) plus the increase in U which is Mgh plus F_f times the distance the sled is pulled. Thus Eq. 5-9 gives

$$(\text{work done}) = 0 + Mgh + F_f \cdot s$$

where $s = 100$ m, $h = 50$ m, and $F_f = 50$ newtons. So

$$(\text{work done}) = (20 \times 9.8 \times 50 + 50 \times 100) \text{ joules}$$
$$= (9800 + 5000) \text{ joules}$$
$$= 14,800 \text{ joules}$$

In part (b), the work done is zero and the change in potential energy is just the reverse of the above. The heat energy generated (5000 joules) must be the same in both cases. Thus for the downhill run Eq. 5-9 becomes

$$0 = \tfrac{1}{2}Mv^2 - 9800 \text{ joules} + 5000 \text{ joules}$$

$\tfrac{1}{2}Mv^2 = 4800$ joules is the kinetic energy

The velocity is

$$v = \sqrt{\frac{2 \times 4800 \text{ joules}}{M}}$$

$v = 21.9$ m/sec

Example 10

A 1-kg earth satellite is circling the earth at a height of 30 km at

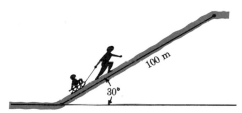

Fig. 5-14. Work is done in pulling sled against gravity and friction.

100 m

30°

the time of re-entry. How much heat energy will be generated during re-entry?

In this example v_0 is the circular velocity of 8 km/sec and $(U - U_0) = -Mgh$ where $h = 30$ km. Thus Eq. 5-9 becomes

$$0 = (0 - \tfrac{1}{2}Mv_0^2) - Mgh + \text{(heat energy)}$$

or

$$\text{(heat energy)} = \tfrac{1}{2}Mv_0^2 + Mgh$$

As we would expect, the law of conservation of energy says that the initial kinetic plus potential energy get converted into heat energy. Putting in the numbers we obtain

$$\text{(heat energy)} = \tfrac{1}{2}(8 \times 10^3) + 9.8 \times 3 \times 10^4 \text{ joules}$$

$$= 32 \times 10^6 + 0.29 \times 10^6 \text{ joules}$$

$$= 32.3 \times 10^6 \text{ joules}$$

This example illustrates the main problem of re-entry—the heating up of the satellite. For example, 1 kg of aluminum heats up $1°$C for every 10^3 joule increase in heat energy. We see that if even just 1% of the heat energy generated in re-entry stays with an aluminum satellite, its temperature will increase by about $300°$C.

5-11   Equivalence of Mass and Energy

The vast, but unobtainable energy in a pinch of sand

In the last two sections of this chapter we shall see that there is an unlimited supply of energy all around us, but none of it can ever be utilized.

As noted in the discussion of Eq. 3-6, the theory of relativity requires that if the velocity or energy of a particle increases, so does its mass. The relation between mass increase and energy increase turns out to be a particularly simple one. Einstein found that relativity theory required that

$$\Delta M = \frac{\Delta W}{c^2}$$

where ΔM is the increase in mass corresponding to an energy increase ΔW. Einstein proposed that the entire energy of a mass M must be

Einstein mass-energy relation $$W = Mc^2 \tag{5-10}$$

where $c = 3.0 \times 10^8$ m/sec is the speed of light. This says that 1 kg of sand must contain 1 kg $\times$ $(3.0 \times 10^8$ m/sec$)^2$ or 9×10^{16} joules of energy. This is almost twice the entire electrical energy consumed by the United States in a week.

But even though such unlimited energy is there in the sand or any other matter, it is inaccessible. Obtaining useful energy from mass was strictly forbidden by the old law of conservation of mass. This law states that matter can neither be created or destroyed. However, we now know this is not quite true. The modern version of the law of conservation of mass which forbids the extraction of energy from sand is discussed in the next section. There are, however, examples of the conversion of rest mass into energy and vice versa. (The rest mass of a particle is defined as the mass of the particle when its velocity is zero.) For example, when matter is bombarded by high energy protons, new elementary particles are copiously produced (see Chapter 16). In such a process the kinetic energy of the high energy protons is directly converted into the rest mass of these particles. As an example of the reverse process, some of these particles decay spontaneously, thus converting their rest mass back into the kinetic energy of their decay products. We should note, however, that according to Eq. 5-10, kinetic energy must also have mass and that strictly speaking it is not correct to say mass can be converted to energy or that energy can be converted to mass. (This is contrary to what appears in some popular textbooks.) Because the total amount of energy in the universe must be conserved, so, according to Eq. 5-10, must the total inertial mass in the universe remain constant. The mass that is not conserved is the rest mass. It is rest mass that can be converted into kinetic energy and vice versa, and even here there is a severe constraint that is discussed in the next section.

5-12 Conservation of Heavy Particles

Protons and neutrons forever

Ordinary matter is made up of atoms. Atoms are made up of atomic nuclei surrounded by orbital electrons. The atomic nuclei are each made up of protons and neutrons which are roughly 1800 times heavier than electrons. For this reason protons and neutrons are called heavy particles. The law of conservation of heavy particles states that the total number of protons and neutrons in any closed system must remain constant. As we shall learn in Chapters 15 and 16 these protons and neutrons can still interact violently with each other by such processes as fusion, fission, and nuclear transmutation. In fact, neutrons can change into protons and vice versa. However, in all these possible nuclear interactions the total number of protons plus the number of neutrons are always observed to remain constant.* No one has ever observed the rest mass of a proton or neutron convert into other forms of energy. The law of conservation of heavy particles rests on a great deal of experimental evidence. According to this law we can never utilize the vast energy contained in a handful of sand.

On the other hand, it is possible to obtain a limited amount of nuclear energy. In atomic nuclei, protons and neutrons attract each other with very strong forces. These nuclear forces contribute a large amount of potential energy to each nucleus. In processes such as nuclear fission and fusion some of this potential energy can be converted into kinetic energy. This is the energy source of nuclear reactors and nuclear bombs. These subjects and the application of the law of conservation of heavy particles to antimatter and other heavy elementary particles are discussed in Chapters 15 and 16.

Problems

1. Suppose the wrench of Fig. 5-3 had been thrown up in the air at an angle of 45° to the horizontal. What, then, would be the path of its center-of-mass?

2. The steady-state theory of the universe postulates that neutrons are spontaneously produced (at a very low rate). Which conservation laws does this theory violate?

3. A cat when released from an upside-down position always lands upright. Is this a violation of the conservation of angular momentum?

4. A spinning top if grabbed by hand stops spinning. Where does the angular momentum go? (*Hint:* Suppose this was done by a man standing on a turntable.)

---

*An antiproton is an anti-heavy particle. In other words a proton plus an antiproton is a total of zero heavy particles. Hence in counting heavy particles we count each proton or neutron as $+1$ and each antiproton or antineutron as -1.

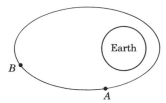

5. An earth satellite is in an elliptical orbit as shown.
   (a) Is the kinetic energy greater at A or B?
   (b) Is the potential energy greater at A or B?
   (c) Is the total energy greater at A or B?
   (d) Is the angular momentum greater at A or B?

6. A uniform meter stick has a mass of 100 gm. A 50-gm mass is attached to the 100-cm reading. What will be the reading of the center-of-mass?

7. An ice cube is released at the top of a hemispherical bowl and without friction rocks back and forth between the two edges. The mass of the cube is 30 gm; the radius of the bowl, 20 cm.

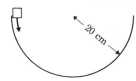

   (a) When the cube passes the bottom of the bowl, what is its velocity?
   (b) At the bottom what force does the bowl exert on the cube?
   (c) If the cube is allowed to oscillate through a small arc at the bottom of the bowl, what is its period of oscillation?

8. The rim of a 0.8 m diameter bicycle wheel has a mass of 1.5 kg. What is the angular momentum of the bicycle wheel when the bicycle has a velocity of 3 m/sec? Ignore the mass of the spokes.

9. A car is traveling at 60 mph. How much faster must it go in order to double its kinetic energy?

10. A lightweight meter stick has a heavy mass fixed at the top end and is pivoted at the bottom end. When released the mass will follow the circumference of a 2 m diameter circle. What will be the maximum velocity and maximum centripetal acceleration of the mass?

11. A 5-gm mass is moved from point A to point B. Assume the mass experiences a constant electrostatic force of 2 dynes pointing to the left over this entire region. How much work must be done to move the mass from A to B. Did its potential energy increase, decrease, or remain the same?

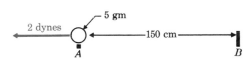

12. (a) How much work is required to push a 10-gm mass up a frictionless inclined plane whose length is 3 m and whose height is 0.5 m?
   (b) Now assume of force of friction between the mass and the inclined plane of 700 dynes. How much work must be done to move the mass up the plane?
   (c) Suppose an applied force of 3000 dynes is used to push the mass up the incline (assume the same frictional force as (b)). What would be the velocity of the mass when it reached the top of the incline?

13. A firecracker is moving with a velocity of 10 ft/sec. Suppose it explodes into just two pieces of equal mass. If one of the two pieces has zero velocity just after the explosion, what is the ratio of the final kinetic energy to the initial kinetic energy of the firecracker?

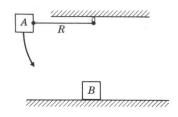

14. Block B is at rest on a frictionless surface. An identical block A is attached to one end of a string of length R. Block A is released from the horizontal position and collides with B. The two blocks stick to each other and move together after the impact.

    (a) What is the velocity of the two blocks immediately after impact?

    (b) How far will they rise above the surface?

15. An 800-gm toy train is being pulled with a constant force of 100 dynes. At first the train speeds up from rest and then reaches a constant velocity v_0. If 2×10^3 ergs of work are done every second, what is v_0? (How far must the train go in order to perform each 2000 ergs of work?)

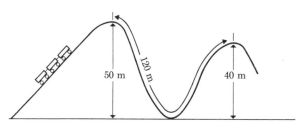

16. A certain roller-coaster rises 50 m. If the next rise is 40 m after 120 m of track, what is the maximum permissible value of the frictional force on a 500-kg car? (If F_f were any larger, the car would not be able to reach the top of the second rise.)

17. A 1-kg block slides down an inclined plane starting from rest at the top of the plane. The velocity of the block at the bottom of the plane is 100 cm/sec.

    (a) What is the work done by this frictional force?

    (b) What is the constant force due to friction?

    (c) If the inclined plane is given a coat of oil and the frictional force is reduced to one-tenth of its previous value, what would be the new velocity of the block at the bottom of the plane?

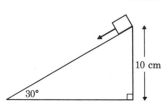

18. In Problem 11, Chapter 3, how much KE was lost in the collision? Where did the energy go?

19. In Problem 14, Chapter 3, what is the velocity of M_1 after it has moved up 50 cm? (*Hint*: determine initial and final potential energies of both masses.)

20. In the three-dumbbell example on page 95 what is the initial and final kinetic energy of the dumbbells? Where did this additional energy come from? The mass of each dumbbell is 20 kg.

21. A 15-kg sign is supported as shown by a wire. What is the tension in the wire and what is the force of the sign against the wall?

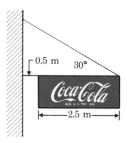

22. Consider a car of mass 1000 kg. When on a level road it always takes a force of 500 newtons to push it slowly at a constant velocity.

    (a) What is the frictional force in newtons?

    (b) If a force of 1000 newtons is applied, what will be the acceleration of the car?

    (c) If the car is parked on the side of a hill and the brakes give out, how far will the car coast before coming to rest if its vertical height on the hill was 10 m above the level ground?

23. All examples of simple harmonic motion have $U = \frac{1}{2}kx^2$ where $k = F/x$. What is the maximum potential energy of a simple pendulum in terms of M, g, L, and x_0? What is the maximum velocity in terms of g, L, and x_0 where x_0 is the maximum displacement?

Prob. 27

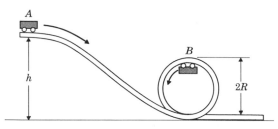

Prob. 28

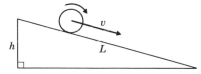

Prob. 29

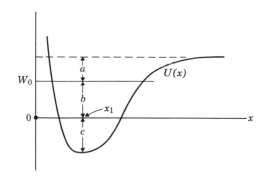

24. If the mass of the ladder in Example 2 is 15 kg, how far can the man climb before the ladder slips?

25. A 10-kg ladder is resting at 45° against a frictionless wall. What is the force of the ladder against the wall?

26. A certain uranium nucleus containing 92 protons and 143 neutrons absorbs a neutron and then fissions into 2 fission products and 4 neutrons. If one of the fission products is Cs^{137} (55 protons and 82 neutrons), how many protons and neutrons will there be in the other fission product?

27. A toy roller coaster works as shown. The car is given a gentle push at position A so that it starts out with essentially zero velocity. It slides down the frictionless track and travels around on the inside of the circular loop of radius R. The height h is such that the car can just barely make the trip around the loop without losing contact with the tracks. What is h in terms of R? What force does the track exert on the car at point B?

28. Consider a solid sphere starting from rest and rolling down an incline. Assume that the kinetic energy of rotation is always equal to the kinetic energy of translation ($\frac{1}{2}Mv^2$) where v is the velocity of the center of mass. It can be shown that the total kinetic energy must always be the sum of the two.

(a) What will be the total kinetic energy of the sphere when at the bottom? Give answer in terms of M, g, and h.
(b) What will be v when at the bottom?
(c) What is the acceleration of the center of mass in terms of v and L?

29. The figure represents the potential energy diagram for an atom in a molecule. The quantities a, b, and c are amounts of energy (they are positive). $W_0 = KE + U$. Express the following quantities in terms of a, b, and c when the atom is at position x_1. Do not ignore minus signs in your answer.

(a) What is the potential energy?
(b) What is the kinetic energy?
(c) What is the total energy (excluding rest energy)?
(d) What is the energy of dissociation (the amount of additional energy needed to remove the atom from the molecule)?

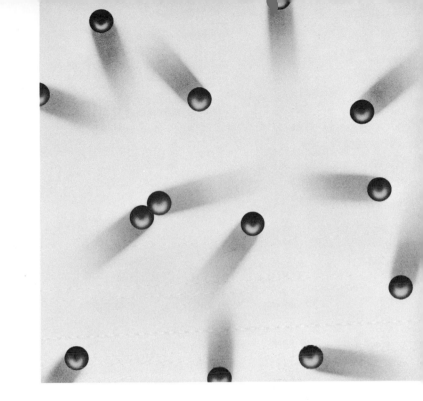

The kinetic theory

Chapter 6

The kinetic theory

The kinetic theory explains the macroscopic properties of gases in terms of the laws of mechanics applied to the gas molecules. Before presenting the kinetic theory, we must first introduce some of the macroscopic concepts such as density and pressure.

6-1 Density

Water is one

The density of a body is defined as the quotient M/V, where M is its mass and V is its volume.

$$D = \frac{M}{V} \tag{6-1}$$

We recall that at the time the metric system was being devised, it was necessary to decide on a standard unit of mass. This unit, the gram, was defined to be the amount of mass in 1 cc of water. Thus the density of water is conveniently 1 gm/cm³. In the MKS system the density of water is

$$D_{\text{water}} = \frac{1 \text{ gm}}{(1 \text{ cm})^3} = \frac{10^{-3} \text{ kg}}{(10^{-2} \text{ m})^3} = 10^3 \frac{\text{kg}}{\text{m}^3}$$

Table 6-1 lists the densities of various substances.

6-2 Pressure

Force on a surface

When a fluid (liquid or gas) is contained in a container, the fluid will exert a force on each element of area of the container. As an example consider the gas inside a balloon. If we continue blowing air into the balloon the pressure keeps increasing until the force exerted on the rubber is so great that the rubber breaks. *The pressure P in a fluid is defined as the force per unit area exerted on the container.*

Pressure

$$P = \frac{F}{A} \tag{6-2}$$

In the CGS system pressure is measured in dynes/cm², and in the MKS system it is newtons/m². The direction of the force is always perpendicular to the confining surface no matter how the surface is oriented. If we could construct a

TABLE 6-1 TABLE OF DENSITIES

Substance	Density in gm/cm³
Nuclear matter	2×10^{14}
Center of sun	100
Platinum	21.4
Gold	19.3
Mercury	13.6
Lead	11.3
Iron	7.8
The earth	5
Aluminum	2.7
Water	1.0
Ice	0.92
Balsa Wood	0.13
Air	0.0013
Best man-made vacuum	10^{-19}
Interstellar space	10^{-24}
Interglactic space	10^{-29}

6

Fig. 6-1. Forces on cube immersed in fluid under pressure P.

Fig. 6-2. Forces on two pistons of a hydraulic pump.

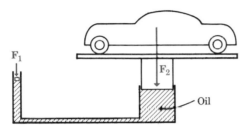

Fig. 6-3. Column of liquid of height h and area A.

small, hollow cube with thin metal surfaces of area A each, and immerse this empty cube into a fluid that is under pressure P, the force F on each surface of our hollow cube will be $F = PA$ no matter how the cube is oriented. In any small region of a fluid at rest, the pressure will be the same in all directions; otherwise there would be a net force on a small cube of the fluid and it would move. Ignoring gravity, the pressure exerted on a container must be the same at all points on the container, independent of its shape.

Example 1

An automobile rack is lifted by a hydraulic pump which consists of two pistons connected by a pipe as shown in Fig. 6-2. The large piston is 1 m in diameter and the small piston is 10 cm in diameter. If the weight of the car is F_G, how much smaller a force is needed on the small piston to lift the car?

Both pistons are walls of the same container and hence experience the same pressure. Let $P_1 = F_1/A_1$ be the pressure on the small piston, and $P_2 = F_2/A_2$ be the pressure on the large piston. Because $P_1 = P_2$, we have

$$\frac{F_1}{A_1} = \frac{F_2}{A_2}$$

$$F_1 = F_G \frac{A_1}{A_2}$$

The ratio A_1/A_2 is $1/100$ because the areas go as the square of the diameters. Thus F_1 only need be 1% of the weight of the car.

6-3 Hydrostatics

Liquid at rest

When a volume of liquid is under the influence of gravity, the weight of the liquid on top is exerting an external force on the liquid below. For this reason underwater pressure increases with depth. Consider a cylinder of area A holding a liquid of density D and height h as in Fig. 6-3. Then the force exerted on the bottom of the cylinder is the mass of the liquid M_{liq} times g.

$$F = (M_{\text{liq}})g$$

According to Eq. 6-1, the mass of the liquid is D times the volume (Ah). Thus

$$F = (DAh)g$$

The pressure is obtained by dividing both sides of the above by A:

$$\frac{F}{A} = \frac{DAhg}{A}$$

or

$$P = Dgh \qquad (6\text{-}3)$$

The above formula does not depend on A or the shape of the container. It tells us the pressure at a depth h due to the weight of the liquid. No matter what the shape of the vessel, the pressure will be the same at any two points with the same height.

Atmospheric pressure

The height of the earth's atmosphere is a few hundred kilometers. According to Eq. 6-3 there should be a pressure P_0 at the surface of the earth equal to the height of the atmosphere times g times the density of the air averaged over the height of the atmosphere. The numerical result is

Atmospheric pressure

$$P_0 = 1.01 \times 10^6 \text{ dynes/cm}^2$$

The barometer

Suppose we take a tube of mercury ($D = 13.6$ gm/cm^3) and invert it over a beaker of mercury as shown in Fig. 6-4. The pressures at points A and B must be the same because these two points are at the same height. According to Eq. 6-3, $P_A = Dgh$ where h is the height of the mercury column. The pressure on the mercury–air surface must be $P_B = P_0$, the atmospheric pressure. Thus

$$Dgh = P_0$$

$$h = \frac{P_0}{Dg}$$

$$h = \frac{1.01 \times 10^6}{13.6 \times 980} \text{ cm} = 76.0 \text{ cm} \qquad (6\text{-}4)$$

The height of the column of mercury is proportional to the atmospheric pressure. A device, called the barometer, is used to measure the value of atmospheric pressure.

Fig. 6-4. The mercury barometer.

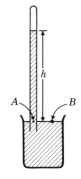

Fig. 6-5. How not to get water from a deep well.

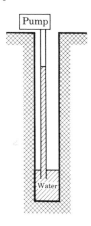

Fig. 6-6. Block of volume AL submerged in liquid of density D.

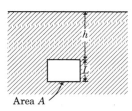

Area A

Example 2

The water level in a certain well is 50 ft below ground. A pipe is inserted into this well and a pump turned on at the top end until all the air in the pipe is evacuated (Fig. 6-5). Will the water ever reach the pump? This is exactly the same physical situation as that of the mercury barometer shown in Fig. 6-4. Thus we can use Eq. 6-4 to solve for the maximum possible height of the water column.

$$h = \frac{P_0}{Dg} = \frac{1.01 \times 10^6}{1 \times 980} = 1.03 \times 10^3 \text{ cm}$$

We see that the water will rise only 10.3 m or about 33 ft. Those of you who have seen deep wells know that the trick is to put the pump down in the well. Then it can pump on water rather than air.

Archimedes' principle

Suppose a block of height L and area A as shown in Fig. 6-6 is submerged a distance h in a liquid of density D. The force on the bottom surface of the block will be pointing up and will be of magnitude

$$F_{\text{up}} = P \cdot A = Dg(h + L) \cdot A$$

The force on the top surface will be

$$F_{\text{down}} = (Dgh) \cdot A$$

The resultant force of the liquid on the block will be

$$F_{\text{up}} - F_{\text{down}} = DgL \cdot A = (M_{\text{liq}})g$$

where $M_{\text{liq}} = DLA$ is the mass of the liquid displaced by the block. Thus the block experiences an upward force equal to the weight of the water displaced. *Archimedes' principle states that a body immersed in a fluid is buoyed up by a force equal to the weight of the fluid displaced.* This principle when applied to the special case of a floating body states that a floating body must displace its own weight in water.

Example 3

A popular thought question is: What happens to the water level in a glass of ice water as the ice melts? Does the melted ice raise or lower the water level?

The answer is that the level remains the same, assuming that the ice had originally been floating in the water. Because an ice cube displaces its own weight in water, it will exactly fill up its own space in the water when it becomes water itself.

Q.1: Assume a bubble of air got into the glass tube mercury barometer in Fig. 6-4. If the air pressure in the tube above the mercury column is $\frac{1}{10}$ of an atmosphere, how high is the mercury column? $h = $ _____cm.

6-4 Atoms and Molecules

The "invisible" building blocks

The structure of matter is treated in detail in Chapters 13 and 14. However, before we can proceed with the kinetic theory, we must review the concepts of atoms and molecules.

About 400 B.C. Democritus, a Greek philosopher, proposed that all matter is made up of particles called atoms and that the space between atoms is completely empty (a vacuum). The Greek word "atom" means indivisible. We now know that there are about 92 different kinds of atoms that occur naturally on the earth. Substances made up entirely of one kind of atom are called elements. Hydrogen, carbon, oxygen, and copper are examples of some of these 92 elements. A table of all the elements appears in the Appendix. They are listed in the order of their atomic masses (with two minor exceptions). The first element on the list is hydrogen, the second is helium, and the third is lithium. The numerical position on the list is called Z, the atomic number. Thus $Z = 1$ for hydrogen and $Z = 3$ for lithium.

The structure of materials is explained by the fact that the forces between some atoms are strongly attractive when the atoms are almost close enough to touch each other. Even certain unlike atoms exert strong attractive forces on each other. For example, an oxygen atom has a strong affinity for one or even two hydrogen atoms. The combined unit of two hydrogen atoms firmly attached to an oxygen atom is called a water molecule. Water is made up entirely of these H_2O molecules. If a sample of water vapor is subdivided down to the smallest particle that still has the chemical properties of water vapor, the result is a H_2O molecule. Any substance made up out of identical molecules containing dissimilar atoms is called a compound.

The structure of atoms and a calculation of their sizes are given in Chapter 13 along with a display of these sizes in Fig. 13-21. Most atoms are about 10^{-8} cm in diameter. A wavelength of light is about 5×10^{-5} cm or thousands of times larger; hence individual atoms and molecules cannot be seen by the eye or even by high-powered optical microscopes. However, it is possible to "see" molecular structure using modern x-ray techniques. Figure 6-7 is a "photograph"

Ans. 1: The mercury column must now supply a pressure $P = 0.9P_0$, hence, $h = 68.4$ cm.

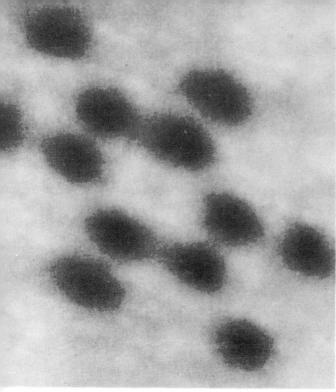

of a hexamethylbenzene molecule using special x-ray techniques. Another technique for "seeing" individual atoms is the field ion microscope invented by E. W. Muller. In this device, atoms on the point of a sharp pin give rise to spots on a flourescent screen as shown in Fig. 6-8. The molecules of a solid or liquid are in fairly close contact with each other. For this reason solids and liquids are almost incompressible. Gases, on the other hand, are usually about 1000 times less dense and thus consist of well-separated molecules. In a gas, a molecule usually moves a distance corresponding to many molecular diameters before it collides with another molecule.

6-5   The Ideal Gas Law

A continual bombardment of molecules

Experiments show that if the volume of a given quantity of gas is decreased, the pressure of the gas will increase provided the temperature remains unchanged. About 300 years ago Charles Boyle observed that for most gases the change in pressure bears a simple relationship to the change in volume. If the initial pressure and volume are P_1 and V_1, and the final pressure and volume are P_2 and V_2, then Boyle's law states

$$P_1 V_1 = P_2 V_2 \qquad (6-5)$$

provided the temperature is kept constant and the same amount of gas is used.

Example 4

The volume of an air bubble triples in volume as it rises from the bottom of a lake to the surface. How deep is the lake?

Let P_1 be the pressure and V_1 the volume at the bottom of the lake. Then $P_2 = P_0$ the atmospheric pressure, and $V_2 = 3V$. According

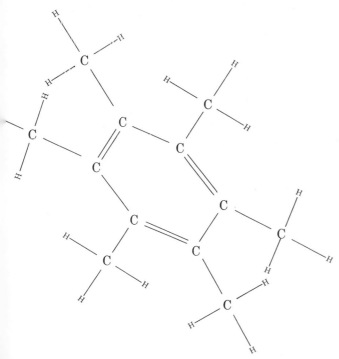

◀ **Fig. 6-7.** "Photograph" of the hexamethylbenzene molecule. Structural diagram shown below. The "photograph" is not taken directly, but is obtained by converting x-ray signal into a visual signal in a manner somewhat analogous to conversion of television electronic signal to a visual signal. Magnification factor is 10^8. (Courtesy Dr. M. L Higgins who performed the x-ray to visual transformation at Kodak Research Laboratory.)

The kinetic theory | 127

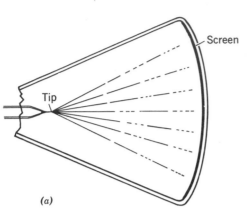

(a)

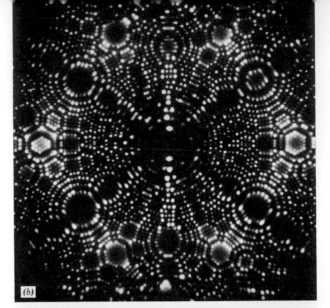

(b)

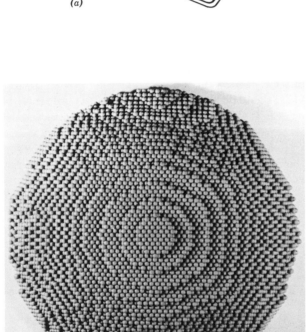

(c)

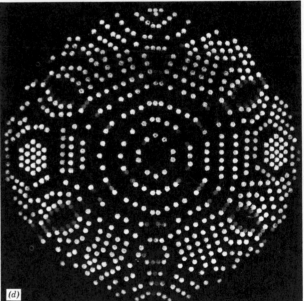

(d)

Fig. 6-8. The field-ion microscope. (a) is a greatly simplified drawing of the microscope. The positively charged tip of the tungsten needle repels positively charged helium atoms (helium ions) in its immediate vicinity. More ions leave from a point near an atom on the surface than from a point between two atoms. The ions travel radially outward in straight lines to the fluorescent screen, where they produce a pattern of illumination which reproduces the arrangement of atoms on the tip. (b) is a photograph of the fluorescent screen and each bright dot is caused by an atom on the tip (or possibly, in some cases, by a group of two or three atoms).

(c) is a model of the tip of the needle made by stacking cork balls, each of which represents a tungsten atom. The simulated needle points directly toward the camera, and its tip is in the center of the photograph. The cork balls that represent those atoms that are particularly effective in repelling ions were coated with luminescent paint and are seen in (d), which is a photograph of (c) made in the dark. The similarity between (b) and (d) confirms the hypothesis that we are seeing individual atoms. (Professor E. W. Müller, Pennsylvania State University.)

to Boyle's law,

$$P_1V_1 = P_2 \times V_2 = P_0 \times (3V_1)$$
$$P_1 = 3P_0$$

The increase in pressure at the bottom of the lake due to the weight of water above is $P_1 - P_0$ or $2P_0$. According to Eq. 6-3, the depth h of the lake is given by

$$2P_0 = Dgh$$

$$h = \frac{2P_0}{Dg}$$

$$h = 20.6 \text{ m}$$

Now that we know that a gas consists of particles colliding against the walls of its container, we should be able to derive Boyle's law using our knowledge of Newtonian mechanics. We know that whenever a particle bounces against a wall, the particle exerts a force on the wall. Thus the pressure of a gas on a wall must be due to the average force of many molecules colliding with the wall. In fact, now that we know that a gas is just a dilute collection of molecules, the above mechanism is the only possible source of force on a wall of the container.

First we shall calculate the average pressure on a wall due to the collisions of just one particle of mass m traveling with velocity v_x in the x-direction as shown in Fig. 6-9. Let this particle be in a box of length L in the x-direction with walls of area A perpendicular to the x-direction. The time between successive collisions with wall A is

$$\Delta t = \frac{2L}{v_x}$$

The change of momentum ΔP_x of the particle in each collision is

$$\Delta P_x = mv_x - m(-v_x) = 2mv_x$$

According to Newton's second law the average force the wall exerts on the molecule is

$$F = \frac{\Delta P_x}{\Delta t}$$

Now we eliminate ΔP_x and Δt by substituting the above two

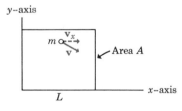

Fig. 6-9. Single particle of mass m bouncing back and forth in a box of volume AL.

Q.2: Can Boyle's law be used to compare two different ideal gases?

expressions to obtain

$$F = \frac{(2mv_x)}{\left(\frac{2L}{v_x}\right)} = \frac{mv_x^2}{L}$$

The average pressure due to this particle is

$$P = \frac{F}{A} = \frac{mv_x^2}{AL} = \frac{mv_x^2}{V}$$

where $V = AL$ is the volume of the box.

The pressure due to N particles in the box is then

$$P = \frac{Nm\overline{v_x^2}}{V} \tag{6-6}$$

where $\overline{v_x^2}$ is the average value of v_x^2 for the N molecules. Gases that obey Eq. 6-6 are called ideal gases. Any gas that is sufficiently dilute (N not too large) behaves as an ideal gas.

The average v_x^2 can be simply related to $\overline{v^2}$ by noting that $v^2 = v_x^2 + v_y^2 + v_z^2$ (Pythagorean theorem in three dimensions). Taking the average of both sides,

$$\overline{v^2} = \overline{v_x^2} + \overline{v_y^2} + \overline{v_z^2}$$

Since the molecules are all moving in random directions the average $\overline{v_x^2} = \overline{v_y^2} = \overline{v_z^2}$. Hence

$$\overline{v^2} = 3\overline{v_x^2}$$

or

$$\overline{v_x^2} = \frac{\overline{v^2}}{3}$$

If this is substituted into Eq. 6-6, we obtain

$$PV = Nm\frac{\overline{v^2}}{3} \tag{6-7}$$

Note that the left-hand side is the same term that appears in Boyle's law. Let us consider the same quantity of gas (N molecules) under two different conditions. According to Eq. 6-7

$$P_1V_1 = \frac{Nm}{3}\overline{v_1^2} \quad \text{(Condition 1)}$$

and

$$P_2V_2 = \frac{Nm}{3}\overline{v_2^2} \quad \text{(Condition 2)}$$

Ans. 2: Not as stated on p. 127. However, if both gases are at the same temperature and have the same number of molecules, then as we shall see on p. 133, $P_1V_1 = P_2V_2$.

As we shall see in the next section, $\overline{v_1^2} = \overline{v_2^2}$ when the temperatures $T_1 = T_2$. Thus the two left-hand sides are both equal to the same quantity

$$\left(\frac{Nm}{3}\overline{v_1^2}\right)$$

when at the same temperature, and

$$P_1 V_1 = P_2 V_2$$

6-6 Temperature

Microscopic kinetic energy

Everybody is familiar with the concept of temperature. We are used to measuring temperature by observing the expansion of a column of mercury or red-colored alcohol. Although it is not so convenient, temperature can be measured by observing the expansion of a column of gas. A simple version of a gas thermometer is shown in Fig. 6-10. The gas is prevented from escaping by a small drop of mercury at the top of the gas column. The pressure on the confined volume of gas will always be atmospheric pressure P_0. According to Eq. 6-7 the volume will be

$$V = \frac{Nm\overline{v^2}}{3P_0}$$

Note that the quantity $m\overline{v^2} = 2(\overline{KE})$, where $(\overline{KE})$ is what is called the average translational kinetic energy per molecule. We shall use the symbol $(\overline{KE})$ to stand for the kinetic energy due to the translational motion of the molecule only. Kinetic energy of a molecule due to rotational and vibrational motion must not be included. Substitution of the quantity $2(\overline{KE})$ for $m\overline{v^2}$ in the above equation gives

$$V = \frac{2}{3}\frac{N}{P_0}(\overline{KE}) \qquad (6\text{-}8)$$

Thus the temperature reading of a gas thermometer is proportional to the average translational kinetic energy per molecule. For an ideal gas this relationship between temperature T and $(\overline{KE})$ can be expressed using a proportionality constant $(\frac{3}{2}k)$ where

$$(\overline{KE}) = \tfrac{3}{2}kT \qquad (6\text{-}9)$$

Fig. 6-10. Simple gas thermometer. The shaded area shows confined gas under constant pressure. Height of mercury drop is proportional to temperature reading.

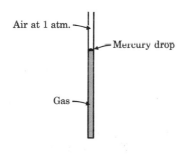

Air at 1 atm.

Mercury drop

Gas

Q.3: If the average velocity of molecules in a box is doubled, what happens to the pressure?

Or, according to Eq. 6-8,

$$T = \frac{P_0}{Nk} V$$

In this book, the *definition of temperature is given by the above equation.* The proportionality constant k is called the Boltzmann constant and must be determined experimentally by measuring P and V at a known temperature. The result is

Boltzmann constant

$$k = 1.38 \times 10^{-16} \text{ erg/degree Centigrade}$$

In the metric system the unit for temperature is the degree centigrade. The centigrade scale was originally established by defining $0°C$ as the temperature of freezing water and $100°C$ as the temperature of boiling water (at atmospheric pressure). In the English system freezing water is $32°F$ (Fahrenheit) and boiling water is $212°F$. Thus a temperature difference of $1°C$ corresponds to $1.8°F$. A comparison of the Fahrenheit and centigrade scales is shown in Fig. 6-11.

Fig. 6-11. Comparison of Fahrenheit, centigrade, and absolute (or Kelvin) temperature scales.

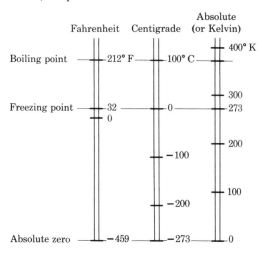

Absolute zero

We note that zero temperature as defined in Eq. 6-9 does not correspond to the zero on the centigrade scale. Equation 6-9 says that $T = 0$ when the kinetic energy of all the molecules becomes zero, which is a much lower temperature than that of freezing water. In order to use Eq. 6-9 we must "redefine" the zero of the centigrade scale. This redefined centigrade scale is called the absolute scale or Kelvin scale of temperature. Experiments show that water freezes at $T = 273°K$ (Kelvin). Thus the zero of the Kelvin scale is at $-273°C$ on the centigrade scale. At this temperature all mechanical motion of the molecules should stop. (According to the quantum theory there will still be a "zero point energy.") This special temperature is called absolute zero. One of the consequences of defining temperature as a measure of the molecular motion is that there can be no temperatures lower than absolute zero. Thus far laboratory experiments have reached temperatures lower than $0.0001°K$. Note that a temperature difference of $1°K$ is also $1°C$. The Kelvin scale is a centigrade scale with its zero shifted by $273°$.

Ans. 3: The average velocity squared would increase by a factor of 4 and so would the pressure which is proportional to $\overline{v^2}$.

We can now write the ideal gas law in its most complete form by substituting the right-hand side of Eq. 6-9 into Eq. 6-8. Then

$$V = \frac{2}{3} \frac{N}{P} \left(\frac{3}{2} kT \right)$$

or

Ideal gas law

$$PV = NkT \qquad (6\text{-}10)$$

Actually gases will condense and liquefy at temperatures somewhat above absolute zero. Obviously Eq. 6-10 does not hold in the temperature region where the gas changes into a liquid. Strictly speaking Eq. 6-10 is exactly true only for what is called an ideal gas. In an ideal gas the total volume of the molecules must be much less than V and the molecules must behave as hard spheres that exert no force on each other except during collisions. So far temperature is just a mathematical definition. It will be of no practical use to us unless we can demonstrate what is called the Zeroth Law of Thermodynamics. This law states that any two substances in statistical equilibrium (thermal equilibrium) will be at the same temperature. That is, if two different ideal gases having molecules of mass m_1 and m_2 are mixed together in the same volume, both gases will come to the same temperature after such close contact. Equation 6-9 defines the temperature of gas 1 as $T_1 = 2(\overline{\text{KE}})_1/3k$ and the temperature of gas 2 as $T_2 = 2((\overline{\text{KE}})_2/3k$. In order to prove the Zeroth Law of Thermodynamics, we must show by using Newtonian mechanics that $(\overline{\text{KE}})_1$ will equal $(\overline{\text{KE}})_2$. We must prove that

$$\tfrac{1}{2}m_1\overline{v_1^2} = \tfrac{1}{2}m_2\overline{v_2^2}$$

or

$$\frac{\overline{v_1^2}}{\overline{v_2^2}} = \frac{m_2}{m_1}$$

The rigorous proof uses mathematics beyond the scope of this book. However, we can get a feeling for how the proof goes by examining what happens when m_1 and m_2 start out with the same velocity and then have a head-on collision. Suppose m_1 is the larger mass. Then m_1 will be slowed down by the collision. The collision of m_2 with a heavier moving object is like that of a tennis ball with a moving tennis racket. The speed of m_2 will be increased. After many successive

Q.4: At what temperature would molecules have half the $\overline{\text{KE}}$ that they have at room temperature? $T =$ _____°C.

collisions the average $m_1v_1{}^2$ will equal the average $m_2v_2{}^2$. Thus any two substances when at the same temperature will have the same average kinetic energy of translational motion.

The Zeroth Law of Thermodynamics holds for all substances, not just ideal gases. The two ideal gases having molecules of mass m_1 and m_2 need not be mixed in the same container, they could be in separate containers with a glass wall between them. Suppose we start out with gas 1 hotter than everything else. The collisions of gas 1 molecules would transfer energy to the glass atoms until an equilibrium was reached. Now that the glass atoms are moving faster, they will transfer energy to the gas 2 molecules until they reach the same average kinetic energy per molecule as that of gas 1. Our gas thermometer will read the same whether it is inserted into gas 1, gas 2, or just the glass wall.

To conclude our discussion of temperature, we ask a thought question. A box contains N identical molecules at a temperature T. If the number of molecules is doubled, but the average kinetic energy per molecule is kept the same, what happens to the temperature? Anyone who has trouble with this thought question does not fully understand temperature. The answer is obtained by looking at our definition of temperature (Eq. 6-9) and noting that it is independent of N.

Example 5

If the mass of a hydrogen atom is 1.67×10^{-24} gm, what is the density of hydrogen gas at atmospheric pressure and $0°C$?

The density D will be the mass Nm divided by the volume V.

$$D = \frac{Nm}{V}$$

Now solve Eq. 6-10 for V and substitute in the above:

$$D = \frac{Nm}{\left(\dfrac{NkT}{P}\right)}$$

$$D = \frac{mP}{kT}$$

The quantity m will be twice 1.67×10^{-24} gm since there are two atoms in each hydrogen molecule. The value for P is 1.01×10^6 dynes/cm^2 and the value for T is $273°K$. Inserting these values gives

Ans. 4: On the absolute scale room temperature is about $300°K$, and one half of this is $150°K$ or $T = -123°C$.

$$D = \frac{2 \times 1.67 \times 10^{-24} \times 1.01 \times 10^6}{1.38 \times 10^{-16} \times 273} \text{ gm/cm}^3$$

$$= 8.9 \times 10^{-5} \text{ gm/cm}^3$$

6-7 Avogadro's Law

It is the number of particles that counts

Avogadro's law states that any two gases, which have identical pressures, volumes, and temperatures, will contain the same number of molecules. We can show that Avogadro's law is a consequence of the kinetic theory by solving Eq. 6-10 for N. For gas 1,

$$N_1 = \frac{P_1 V_1}{k T_1}$$

For gas 2,

$$N_2 = \frac{P_2 V_2}{k T_2}$$

Now if $P_1 = P_2$, $V_1 = V_2$, and $T_1 = T_2$, we have

$$N_1 = N_2$$

Thus one liter of any gas at room temperature and atmospheric pressure contains the same number of molecules as a liter of a different gas also at room temperature and atmospheric pressure. The ratio of the masses of the two gases would then be m_1/m_2, the ratio of their molecular masses.

Example 6

How many molecules are contained in 1 cm³ of air at 0°C? According to Eq. 6-10,

$$N = \frac{PV}{kT}$$

$$N = \frac{(1.01 \times 10^6)(1)}{(1.38 \times 10^{-16})(273)} = 2.68 \times 10^{19}$$

Q.5: If box 2 has twice the volume of box 1 and holds a gas at twice the temperature and pressure, what is the ratio of particles in the boxes? $\dfrac{N_2}{N_1} =$ _____ .

The mole

The ratio of the atomic masses of oxygen to hydrogen are 16.00 to 1.008. In order to have a standardized table of relative atomic masses, chemists have defined the atomic "weight" of naturally occurring oxygen to be exactly 16.* Then the molecular "weight" of an oxygen molecule (O_2) is 32 and that of a hydrogen molecule (H_2) is 2.016. *The term mole is defined as the molecular "weight" in grams.* Thus one mole of H_2 is 2.016 gm of hydrogen gas and one mole of H_2O is 18.016 gm of water. The number of molecules in one mole is defined as Avogadro's number N_0 and can be calculated if the mass of an individual molecule is known. By x-ray techniques that accurately determine the interatomic spacing in solid crystals, the mass of the hydrogen atom is known to be 1.67×10^{-24} gm. The number of molecules in a mole is then 2.016 gm of hydrogen gas divided by $2 \times 1.67 \times 10^{-24}$ gm (mass of one H_2 molecule). The result is

Avogadro's number

$$N_0 = 6.02 \times 10^{23}$$

According to Avogadro's law, one mole of any gas at atmospheric pressure and $0°C$ should occupy the same volume as any other gas also at atmospheric pressure and $0°C$. The volume occupied by one mole of gas under these conditions can be obtained from Eq. 6-10.

$$V = \frac{N_0 kT}{P_0}$$

$$= \frac{(6.02 \times 10^{23}) \times (1.38 \times 10^{-16}) \times 273}{1.01 \times 10^6} \text{ cm}^3$$

$$= 22.4 \times 10^3 \text{ cm}^3$$

$$= 22.4 \text{ liters}$$

Example 7

Carbon has atomic "weight" 12. If each carbon atom contains 6 electrons, how many electrons are there in 1 gm of carbon?

One mole or 12 gm of carbon contains N_0 carbon atoms or $6N_0$ electrons. Thus 1 gm of carbon contains $6N_0/12$ or 3.01×10^{23} electrons.

---

*Physicists did not like this definition because there are three isotopes of oxygen occurring in nature of which oxygen 16 is most abundant. In 1961 physicists and chemists got together and agreed on an atomic mass scale which is based upon a single isotope of carbon, carbon-12. Now it is the atom of carbon-12 which has exactly 12 mass units. This changes the previous scale by less than one part in 10^4, so we will not worry about it in this book.

Ans. 5: Since $\dfrac{P_2 V_2}{T_2} = 2\dfrac{P_1 V_1}{T_1}$, $N_2 = 2N_1$

6-8 The Kinetic Theory of Heat

A microscopic view of macroscopic energy

The unit of heat in the metric system is the calorie. One calorie is defined as the amount of heat required to raise the temperature of 1 gm of water by 1°C. As discussed in Section 5-10 a given amount of heat can be obtained from an equivalent amount of mechanical energy. For example, a known amount of work can be done on a known amount of water by rotating a paddle wheel inside the water. The temperature of the water is observed to increase proportionally to the amount of work done. All such experiments give the result that one calorie of heat is equivalent to 4.18 joules of energy.

Mechanical equivalent of heat

1 calorie = 4.18×10^7 ergs

Instead of heating up water, let us heat up one mole of monatomic (one atom to the molecule) gas. The amount of heat required to raise the temperature by one degree on one mole of a gas kept at a constant volume is called the specific heat C_v (the subscript indicates the condition of constant volume). Experiments show that the specific heat of helium is 2.98 calories/mole–degree; that is, it takes 2.98 calories, or 12.5×10^7 ergs of work to increase the temperature of 4 gm of helium by 1°C. Where did the energy go? At this state we suspect that the energy completely went into increasing the kinetic energy of the helium molecules. Let us see now if the experimental value of 12.5×10^7 ergs checks with the prediction of the kinetic theory. According to the kinetic theory, the kinetic energy of one mole of helium molecules is given by multiplying Eq. 6-9 by N_0:

$$\text{KE/mole} = N_0 \times \tfrac{3}{2}kT$$

The increase in kinetic energy per degree centigrade is

$$\text{KE/mole-degree} = \tfrac{3}{2}N_0 k$$

$$C_v = \tfrac{3}{2} \times 6.02 \times 10^{23} \times 1.38 \times 10^{-16}$$
$$\text{erg/mole-degree}$$
$$C_v = 12.5 \times 10^7 \text{ ergs/mole-degree}$$

We see that the kinetic theory gives a complete physical understanding of what this "mysterious" quantity heat really is.

Q.6: How many liters of hydrogen combines with how many liters of oxygen at 0°C and atmospheric pressure to form 18 gm (1 mole) of water? V_{H_2} = _____ liters. V_{O_2} = _____ liters.

For diatomic molecules (two atoms to the molecule), such as H_2, O_2, and N_2, the heat not only goes into kinetic energy of translational motion but also into kinetic energy of rotational and vibrational motion. Hence the specific heat of a diatomic gas is somewhat greater than the specific heat of a monatomic gas.

Newtonian or classical mechanics makes definite predictions of just how much energy should go into rotational and vibrational motion of molecules. In general, the measured specific heats of gases (heat required to raise the temperature by $1°C$) do not agree with the predictions of classical mechanics. However, there is excellent agreement between specific heat measurements and the newer quantum theory (see Chapter 12).

As an example of the failure of classical mechanics we need only consider the rotational motion of molecules in monatomic gases. Each atom has three degrees of freedom of rotation in addition to its three degrees of freedom of translational motion. Using Newton's laws and higher mathematics, one can derive the law of equipartition of energy which predicts each degree of freedom of a molecule will take up an average kinetic energy of $\frac{1}{2}kT$. Hence the average kinetic energy per atom for a monatomic gas should be $6 \times \frac{1}{2}kT = 3kT$, whereas the experimental value at room temperature is close to $\frac{3}{2}kT$. This discrepancy was explained away by assuming that atoms were so "smooth" that no kind of a collision could start them rotating. But even this could not explain the discrepancy in the case of diatomic molecules. A diatomic molecule has two rotational degrees of freedom plus one vibrational degree of freedom. The average kinetic energy per molecule will then be $\frac{3}{2}kT$ for the translational motion plus $\frac{2}{2}kT$ for the rotational motion plus $\frac{1}{2}kT$ for the vibrational kinetic energy. Also since the average potential energy of an oscillator in simple harmonic motion is equal to the average kinetic energy, there must be an additional $\frac{1}{2}kT$ that goes into the potential energy of vibrational motion. Hence

$$\text{KE/mole} = N_0 \times \tfrac{7}{2}kT$$

$$\text{KE/mole-degree} = 6.02 \times 10^{23} \times \tfrac{7}{2} \times 1.38 \times 10^{-16}$$

$$C_v = 29.1 \times 10^7 \text{ ergs/mole-degree}$$

Ans. 6: The chemical reaction is

$$H_2 + \tfrac{1}{2}O_2 \rightarrow H_2O,$$

or one mole of molecular hydrogen plus $\frac{1}{2}$ mole of molecular oxygen gives one mole of H_2O; hence, $V_{H_2} = 22.4$ liters, $V_{O_2} = 11.2$ liters.

However, at room temperature the observed value is 20.8×10^7 ergs/mole-degree. The diatomic molecule behaves as if only its rotational degrees of freedom and none of its vibrational motion can be excited. But at much higher temperatures the vibrational motion finally shows up. This "strange" temperature dependence of the specific heat is accurately predicted by the quantum theory. Classically, the specific heat should be independent of temperature.

6-9 Changes of State

Imprisonment versus freedom (of molecules)

A collection of molecules can occur either in the solid, liquid, or gaseous state, depending on the strength of the intermolecular forces and the average kinetic energy per molecule or temperature. In a solid at room temperature all the molecules are vibrating about their equilibrium positions with an average kinetic energy of $\frac{3}{2}kT$ (neglecting quantum effects). However, the intermolecular forces are so strong that it takes a kinetic energy much greater than $\frac{3}{2}kT$ for a molecule to free itself from its position in the solid. As the temperature is increased, the molecules do finally reach a temperature where their relative positions are no longer fixed but where the forces are still strong enough to hold them together. This temperature is known as the melting point. A collection of molecules that are held together, although their relative positions are not fixed, is the microscopic description of a liquid. In a liquid a single molecule will have great difficulty in breaking away from the main body of molecules.

At an even higher temperature (the boiling point), most of the molecules do have enough energy to break away completely from the intermolecular forces. Even at temperatures below the boiling point, an occasional molecule near the liquid surface that happens to be enough above average in kinetic energy can escape from the surface. This process wherein the highest velocity molecules escape from the surface is called evaporation. The escape of the highest velocity molecules will tend to lower the average kinetic energy or temperature of the remaining molecules. Hence, in order to convert a given amount of liquid from the liquid state to the gaseous state at the same temperature, a certain quantity of

Q.7: If in a solid each atom has 3 vibrational degrees of freedom, what would be the specific heat per mole?

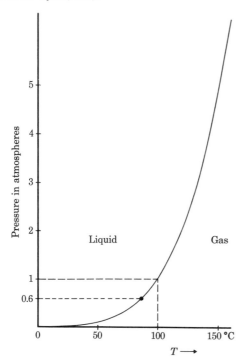

Fig. 6-12. Vapor pressure curve of water in the temperature region 0 to 150°C. The pressure scale is in units of atmospheres (1.01×10^6 dynes/cm²).

heat must be supplied. This quantity of heat per gram of liquid is called the latent heat of evaporation. The latent heat of evaporation for H_2O at 100°C is 540 calories/gm. We see that it takes a significant amount of energy to boil away a pot of water. For similar type reasons it takes 80 calories/gm to melt a gram of ice at 0°C. This is called the latent heat of fusion. In general, a change of state of any substance will have an associated latent heat.

Vapor pressure

Suppose a certain quantity of water is inserted into a closed evacuated box. The water will continue to evaporate until the number of molecules leaving the surface per second equals the number returning to the surface each second. At any given temperature T this equilibrium condition will determine a unique pressure P of the gas. This particular value of P corresponding to the particular value of T is called the vapor pressure. The gas when in equilibrium with its liquid is usually called a vapor. The curve in Fig. 6-12 shows the particular value of P for water vapor corresponding to each value of T. Note that for $T = 100°C, P = 1$ atmosphere. This is why water boils at 100°C when at sea level. Cold water inserted into a vacuum chamber will boil just as well.

Example 8

On a 14,000 ft mountain top the air pressure is 40% less than at sea level. At what temperature will water boil when at 14,000 ft?

According to Fig. 6-12 a pressure of 0.6 atmosphere corresponds to a temperature of 86°C. At this temperature the vapor pressure of water would equal the atmospheric pressure on the water and the water would then boil.

If the pressure on a sample of H_2O is greater than that given by the curve in Fig. 6-12 all of the H_2O will be in the liquid state. If the pressure is less than that given by the curve, it will all be in the gaseous state. At the special value of P given by the curve, both liquid and gas can exist simultaneously in equilibrium.

Ans. 7: For each atom there would be $2(\tfrac{1}{2}kT)$ of energy for each vibrational degree of freedom. Altogether we have $\tfrac{6}{2}kT$ per atom or $3N_0kT$ per mole. Hence the specific heat per mole of a solid should be $C = 3N_0k$. This law works quite well, especially at high temperatures. It is called the law of Dulong and Petit.

Fig. 6-13. Vapor pressure curve for a bubble chamber. Before expansion conditions of liquid are represented by point A. Point B corresponds to conditions after expansion.

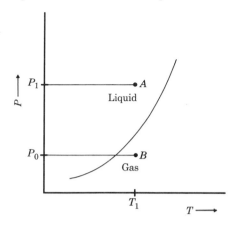

6-10   The Bubble Chamber

A liquid trying to boil faster than it can

Suppose a certain volume of liquid is being pushed by a piston. Let the force on the piston be such that the pressure is P_1 at a temperature T_1 corresponding to point A in Fig. 6-13a. Now suddenly let go of the piston. The pressure on the liquid drops to P_0, atmospheric pressure, corresponding to point B on Fig. 6-13. The liquid is suddenly put under pressure and temperature conditions in which it should be a gas instead. Under such special conditions a liquid is called superheated and is unstable. As soon as the pressure on the piston is released, many liquids will begin boiling on metal surfaces, but not on glass surfaces. But, if a charged particle like an electron or proton passes through such a superheated liquid, it will start boiling at various points along the track. The theory of where a liquid should first start boiling is not well understood. If one chooses a liquid that will not start boiling on a glass surface, but will start boiling along the path of a charged particle, a photograph of the liquid taken through a glass window will show the tracks of charged particles having traveled in the liquid. Photographs of bubble chamber tracks are shown in Figs. 3-2, 8-12, 16-4, 16-6, and 16-7. The first bubble chamber ever built with metal walls, which is shown in Fig. 6-14 was 2 in. in diameter. This bubble chamber was built in 1954 by D. A. Glaser, its inventor. By 1959 a bubble chamber in the shape of a 6-ft coffin had been built by a group under the direction of L. W. Alvarez at the Lawrence Radiation Laboratory at the University of California. This liquid hydrogen bubble chamber along with its auxiliary apparatus shown in Fig. 6-15 is so heavy that it cannot be supported by wheels. Instead it moves itself around by lifting itself up and down on 4 giant feet. A 14-ft. diameter liquid hydrogen bubble chamber is now under design at Brookhaven National Laboratory.

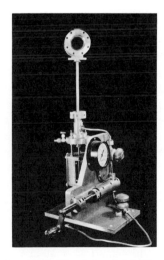

Fig. 6-14. First bubble chamber ever built using metal walls. The chamber which is 2 in. diameter was built in 1954. (Courtesy Professor D. A. Glaser.)

6-11   Statistical Mechanics

"All the kings horses and all the kings men, couldn't put Humpty Dumpty together again."

Any sample of matter contains such an enormous number of molecules that it takes mathematical methods called

Fig. 6-15. Stainless steel "coffin" for 72-in. liquid hydrogen bubble chamber. This and the following photograph are the courtesy of the 72-in. Bubble Chamber Group of the Lawrence Radiation Laboratory, University of California.

statistics to describe its physical properties. As an example let us consider two boxes, each 1 cm³, with a partition between them as shown in Fig. 6-17. If the pressure in Box 1 is 1 atmosphere, the number of particles in the box will be 6.02×10^{23} divided by the number of cubic centimeters in 22.4 liters, or 2.7×10^{19} particles/cm³. At the start Box 2 is empty. We now open up a hole in the partition, and within a short time we shall find half of the particles in Box 2. The gas expanded into the vacuum. No matter how long we wait, the reverse of this process will never occur. Actually the number of particles in Box 2 will fluctuate slightly. Mathematical statistics tell us that about 70% of the time the number of particles in a given volume will lie between $N - \sqrt{N}$ and $N + \sqrt{N}$, where N is the average number. In this case we have

$$(1.35 \times 10^{19} \pm \sqrt{1.35 \times 10^{19}}) =$$
$$(1.35 \pm 0.00000000037) \times 10^{19}.$$

We see that the fluctuations are so small that they are in-

Fig. 6-16. Seventy-two inch bubble chamber with magnet assembly "walking" from one building to another at the Lawrence Radiation Laboratory.

Fig. 6-17. Irreversible expansion of gas into a vacuum. Gas from Box 1 will expand into empty Box 2 when hole in partition is opened.

| Box 1 | Box 2 |

capable of detection and that it is virtually impossible to have a fluctuation so large that no particles are left in Box 2.

Suppose, however, that after opening up the hole, and after half the particles have escaped from Box 1 to Box 2, time was suddenly stopped and made to run backwards. Physically time can never run backwards, but we can observe what it would be like by taking a movie of the experiment and running the film backwards in the projector. Box 2 would then spontaneously empty and produce a vacuum. We are now faced with a paradox. We know that in nature a box would never spontaneously empty and produce a vacuum when open to the outside air; yet, in the movie film run backwards, none of Newton's laws were violated. In fact, the movie film told us of a specific configuration of particle positions and velocities in Box 2 that would require the particles there to move and collide in such a way that they all would leave the box. During this process there would be no violation of any law of physics. The paradox is resolved by noting that for this one special configuration of particles in Box 2 there are virtually an infinite number of configurations where the particles will stay almost equally in both boxes. Thus in practice this configuration which allows all the particles to leave Box 2, although permitted, never occurs. Hence the process of a gas expanding into a vacuum is irreversible, even though in principle it is possible to have a situation where the vacuum is "spontaneously produced."

Another example of irreversibility because of statistics is the flow of heat from a higher temperature to a lower temperature. If a hot and a cold piece of metal are in contact, heat will never flow from the cold metal to the hot metal. If this were ever possible one could extract heat from a piece of metal or from the ocean and use it to run a steam engine. There is enough heat energy contained in the ocean to satisfy all of man's power requirements for millions of years. However, it is impossible statistically to convert this disordered kinetic energy of the water molecules into the ordered motion of a machine. A machine run by heat energy extracted from a single heat reservoir is called a perpetual motion machine of the second kind. Such a machine would not violate the law of conservation of energy. All that is required by the conservation of energy is that, as such a machine does work, the temperature of the heat reservoir must decrease. The conservation of energy applied to problems of heat flow is called the first law of thermodynamics.

Second law of thermodynamics

There must be some new law that forbids us to get useful energy out of the heat in the ocean and that forbids the spontaneous formation of a vacuum, to name only two examples. Humpty Dumpty is a third example. This law is the second law of thermodynamics and is merely a consequence of the statistics or mathematics of large numbers. In this sense there is no new fundamental physical principle involved here—it is a mathematical consequence of having a very large number of particles. One way of stating the second law of thermodynamics is that a perpetual motion machine of the second kind is impossible in practice. Another way is that the total amount of disorder (also called entropy) in the universe can increase, but it cannot decrease.

Time reversal

We see that the second law of thermodynamics when expressed as a mathematical equation predicts a different physical result, depending whether time flows forward or back. The preceding discussion has attempted to point out that this is purely a consequence of statistics and that in this sense the second law of thermodynamics is really not a new law or fundamental principle of physics at all. It is a statistical macroscopic description that works only when there are a large number of particles. It can be deduced from Newton's laws and the kinetic theory using mathematics (statistics). On the other hand, the application of mathematical statistics to Newton's laws opens up a whole new field of physics and provides us with useful physical concepts such as temperature.

So far all the truly fundamental laws of nature we have encountered are time reversible. Time reversible means that if one reverses the directions of motion of all the particles (including their rotations), then the same equations or laws of physics (with time running forward) still hold. This very fundamental symmetry principle of nature has recently been checked by special experiments designed to show up possible violations. In 1964 a violation was found in the weak interactions. If the violation is confined to the weak interaction, it will not affect strong and electromagnetic interactions which determine nuclear and atomic physics. In addition, violations were found of two other very basic symmetry principles (conservation of parity and antiparticle symmetry) that were checked for the same reasons as time reversibility. The recent overthrow of these three "sacred" symmetry principles is discussed in Chapter 16.

Problems

1. How far below the surface of a lake must you dive to find the pressure on you twice that at the surface?

2. At a temperature of 0°C and a pressure of 1 atm, the average velocity of hydrogen molecules is 1.8×10^5 cm/sec. Suppose a sample of hydrogen at the above temperature and pressure is heated with the pressure remaining constant until the volume is three times as great.

(a) What would be the temperature of the gas at the new volume?

(b) What would be the average velocity under the new conditions?

Prob. 3

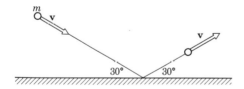

3. A molecule of mass M strikes a surface at an angle of 30° as shown and then bounces off at an angle of 30° with no change in magnitude of velocity. What is the change in momentum of this molecule?

4. A certain gas has a molecular weight of 39. What will be the mass of 6.02×10^{23} molecules of this gas?

5. When New York State had an elaborate canal system, there was a canal that crossed over a river on a bridge. What is the increased load on this bridge when it carried a boat of mass 10^5 kg? The average density of the boat is 0.8 gm/cm³.

6. Can heat be extracted from water when it is at absolute zero? Explain.

7. The area of contact between a phonograph needle and an LP record is a circle of 10^{-3} in. diameter. If the needle exerts a force of 5 gm times g, what is the pressure in dynes/cm²?

8. A hydraulic jack has two pistons of diameters 1 and 5 cm.

(a) What is the force required on the small piston to enable the large piston to lift an object weighing 10 newtons?

(b) How far will the object be raised when the small piston is moved 0.1 m?

9. A weather balloon has a mass of 50 kg (including the enclosed gas) and a volume of 110 m³. It is tied to the earth with a rope. Density of air = 1.3 kg/m³.

    (a) If the rope stands vertical, what is the tension in it?

    (b) If there is a wind causing the rope to assume a direction at an angle of 30° to the vertical, what is the tension in the rope?

10. An object when weighed under normal conditions weighs 300 newtons. When weighed in water this same object weighs 200 newtons.

    (a) What is its density?

    (b) What is its volume?

11. The temperature of a certain quantity of gas at a pressure of 1 atm falls from 100°C to 0°C. What change in pressure (in atmospheres) is required to keep the gas volume a constant?

12. The density of a 1 cm³ ice cube is measured to be 0.90 gm/cm³. How high will the top of the cube float above the surface of the water?

13. A block of lead of density 11.5 gm/cm³ floats on a pool of mercury of density 13.6 gm/cm³.

    (a) What fraction of the block is submerged?

    (b) If the block has a mass of 2 kg, what force must be applied to it to keep it totally submerged?

14. One liter of a certain gas has a mass of 0.1 gm at 0°C and a pressure of 1 atm. What is the gas?

15. A piece of wood whose density is 0.8 gm/cm³ floats in a liquid of density 1.2 gm/cm³. The total volume of the wood is 36 cm³.

    (a) What is the mass of the wood?

    (b) What is the mass of the displaced liquid?

    (c) What volume of wood appears above the surface of the liquid?

Prob. 16

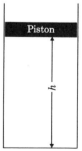

16. A sample of gas is contained in a cylinder which is sealed with a piston that is free to move without friction. The volume of the gas sample is 0.5 m³ and the height h is 1.0 m. The piston weighs 5×10^4 newtons. Atmospheric pressure is 10^5 newtons/m².

    (a) What is the pressure in the gas?

    (b) What additional force must be applied to the piston to reduce h to 0.6 m, keeping the temperature constant?

17. A 1 cm cube of density 0.8 is floating in water. The temperature is increased 50°C and the cube expands 10% in its linear dimensions. What is the increase in volume of water displaced? (Assume the density of water is 1 gm/cm³ at both temperatures.)

18. An underwater bubble of 1 gm of H_2 gas takes up a volume of 5.6×10^3 cm³ at room temperature (21°C).

    (a) What is the pressure in the gas?

    (b) How far below the water surface is the bubble? The atmospheric pressure above the water surface is 10^6 dynes/cm².

19. In a box of volume V we have N particles where the average KE per particle is ϵ. Give all answers in terms of V, N, ϵ, and k.
 (a) What is the total KE in the box?
 (b) What is the temperature in the box?
 (c) What is the pressure in the box?
 (d) If the volume is doubled by connecting to an empty box of the same volume, what happens to the temperature and pressure?

20. The best vacuum man can achieve is on the order of 10^{-14} cm of Hg. How many molecules still remain in 1 cm^3 in this so-called vacuum? The vacuum of interstellar space contains about one proton per cubic centimeter.

21. During an H-bomb explosion, the temperature is 10^8 degrees centigrade. In this very hot gas there will be free hydrogen nuclei or protons and also many deuterons (twice the mass of a proton).
 (a) What is the average velocity of the protons?
 (b) What would be the ratio of the average kinetic energy per proton to the average kinetic energy per deuteron assuming thermal equilibrium?

22. A sample of an ideal gas occupies a volume V at a pressure P and absolute temperature T. The mass of each molecule is m. Which of the following expressions gives the number of molecules in the sample: PV/m, k/PVT, m/k, kT/V, PV/kT? Which of the following expressions gives the density of the gas: mkT, m/V, Pm/kT, P/kTV, or P/kT?

23. An ideal gas is contained in a vessel of fixed volume at $0°C$ and atmospheric pressure.
 (a) If the average velocity per molecule is doubled, what will be the new temperature?
 (b) If the average velocity per molecule is doubled, what will be the new pressure?
 (c) If the volume of the vessel is one liter, how many molecules does it contain?

24. Consider two gases where the molecules of Gas 2 have $\frac{1}{2}$ the mass of the molecules of Gas 1 ($m_2 = \frac{1}{2}m_1$). When Gas 1 is at a temperature of $27°C$ its molecules have an average velocity of 10^5 cm/sec. What is the temperature of Gas 2 when its molecules have an average velocity of 2×10^5 cm/sec?

25. The atomic weight of oxygen is 16. Eight grams of O_2 is contained in an 8-liter vessel (1 liter = 10^3 cm^3). The pressure in the gas is 1 atm (10^6 dynes/cm^2).
 (a) How many moles of oxygen gas O_2 are in the vessel?
 (b) How many molecules of O_2 are in the vessel?
 (c) What is the temperature?
 (d) Assuming the O_2 behaves as an ideal gas at all temperatures, what is the total translational KE in the vessel?

26. An ideal gas in a box has N molecules. Now double the number of molecules in the same box, while keeping the total kinetic or heat energy of the gas the same as before (the total energy content of the new amount of gas is the same as that of the original amount of gas).

(a) What will be the ratio of the new pressure to the original pressure?

(b) What will be the ratio of the new temperature to the original temperature?

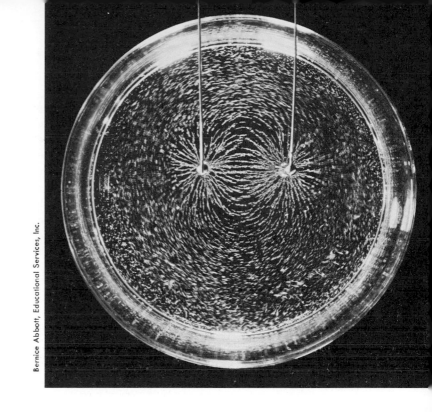

Bernice Abbott, Educational Services, Inc.

Electrostatics

Chapter 7

Electrostatics

7-1 Electronic Structure of Matter
Electrons everywhere

The study of electricity and its many applications comprises a major part of physics. One of the applications, as we shall see, is the electromagnetic theory of light. Thus the entire subject of light could be considered as a side topic in electricity. Another subject of great importance, the theory of relativity, follows deductively from the theory of electricity if we assume that the equations of electricity must have the same form in all reference systems. In the study of atomic physics, solid state physics, molecular theory, and chemistry we deal with the electromagnetic electron-electron and electron-nucleus interactions. The forces between atoms and molecules result from the electromagnetic interaction. Thus nearly all of this book involves the vast subject of electricity in one way or another.

This book assumes that the reader has a general idea of atomic structure. Atomic symbols showing a nucleus encircled by revolving electrons appear every day in newspapers, magazines, television, children's toys, breakfast cereal containers, etc. An atom is approximately 10^{-8} cm in diameter and consists of a small, heavy nucleus of positive charge encircled by lighter orbital electrons of negative charge. Notice that the terms positive and negative charge have just been used without first defining them. The entire next section is devoted to presenting this important new concept of electric charge.

The familiar atomic symbols are based on a theory of atomic structure proposed by Niels Bohr in 1914. Figure 7-1 represents the lithium atom according to the Bohr model. Lithium has an atomic weight of 7 and atomic number $Z = 3$. The lithium nucleus is made up of three positive protons and four neutral neutrons. Suppose electrons are attracted to a nucleus by an inverse square force. Then, as Newton has shown, a single atom would be like a miniature solar system with each electron in a planetary orbit revolving about the nucleus. At first one might think that it is the gravitational force which holds an atom together. However, if this were so, atoms would always attract each other no matter how close together. Matter could then be compressed until it was all

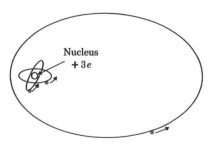

Fig. 7-1. The lithium atom according to the Bohr model. Three electrons, each of charge $-e$, revolve in planetary orbits around the nucleus of charge $+3e$.

Nucleus
$+3e$

one giant nucleus of protons, neutrons, and stationary electrons. Because of the great difficulty in compressing liquids or solids, we know that atoms must exert strong repulsive forces on each other when compressed. Thus, there must exist a new type of repulsive force which is stronger than the attractive gravitational force. The explanation of this new force requires what is called the electrostatic force or Coulomb force, which turns out also to be an inverse square force. One difference between electrostatic force and gravitational force is that there are both positive and negative electric charges and the electrostatic force is repulsive between bodies of like charge and attractive for bodies of unlike charge. Protons and atomic nuclei repel each other, but electrons are attracted to nuclei by the electrostatic force. Furthermore, the electrostatic repulsive force between two electrons is 4.17×10^{42} times as strong as the attractive gravitational force! The reason why gravitational forces seem stronger to ordinary people than electrical forces is that people and objects around them have equal parts of positive and negative charge. If a person had just a small excess of one kind of charge, a lightning bolt would immediately strike him down! Then he would be convinced that electricity is a stronger effect than gravity.

Actually the picture of electrons in planetary orbits about a nucleus is not correct. It conflicts with the more modern theory of quantum mechanics which is discussed starting with Chapter 12. However, when the quantum theory is taken into account, the electrostatic force does play the role of keeping the "orbital" electrons bound to the nucleus. The modern electron wave picture of the lithium atom which replaces Fig. 7-1 is on page 150.

7-2 Concept of Charge

It's all in the mind

We cannot give a quantitative definition of charge until we write down Coulomb's law which describes all electrostatic forces. Before doing that it will be helpful to discuss certain properties of charge. For example, is charge an additive quantity? Can it be destroyed?

Q.1: Which is stronger, the electrostatic force between two electrons, or the electrostatic force between two protons?

The effects of charge were first noticed by ancient man in the same way that modern man cannot help but notice electrical sparks when he walks across a carpet on a dry day, or when he takes off a coat or shirt. These effects are called electrification and result from the contact of two dissimilar substances. Effects of electrification were studied by the ancient Greeks.

Suppose substance A is rubbed against substance B. Then some of the orbital electrons of A will interact with electrons of B. Some of them will leave substance A and attach to B, and vice versa. If the outer electrons of A are not bound as tightly as those of B, a few of the electrons of A will be missing when the two materials are separated. The object A would be left with a net positive charge and B with a corresponding negative charge. The terms positive and negative are arbitrary when applied to charge. It was Benjamin Franklin who established the present-day convention that electrons are negative. He arbitrarily defined the charge left on glass after being rubbed by silk as positive. In the following chapters, when we study currents in conductors and in vacuum tubes, we will wish Franklin had chosen the reverse definition. According to Franklin's definition positive current will have the direction opposite to that of the actual moving charge. In conductors and vacuum tubes the current is due to the motion of the negative electrons. We must not blame Franklin for adding to the confusion because his theory of electric charge preceded the discovery of the electron by about one hundred years.

The concept of charge has similarities to the concept of mass. Just as every body or particle has the abstract attribute of mass assigned to it, so has every body or particle an inherent charge which may be positive, negative, or zero. We found that the calculations of interacting bodies (their relative accelerations, etc.) were greatly simplified by the introduction of the purely abstract concept of mass. Similarly, the introduction of the concept of charge gives us a simple representation of this new type of force which can hold one atom together and at the same time keep two atoms apart. The following quotation helps explain the reality of charge: "When I say that an electron has a certain

Ans. 1: Neither, they are the same.

amount of negative charge, I mean merely that it behaves in a certain way. Charge is not like red paint, a substance which can be put on to the electron and taken off again, it is merely a convenient name for certain physical laws."[*]

The question arises: Just what kind of a quantity is charge? If a first charge Q_1 and then a second charge Q_2 is applied to a body, will the resultant charge be $Q_1 + Q_2$? Is there any method by which charge can be created or destroyed? For example, rest mass can be created in pair production of an electron and positron (positive electron) by a photon of zero rest mass. The answer is that conservation of charge is postulated to be a strictly obeyed law of physics. Like other basic laws of physics, there is no absolute proof. The only proof we have for conservation laws or other laws of nature is that they continue to withstand repeated testing. So far in this book we have accumulated five conservation laws which are considered to be exact (no violation of any kind permitted). They apply only to closed systems (no outside agents). Our list of conservation laws now reads:

1. *Conservation of total energy*

One must use relativistic energy which includes the rest energy $W_0 = M_0 c^2$.

2. *Conservation of total linear momentum*

To be exact one must use relativistic momentum which is relativistic mass times velocity.

3. *Conservation of total angular momentum*

4. *Conservation of heavy particles*

Rest mass can be converted to energy, but with the constraint that the total number of protons, neutrons, and certain other heavy particles (see Chapter 16) must forever remain constant.

5. *Conservation of charge*

The algebraic sum of the charge in a closed system must always remain the same.

All electrons and protons have the same charge, but with opposite sign. This quantity of charge is called the electronic charge e and is the smallest amount of charge any body or

Q.2: Is the electric charge of an atomic nucleus directly proportional to its mass?

[*]Bertrand Russell, *ABC of Atoms,* E. P. Dutton, 1923.

particle can have. We say that charge is quantized; that is, it can only have values 0, $\pm e$, $\pm 2e$, $\pm 3e$, etc., where e is the electronic charge. In fact, according to the quantum theory, other physical quantities such as energy, momentum, and angular momentum are quantized. For the rest of this section on electricity we shall continue with the approximation that energy, momentum, and angular momentum can be continuous variables. This approximation is called classical electrodynamics and is quite adequate for dealing with electrical interactions between macroscopic bodies. For electrical interactions of microscopic bodies (atomic dimensions) quantum theory is an absolute necessity.

The fundamental theory is called quantum electrodynamics. Classical electrodynamics is a limiting case; that is, when quantum electrodynamics is applied to macroscopic bodies it usually gives the same answer as classical electrodynamics.

7-3 Coulomb's Law

The "new" force

Experiments of very great accuracy show that the electrostatic force between two charges is proportional to the product of the charges and varies inversely as the square of the distance in the same manner as the gravitational force. This behavior of the electrostatic force is called Coulomb's law. This law states that the force between two stationary point charges Q_1 and Q_2 is

$$F = K \frac{Q_1 Q_2}{r^2} \tag{7-1}$$

where r is the distance between the two charges and K is a positive proportionality constant which depends on the choice of the unit of charge. If Q_1 and Q_2 are of opposite sign, their product is negative and F is directed opposite to r; that is, F is attractive. The relation between the direction of F and the signs of the charges is illustrated in Fig. 7-2. The electrostatic force can be either repulsive or attractive, depending whether the two charges are of the same or opposite sign.

Fig. 7-2. Dependence of the direction of the electrostatic force on the signs of the two charges.

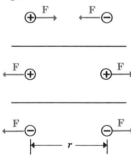

Ans. 2: No. Different atomic nuclei contain different numbers of neutrons which have zero charge.

In the CGS system the units for r and F are centimeter and dyne, respectively. If the charge of the electron were used as the unit for Q, then K as determined by experiment would be an unwieldy 2.3×10^{-19}. Since we have the freedom of define the size of the unit charge, it would be advantageous to define it so that $K = 1$. This is what is done in the CGS system of units. The CGS unit of charge is defined by setting $K = 1$ in Eq. 7-1. This unit of charge is given the name statcoulomb. One statcoulomb is often referred to as one esu (short for electrostatic unit) of charge. To avoid confusion and to make the physics easier to see, the statcoulomb will be the standard unit of charge used throughout this text. Hence, *Coulomb's law* becomes

Coulomb's law
$$F = \frac{Q_1 Q_2}{r^2} \qquad (7\text{-}2)$$

where Q is in statcoulombs. *We see that the unit electric charge (statcoulomb) is defined as that charge which, if placed 1 cm distant from a charge of the same magnitude and sign, will repel the latter with an electrostatic force of 1 dyne.* In these units the electronic charge is

$$e = 4.8 \times 10^{-10} \text{ statcoulomb} \qquad (7\text{-}3)$$

Fig. 7-3. The two charged balls of Example 2.

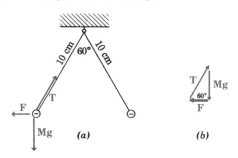

(a)   *(b)*

Q.3: The electrostatic force between any two objects of charges Q_1 and Q_2 is given by Eq. 7-2. True or false?

Example 1

A small carbon ball has a mass of 1 gm. How many electrons are in the ball?

Carbon has $Z = 6$ and atomic weight 12. Hence 12 gm of carbon is one mole and contains 6.02×10^{23} atoms or six times as many electrons. The number of electrons in 1 gm of carbon is then $\frac{1}{12}(6 \times 6.02 \times 10^{23}) = 301 \times 10^{23}$.

Example 2

In Fig. 7-3 two of the above balls are given identical negative charges. They are suspended by 10-cm strings which have an angle of 60° after application of the charge. (a) What is the electrostatic force between the two balls? (b) How many electrons were added to each ball? (c) What is the ratio of electrons to protons in each ball? (d) What is the gravitational force between the two balls?

In Fig. 7-3b the sum of the electrostatic force **F**, the tension in the string **T**, and the downward gravitational force of 980 dynes must be zero. Since in a 30°-60° right triangle the ratio of the short arm to the long arm is $1/\sqrt{3}$, we have $F/980 = 1/\sqrt{3}$ or $F = 565$ dynes.

For Part (b) we must compute how much charge was added to each ball. This can be determined by using Coulomb's law $F = Q^2/r^2$, or $Q = \sqrt{Fr^2}$. Substitution of 565 dynes for F and 10 cm for r gives $Q = \sqrt{565 \times 10^2} = 238$ statcoulombs. The number of electrons added is this value divided by 4.8×10^{-10}, or 4.95×10^{11} electrons.

For part (c) the ratio of added electrons to original electrons (or protons) is the above 4.95×10^{11} electrons divided by the 3.01×10^{23} electrons of Example 1, which is 1.65×10^{-12}. The ratio of electrons to protons is then 1.00000000000165. We note again that very small charge excesses give rise to gross effects.

According to Newton's law of gravitation the mutual gravitational attraction of the two balls is

$$F = G\frac{M^2}{r^2} = 6.67 \times 10^{-8} \times \frac{1^2}{10^2} = 6.67 \times 10^{-6} \text{ dyne}$$

Note that the electrostatic force between the balls is 8.5×10^{11} times as much.

Example 3

According to the Bohr theory of atomic structure the hydrogen atom consists of an electron revolving about a proton in a circular orbit (see Fig. 7-4). The radius of the Bohr electron orbit in hydrogen is 0.53×10^{-8} cm. (a) What is the force between the electron and proton? (b) What is the electron velocity? (c) What is the kinetic energy of the electron?

According to Coulomb's law the force would be

$$F = \frac{e^2}{r^2} = \frac{(4.8 \times 10^{-10})^2}{(0.53 \times 10^{-8})^2} = 8.2 \times 10^{-3} \text{ dyne}$$

Since this is the centripetal force which keeps the electron in orbit we have

$$m\frac{v^2}{r} = F \qquad \text{where} \quad m = 9.1 \times 10^{-28} \text{ gm}$$

is the electron mass.

$$v = \sqrt{\frac{rF}{m}} = \sqrt{\frac{0.53 \times 10^{-8} \times 8.2 \times 10^{-3}}{9.1 \times 10^{-28}}}$$

$$v = 2.18 \times 10^8 \text{ cm/sec}$$

The kinetic energy is

$$KE = \tfrac{1}{2}mv^2 = \tfrac{1}{2}(9.1 \times 10^{-28})(2.18 \times 10^8)^2 = 2.16 \times 10^{-11} \text{ erg}$$

Another common unit of charge that we shall encounter in the next two chapters is the coulomb. For very important reasons which will be discussed in Chapter 8, the coulomb is related to the statcoulomb by the speed of light.

Fig. 7-4. Bohr model of hydrogen atom.

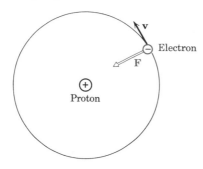

Electron

Proton

Ans. 3: False. Eq. 7-2 applies to each pair of point charges. The net force on an object is the vector sum of the forces on each of its point charges.

$$1 \text{ coulomb} = \frac{c}{10} \frac{\text{statcoul}}{\text{cm/sec}} = 3 \times 10^9 \text{ statcoul}$$

where $c = 3 \times 10^{10}$ cm/sec is the speed of light. Actually there are several different systems of electrical units in use today. One common system used in certain fields of engineering and appearing in some physics textbooks is called the rationalized MKS system. In this system, F is in newtons, r in meters, and Q in coulombs. Then K would be $c^2/10^7$, where c is 3×10^8 m/sec. However, in this system $1/4\pi\epsilon_0$ is used in place of K, so that $\epsilon_0 = 10^7/4\pi c^2$. In the MKS system Coulomb's law becomes

$$F = \frac{1}{\epsilon_0} \frac{Q_1 Q_2}{4\pi r^2} \text{ newtons}$$

where $\epsilon_0 = 8.85 \times 10^{-12}$ coulomb^{2}sec^2/kg m^2. In this book we need not worry about such additional complications.

Another advantage to our system of units is that we will see when, how, and why the speed of light enters into the equations of electricity.

7-4 Electrostatic Induction

Unlimited "production" of charge

One of the easiest ways to charge a glass or plastic rod is to rub it with a cloth. Some of the charge on the rod can then be given to objects such as pith balls by touching them to the charged rod. However, by making use of electrostatic induction it is possible to charge conductors over and over again without giving up any of the original charge of the rod. A conductor has the special property that the outer electrons of each atom are no longer bound to their particular atom. They are free to flow anywhere in the conductor or into a connecting conductor. Thus in Fig. 7-5a when the negatively charged rod is brought near the spherical conductor, the electrons are repelled to the far side. If the sphere is touched by hand (the human body is a conductor), these electrons are now free to be repelled even farther away and leave the sphere completely. If the finger is removed, the sphere is left with a net positive charge. This can be done to as many spheres as desired without losing any of the original charge on the rod.

Fig. 7-5. Charging a conducting sphere by induction.

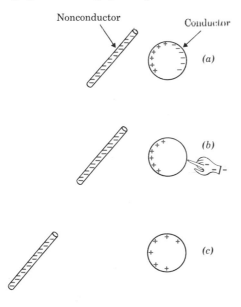

Nonconductor

Conductor

(a)

(b)

(c)

In fact with the help of a larger hollow sphere this original charged rod can be used in principle to build up a giant charge at a potential of millions of volts (potential will formally be defined in Section 7-9). We will see in the next section that a charged conductor must have all its charge on the outside surface and none on the inside. In Fig. 7-6 we can insert the small charged sphere from Fig. 7-5 through a hole into the larger sphere and touch it to the inside surface. Then electrons from the outside surface will immediately run into the small sphere and neutralize its charge. The outside surface will now contain the original positive charge of the small sphere. After repeating this operation many times the total charge on the outside of the big hollow sphere will be the charge on the small sphere times the number of times this process has been repeated. This is the principle of the modern high voltage Van de Graaff generators which are used to produce millions of volts. In the Van de Graaff a continuous belt delivers charge to the inside of a large hollow sphere rather than doing the process by hand and step by step.

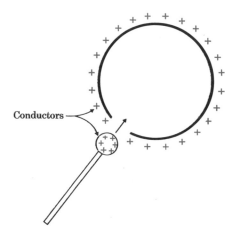

Conductors

7-5 Electric Field

What the force would be everywhere

One of the purposes of using electric and magnetic fields is to avoid the concept of action at a distance. If we have a charge at point P, according to Coulomb's law it is affected by all other charges at their respective distances; that is, force at a distance. But when we use electric field, we say that all these other charges produce some condition in the environment at point P; that is, an electric field at P. If we know the field at P, we can describe the force on a charge at that point without any further reference as to how the field came about. In this sense electric and magnetic fields are mathematical definitions which simplify calculations and should make things easier to understand. The electric field at any position in space is defined as the electric force F on a small stationary test charge q divided by q.

Definition of electric field

$$\mathbf{E} = \frac{\mathbf{F}}{q} \tag{7-4}$$

is the electric field where $\mathbf{F}$ is the electrostatic force on charge q. The units of **E** are dynes/statcoulomb. Hence the electric field at a given point is numerically equal to the force a 1 esu positive charge would experience at that point. Electric field at any point in space is what the force per unit charge would be at that point if we had a test charge available to measure it. The electric field is produced by all the *other* charges in the problem. Although the test charge itself produces an electric field, this contribution must not be used in calculating the force on the test charge.

Next we shall determine the field E produced by one of the "other" charges, a single point charge Q. According to Coulomb's law, the force on a test charge q at a distance r from Q is $F = q(Q/r^2)$. The field E at this position is obtained by dividing by q:

Field produced by a point charge

$$E = \frac{Q}{r^2} \tag{7-5}$$

is the electric field at a distance r from a single point charge Q.

In many physical situations there will be several point charges, or a smooth distribution of charge. Then the resultant field **E** at any given point would be the vector sum of the separate fields due to each charge alone.

Example 4

An electric dipole is defined as two equal and opposite charges Q separated by a distance L as illustrated in Fig. 7-7. What is the field **E** at a point equidistant from each charge of an electric dipole? The distance from the point to each charge is r.

Let E_1 be the field due to charge No. 1. Then

$$E_1 = \frac{Q}{r^2}$$

By similar triangles $E/E_1 = L/r$. Then $E = (L/r)E_1$, or $E = LQ/r^3$.

In the above example note that the force due to a dipole is an inverse cube law rather than the usual inverse square law.

Fig. 7-7. Electric field produced by an electric dipole.

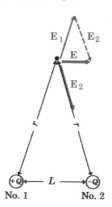

7-6 Lines of Force

A poor man's calculus

The direction of **E** as one travels through space can be indicated by continuous lines such as are shown in Fig. 7-8.

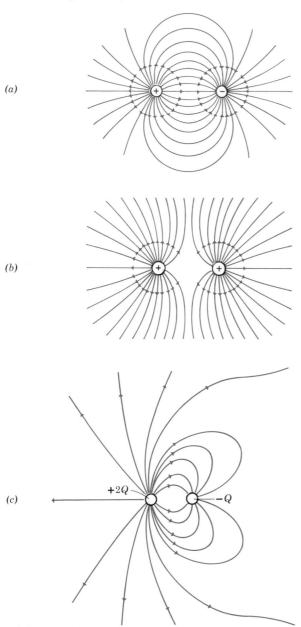

Fig. 7-8. (*a*) Lines of force diagram for two equal but unlike charges. (*b*) Lines of force diagram for two identical charges. (*c*) Lines of force diagram for two charges whose values are $-Q$ and $+2Q$.

(*a*)

(*b*)

(*c*)

$+2Q$ $-Q$

These lines of **E** happen to be called lines of force. Lines of force can be used quantitatively as a powerful and respectable mathematical tool in electrostatics (nothing is moving). However, in problems dealing with moving observers, the lines of force formalism does not hold and must not be used. By using this formalism in electrostatics we will be able to derive and prove many things which normally require integral calculus. For example, we will be able to prove with almost no effort that the gravitational field of the earth behaves as if all the earth's mass were concentrated at the center. As explained in Chapter 3 this was the great stumbling block that caused Newton to withhold his Law of Gravitation for many years. The following is a list of statements that we will verify using lines of force:

1. The field of a uniformly charged sphere is $E = Q/r^2$ outside the sphere.

2. The field inside a uniformly charged spherical or cylindrical shell is zero.

3. The field everywhere inside a conductor is zero.

4. Excess charges can only be on the outside surface of a conductor.

5. The field outside a cylinder or line charge of ρ statcoulomb/cm is $E = 2\rho/r$.

6. The field due to a uniformly charged plane of σ statcoulombs/cm^2 is $E = 2\pi\sigma$.

7. The field between two capacitor plates of area A and charges $+Q$ and $-Q$ is $E = 4\pi(Q/A)$.

For the lines of force to be a quantitative tool, the number of lines passing through each square centimeter must be numerically equal to E as illustrated in Fig. 7-9.* If N is the number of lines passing through an area of A square centimeters and E is perpendicular to A,

$$E = \frac{N}{A}$$

or

$$N = E \cdot A \qquad (7\text{-}6)$$

---

*If $E = 1.5$ dyne/statcoul some square centimeters will contain one line and others two lines. In order to avoid such least count difficulties, we could use a large-scale factor, say divide each line into 10^6 sublines.

Fig. 7-9. Lines of force corresponding to a field $E = 5$ dynes/statcoulomb pointing up. There must be five lines passing through each square centimeter.

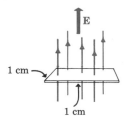

Fig. 7-10. Figure for Example 6.

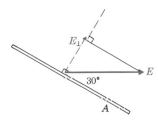

Q.4: Is it ever possible to have a zero electric field produced by several point charges all of the same sign?

If **E** is not perpendicular to A, then $N = E_\perp \cdot A$, where $E_\perp$ is the component of **E**, which is perpendicular to A.

Example 5

Fifteen lines of force cut through an area of 3 cm². If this area is perpendicular to the lines, what is the magnitude of the electric field?

Solving Eq. 7-6 for E gives

$$E = \frac{N}{A} = \frac{15}{3} = 5 \text{ dynes/statcoulomb}$$

Example 6

An area of 3 cm² is in a uniform field of $E = 100$ dynes/statcoul. If the angle between the lines and the surface is 30° as shown in Fig. 7-10, how many lines will cut through the surface?

From Fig. 7-10 we see that $E_\perp = E \sin 30° = 50$ dynes/statcoul. Equation 7-6 gives

$$N = E_\perp \cdot A = 50 \cdot 3 = 150 \text{ lines.}$$

We shall now calculate the total number of lines at a distance R from a point charge Q. The total area at this distance is $A = 4\pi R^2$. According to Eq. 7-6

$$N = E \cdot (4\pi R^2)$$

$$N = \frac{Q}{R^2} \cdot 4\pi R^2$$

$$N = 4\pi Q \tag{7-7}$$

is the total number of lines leaving a point charge Q. Note that this result is independent of R. Thus the lines leaving Q are continuous in space and travel out radially to infinity. If Q is negative, the direction is reversed: the lines travel from infinity and terminate on the charge Q.

Will the lines of force of the resultant field also be continuous when we have two or more charges as shown in Fig. 7-8? The answer is yes, but one must prove this using the quantitative relation Eq. 7-6. We shall prove it for the case of two point charges, although the same derivation can be used for any number of point charges distributed in any way. For two charges the resultant field is

$$\mathbf{E} = \mathbf{E}_1 + \mathbf{E}_2$$

and

$$E_x = (E_1)_x + (E_2)_x \tag{7-8}$$

is the x-component of **E**. Consider any small area A and choose the x-axis perpendicular to A. Let N be the lines of **E**, and N_1 and N_2 be the lines of $\mathbf{E}_1$ and $\mathbf{E}_2$, respectively that pass through A. Then

$$N_1 = (E_1)_x \cdot A, \ N_2 = (E_2)_x \cdot A, \ \text{and} \ N = E_x \cdot A \qquad (7\text{-}9)$$

Multiplying Eq. 7-8 by A gives

$$E_x \cdot A = (E_1)_x \cdot A + (E_2)_x \cdot A$$

Now substitute Eqs. 7-9 in the above equation and we obtain the expected result

$$N = N_1 + N_2$$

Since N_1 and N_2 are each separately continuous, their sum must also be continuous. We have now shown that lines of force can never suddenly start or stop except on charges. Also the total number of lines passing through a surface which completely encloses both Q_1 and Q_2 is $4\pi(Q_1 + Q_2)$. Hence

Gauss's Law

$$N = 4\pi Q_{\text{total}}$$

for any arbitrary shaped body of total charge Q_{total}. We shall find this relation so useful that it deserves a special name. It is named Gauss's law.

For example, if two charges Q_1 and Q_2 are somewhere inside a cubical box of volume $(L \text{ cm})^3$, the number of lines leaving the box will not depend on the area which is $6L^2$. Neither will it depend on the positions of Q_1 and Q_2. They can be anywhere inside the box. Suppose there is a third charge Q_3 just outside the box. Some lines from it will enter the box, but these same lines will also leave the box and their net effect is zero. (An entering line is the negative of a leaving line.) No matter what we do, the net number of lines leaving the box must be $4\pi(Q_1 + Q_2)$.

In summary, lines of force must obey the following rules for charged bodies of any shape and distribution:

1. The total number of lines leaving a body of total charge Q is $4\pi Q$.

2. Lines of force are continuous. They can never start or stop except on charges. Otherwise they travel to infinity.

Ans. 4: Yes. One example is equal charges equally spaced on a circle. Then the field at the center will be zero.

3. Lines of force can never cross. If they did cross, **E** would have two different directions at the point of intersection.

The above formalism follows mathematically from Coulomb's law and may be used in its place. By using this new formalism and principles of symmetry we can now prove the seven statements listed on page 160. This will be done in the following section. For those who may read other books on electricity, it should be pointed out that our definition of lines of force is the same as that of electric flux. The two terms can be used interchangeably.

7-7 Charge Distributions

A quick way to make an infinite number of vector additions

The qualitative lines of force pattern for different charged objects can be displayed by suspending grass seeds in an insulating liquid as shown in Fig. 7-11. An electric field induces equal and opposite charges at the ends of a grass seed, tending to line it up with the field.

Field due to charged sphere

Let us first dispose of the problem that gave Newton so much trouble. First we shall consider a uniformly charged sphere of total charge Q. Because of the spherical symmetry the lines of force leaving the sphere will be of uniform density on all sides and radiate out from the center. According to Eq. 7-6 the field at point P in Fig. 7-12 is $E = \dfrac{N_{\text{total}}}{A_{\text{total}}}$, where $A_{\text{total}} = 4\pi r^2$ is the spherical surface of radius r. According to Gauss's law $N_{\text{total}} = 4\pi Q$. Thus $E = 4\pi Q/4\pi r^2$, or $E = Q/r^2$. According to Eq. 7-5 this is exactly the same result as would be obtained by concentrating the entire charge Q into a point at the center of the sphere. In the gravitational case one uses gravitational field (gravitational force per unit mass) in place of electrostatic field (electrostatic force per unit charge). Since gravitational force is also an inverse square law, the lines of force formalism can also be applied to gravitational fields and we have completed the derivation which took Newton so many years.

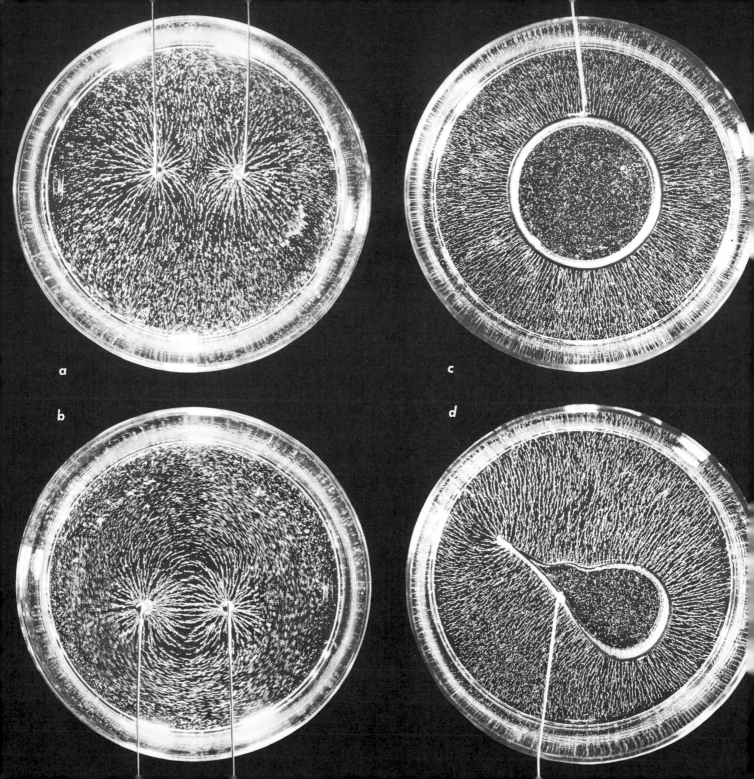

a

c

b

d

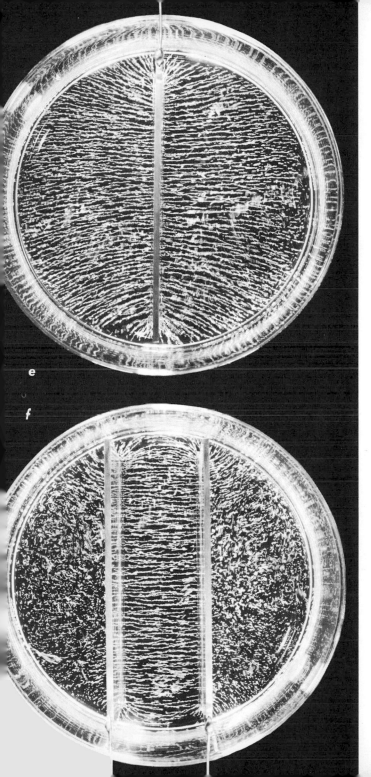

Fig. 7-11. Photographs of lines of force patterns around various charged conductors using grass seeds suspended in an insulating liquid. (Courtesy Berenice Abbott, Educational Services Inc.)

(*a*) Two rods with the same charge.
(*b*) Two rods with equal and opposite charge.
(*c*) A charged cylinder. Note field is zero everywhere inside.
(*d*) A charged conductor of arbitrary shape. Note field is zero inside.
(*e*) A single charged plane.
(*f*) Two charged planes of equal and opposite charge. A capacitor.

Fig. 7-12. Lines of force leaving sphere of charge Q and radius R.

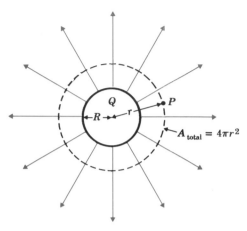

$A_{total} = 4\pi r^2$

Fig.7-13. Uniformly charged cylindrical surface showing lines of force.

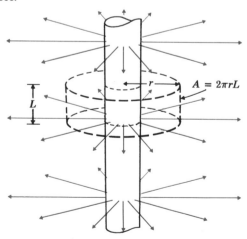

$A = 2\pi rL$

Suppose in Fig. 7-12 that all the charge were concentrated on the surface rather than being uniformly distributed throughout the volume. Then the lines of force would appear just as shown in Fig. 7-12. Any lines of force inside the sphere would either cross each other or else violate the condition of spherical symmetry (there can be no preferred direction away from the center). Thus we see that the field anywhere inside a sphere or cylinder of uniform surface charge must be zero.

Field inside a conductor

We can show that there can be no field inside a current-free solid conductor by considering what would happen if there were a field E inside. The conduction electrons inside would each experience a force, $-eE$, and they would move. But moving electrons constitute an electric current, thus violating our condition that the conductor must be current free (electrostatics is the study of charges at rest). Actually when a charge is initially applied to a solid conductor, there is a field inside, and charges move around quickly readjusting themselves until a static equilibrium is reached.

We can now show that all the excess charge in a conductor must be out on the surface. Suppose there were an excess charge q anywhere inside the conductor. Since $4\pi q$ lines must leave this charge, there would then be lines of force inside the conductor, and this would violate the previous condition that no lines can be inside a conductor holding stationary charges.

Field due to uniformly charged cylinder

The cylinder in Fig. 7-13 has a uniform charge of ρ statcoulomb per centimeter of length along the cylinder.* Since the up direction can have no preference over the down direction, the lines must be perpendicular to the cylinder and extend out radially. Consider a length L centimeters along the cylinder. This length of the cylinder contains a charge ρL and thus radiates $N = 4\pi\rho L$ lines. According to Eq. 7-6 the field is

$$E = \frac{N}{A} = \frac{4\pi\rho L}{2\pi rL}$$

*The symbols ρ and σ are the Greek letters rho and sigma respectively.

Fig. 7-14. Lines of force from a uniformly charged infinite plane (side view).

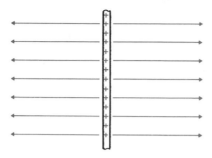

$$E = \frac{2\rho}{r} \qquad (7\text{-}10)$$

outside the cylinder at a distance r from the axis. This result is also true for a charged wire or line charge of ρ statcoulomb/cm.

Example 7
We wish to find a formula for the force per unit length between two parallel charged wires of charges ρ_1 and ρ_2 statcoulomb/cm respectively which are R cm apart. By Newton's third law, the force on a 1 cm length of Wire 1 must be the same as on a 1 cm length of Wire 2.

We can find the force on 1 cm of Wire 2 by finding the electric field at the position of Wire 2 that is produced by Wire 1:

$$E_1(\text{at Wire 2}) = \frac{2\rho_1}{R}$$

Then the force per unit length on Wire 2 is its charge per unit length times E_1:

$$\frac{F_2}{L_2} = \rho_2 E_1 = \rho_2 \left(\frac{2\rho_1}{R}\right) = \frac{2\rho_1\rho_2}{R}$$

Charged planes

Consider the uniformly charged plane in Fig. 7-14 of σ statcoulomb/cm^2. The total number of lines leaving each square centimeter is $4\pi\sigma$. Since there can be no preference of right over left, these lines must divide equally to the right and to the left. Since the field is the number of lines per square centimeter,

$$E = 2\pi\sigma \qquad (7\text{-}11)$$

everywhere in space. We shall next consider the case of two parallel planes of equal and opposite charge density (Fig. 7-15). Since **E** is a vector quantity, the field anywhere in space is the sum of the fields due to each plane separately. In the region outside the planes the field would then be

$$E = 2\pi\sigma + 2\pi(-\sigma) = 0$$

Between the two planes the field would be made up of the $E = 2\pi\rho$ pointing away from the left plane plus the $E = 2\pi\sigma$ pointing toward the right plane. Thus

$$E = 2\pi\sigma + 2\pi\sigma \qquad \text{pointing to the right}$$

Fig. 7-15. Lines of force diagram for two parallel infinite planes of equal and opposite charge.

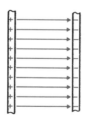

Q.5: What is the field both inside and outside the parallel planes of Fig. 7-15 when the charges on them are equal and of the same sign? $E_{\text{in}} = \underline{\hspace{1cm}}$. $E_{\text{out}} = \underline{\hspace{1cm}}$.

or

$$E = 4\pi\sigma \qquad \text{between the plates} \qquad (7\text{-}12)$$

A common electrical device used in many electrical circuits is the capacitor or condenser. Both words are commonly used for the same thing. The capacitor consists of two plates of area A separated by a small distance d. In most applications the capacitor has a charge $+Q$ on one plate and $-Q$ on the other. The charge density is then $\sigma = Q/A$. According to Eq. 7-12

$$E = 4\pi\frac{Q}{A} \qquad (7\text{-}13)$$

for a parallel plate capacitor.

Actually in many practical capacitors the space between the plates is filled with a dielectric substance. A dielectric has the property that, when inserted in an electric field, induced charges appear on the outside which reduce the strength of the field inside the dielectric. In this instance Q in Eq. 7-13 must be the sum of the original applied charge and the induced charge which is smaller and of opposite sign. The ratio of the field in vacuum to that in the dielectric is defined as the dielectric constant ϵ.

7-8 Electric Potential Energy

A finite amount of work over an infinite distance

It was pointed out in Chapter 5 that there can be many different kinds of potential energy. Potential energy can be thought of as stored-up energy available for further use. If energy can be stored up by lifting a mass against gravity (overcoming gravitational force), or stretching a spring (overcoming the spring force), it can also be stored up by moving a charge against an electric force. Then when the charge is released, the electric force will accelerate it and give back the same amount of work that was done against it. According to the definition of potential energy on page 103, when a charge q is moved from point A to point B, the increase in potential energy will be

$$U_B - U_A = -\overline{F}_x \cdot \Delta x \qquad (7\text{-}14)$$

Ans. 5: Inside the separate fields point in opposite directions, so $E_{\text{in}} = 0$. Outside they add, so $E_{\text{out}} = 4\pi\sigma$.

where $\overline{F}_x$ is the average x-component of the electrostatic force and Δx is the distance from A to B. This contribution to potential energy due to electrostatic forces is called electric potential energy.

As an example we will consider the potential energy of a positive test charge q placed between the charged parallel plates shown in Fig. 7-15. The force on q is pointing to the right and has the value qE. In order to do positive work on q it must be moved to the left. Hence q will have a higher potential energy when at the left-hand plate than when at the right-hand plate. The potential energy difference is qEd, where d is the distance between plates. If the charge is free to move on its own, it will accelerate toward the right-hand plate converting electric potential energy to kinetic energy along the way.

Potential energy of a sphere and a point charge

Suppose the force F_x is due to the electric field produced by a charged sphere of total charge Q, as shown in Fig. 7-16, then $F = Qq/r^2$. Mathematically this is the same problem as the potential energy of a mass m in the gravitational field produced by a sphere of mass M. In the gravitational case $F_G = -GMm/r^2$. Both the electrostatic and gravitational forces are inverse square forces, and one can be transformed to the other by replacing the constant (GMm) with $(-Qq)$. Hence we can obtain the formula for gravitational potential energy by replacing the constant (GMm) in Eq. 5-6 with $(-Qq)$ giving

$$U_r - U_R = -Qq\left(\frac{1}{R} - \frac{1}{r}\right)$$

$$= Qq\left(\frac{1}{r} - \frac{1}{R}\right)$$

We recall that this is the amount of work that must be done in order to move the point charge q from the surface of the sphere to a distance r from the center as shown in Fig. 7-16. If the force is attractive (the product Qq is negative), this potential energy difference will be positive as it was with the gravitational case. But if the electrostatic force is repulsive (Qq is positive), the quantity $U_r - U_R$ will be negative.

Fig. 7-16. Point charge q near a charged sphere of total charge Q.

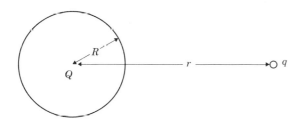

Q.6: The potential energy $(U_r - U_R)$ is the work done in moving the point charge q from r to R. True or false?

Potential energy of two point charges

Since the field E outside a sphere of charge Q is the same as the field produced by a point charge of charge Q, the increase in potential energy in going from point A to point B must also be

$$U_B - U_A = Qq \left(\frac{1}{r_B} - \frac{1}{r_A} \right) \qquad (7\text{-}15)$$

where r_A and r_B are distances from a point charge Q. This is the work done against the electrostatic force in moving q from point A to point B. It is interesting to note that this amount of work must be independent of the path taken in traveling from A to B. If the work was greater by one path than by another, one could have a source of perpetual energy by traveling from A to B over the path of smaller work and always returning by the path of greater work. Since the work done by the electrostatic force on the return trip gets converted into kinetic energy of the moving charge, there would be a net gain of energy on each round trip.

In all our calculations of energy transfer, only the *change* in potential energy occurs. However, it is convenient to speak of *the* potential energy of a test charge q at a distance r from a point charge Q. Such a definition of "absolute" potential energy must be arbitrary. It involves the choice of an arbitrary position or reference level from which to measure potential energy. The convention is to measure the potential energy from the distance $r = \infty$. In Eq. 7-15 let U_A be this reference level ($r_A = \infty$). Then $U_A = 0$ and

$$U_B - 0 = Qq \left(\frac{1}{r_B} - \frac{1}{\infty} \right)$$

or

$$U_B = \frac{Qq}{r_B}$$

This is usually written as

Potential energy of two point charges

$$U = \frac{Qq}{r} \qquad (7\text{-}16)$$

(potential energy of two point charges separated by a distance r). The physical meaning is that the electric potential energy U of two point charges separated by a distance r is

Ans. 6: False. It is the work in moving from R to r.

the amount of work required to bring the charges this close together starting from an infinite separation. If both charges are the same polarity, the electrostatic force will be repulsive and it will take positive work to bring the charges from infinity to a distance r apart. However, if the polarities are opposite, then the work done, or potential energy, will be negative.

As an example, let us consider the potential energy of an electron in the field of a proton. For this example $Q = e$ (charge of proton) and $q = -e$ (charge of electron). Then $U = -e^2/r$. This function is plotted in Fig. 7-17.

Fig. 7-17. Potential diagram of an electron in the field of a proton. The curve is the potential energy of the electron as a function of its distance r from the proton.

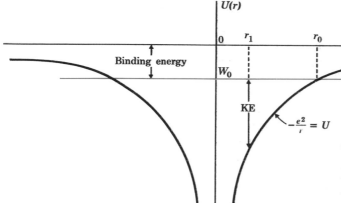

Suppose the electron is initially at rest a distance r_0 from the proton. Then its total mechanical energy is $W_0 = KE + U$, where $KE = 0$ and $U = -e^2/r_0$ with the result that $W_0 = -e^2/r_0$. If left to itself, the electron will accelerate toward the proton, picking up kinetic energy and losing a corresponding amount of potential energy. According to the law of conservation of energy the sum $(KE + U)$ must remain constant and equal to its initial value of W_0. The relationship $W_0 = KE + U$ can be conveniently represented using a potential diagram (see page 105). In Fig. 7-17 the sum $W_0 = KE + U$ has a constant value and is represented by the red horizontal line. Since $KE = W_0 - U$, the kinetic energy is the difference between the height of this line and the height of the curve of U. At $r = r_0$ this difference is zero. Thus r_0 is the maximum possible distance that an electron of total energy W_0 can be from the proton. Note that in this diagram, both W_0 and U are always negative; however, the difference $(W_0 - U)$ is positive whenever r is less than r_0.

Binding energy

The binding energy of the electron is defined as the amount of external energy that must be supplied to the electron in order for it to get to infinity. Physically this is the

Q.7: The potential energy in Eq. 7-16 is the work done in moving q from a distance r to infinity. True or false?

amount of energy needed to pull the electron and proton completely apart. In order for the electron to just barely get to $r = \infty$, its KE $= 0$ when $r = \infty$. Substituting KE $= 0$ and $r = \infty$ into the equation $W = $ KE $+ U$ gives $W = 0 + 0$. If W_0 is negative, the electron energy is lower than $W_0 = 0$ and we must supply an additional positive energy $(-W_0)$ in order to get $W = 0$ so that the electron can go to infinity. By definition this is the binding energy BE; that is,

$$\text{BE} = -W_0$$

Binding energy or

$$\text{BE} = \frac{e^2}{r_0}$$

The binding energy must always be a positive number.

Example 8

An electron has a kinetic energy of 1.6×10^{-11} erg when it is 2.4×10^{-9} cm from a proton. What is its binding energy?

$$W_0 = \text{KE} + U = \text{KE} - \frac{e^2}{r}$$

$$W_0 = 1.6 \times 10^{-11} - \frac{(4.8 \times 10^{-10})^2}{2.4 \times 10^{-9}} \text{ erg}$$

$$W_0 = -7 \times 10^{-11} \text{ erg}$$

Thus the binding energy is BE $= -W_0 = 7 \times 10^{-11}$ erg.

Example 9

In the Bohr model of the hydrogen atom the electron is in a circular orbit of radius $R = 5.3 \times 10^{-9}$ cm. (a) Derive a formula for the binding energy in terms of e and R only. (b) What is the binding energy in ergs?

The binding energy is

$$\text{BE} = -W_0$$

$$= -\left(\frac{1}{2}mv^2 - \frac{e^2}{R}\right)$$

Another expression for v can be obtained by equating the centripetal force to the electrostatic force:

$$\frac{mv^2}{R} = \frac{e^2}{R^2}$$

or

$$mv^2 = \frac{e^2}{R}$$

Ans. 7: False. It is the work done in moving q from infinity to r.

Substituting this in the equation for binding energy gives

$$-W_0 = -\frac{1}{2}\left(\frac{e^2}{R}\right) + \frac{e^2}{R}$$

$$= \frac{1}{2}\left(\frac{e^2}{R}\right) \quad \text{is the answer to Part (a)}$$

$$= \frac{1}{2}\frac{(4.8 \times 10^{-10})^2}{5.3 \times 10^{-9}}$$

$$= 21.8 \times 10^{-12} \text{ erg}$$

7-9 Electric Potential

Where volts come from

Just as electric force per unit charge is a convenient quantity to work with, so is electric potential energy per unit charge also convenient. The force per unit charge was given the name electric field. The electric potential energy per unit charge is given the name electric potential. *By definition, the electric potential at any given point in space is the work that must be done to bring a unit positive charge from infinity to the given point.* The CGS unit for potential is erg/statcoulomb. This unit is given the special name of statvolt. In the MKS system the corresponding unit is joule/coulomb which bears the special name of volt. The conversion factor is

1 statvolt = 300 volts

If a test charge q has an electric potential energy U at a given point in space, the potential V at that point in space is then

$$V = \frac{U}{q} \tag{7-17}$$

The work done on a charge q in moving it a distance Δx against a field E is the force $-qE_x$ times the distance Δx. Thus

$$\Delta U = -qE_x \Delta x$$

is the increase in electric potential energy. Dividing both sides by q gives

$$\Delta V = -E_x \Delta x \tag{7-18}$$

Q.8: Suppose the electric potential is increasing as one moves along the x-axis. What is the direction of E_x?

Example 10

The electric potential energy of an electron is 21.8×10^{-12} ergs. (a) What is the potential in statvolts? (b) What is the potential in volts?

$$V = \frac{U}{q} = \frac{21.8 \times 10^{-12}}{4.8 \times 10^{-10}} = 0.0453 \text{ statvolt}$$

Since 1 statvolt = 300 volts, we multiply by 300 to convert.

$$V = 300 \times 0.0453 = 13.6 \text{ volts}$$

Capacitance

A simple application of Eq. 7-18 is to calculate the potential difference V between two capacitor plates. Since E is constant between the plates, $V = -Ed$ is the potential difference between the plates where d is the plate separation. Substituting the expression for E in Eq. 7-13 gives for the magnitude of V:

$$V = \frac{4\pi d}{A} Q \qquad (7\text{-}19)$$

Note that the potential difference is directly proportional to the charge. The ratio of charge to potential difference is a useful quantity defined as the capacitance or capacity C.

$$C = \frac{Q}{V} \qquad (7\text{-}20)$$

The capacitance of a parallel plate capacitor is then obtained by substituting the right-hand side of Eq. 7-19 into Eq. 7-20.

$$C = \frac{A}{4\pi d} \text{ for a parallel plate capacitor}$$

With a dielectric between the plates, E and thus V is reduced by a factor ϵ for a given applied charge Q. For a parallel plate capacitor with dielectric of strength ϵ, $C = \epsilon(A/4\pi d)$.

Example 11

The spherical capacitor in Fig. 7-18 has a charge $+Q$ on the inner spherical shell and $-Q$ on the outer shell. What is the electrostatic field in regions I, II, and III? What is the capacitance?

In this problem the lines of force will exist only in region II. $4\pi Q$ lines will leave the inner surface and terminate on the outer. Thus E_{I} and E_{III} are zero. The field in region II is $4\pi Q$ (the total number of lines) divided by $4\pi r^2$ (the total area through which these lines

Fig. 7-18. A spherical capacitor consisting of two concentric shells of radii R_A and R_B.

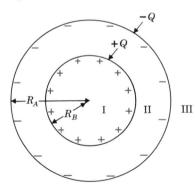

Ans. 8: E_x is in the negative direction and has the value $E_x = -\dfrac{\Delta V}{\Delta x}$.

pass). Thus $E_{II} = Q/r^2$. Since this is the same field as for a point charge Q, Eq. 7-15 can be used to obtain the potential difference

$$V_B - V_A = \frac{Q}{R_B} - \frac{Q}{R_A}$$

$$\frac{Q}{V_B - V_A} = \frac{1}{\dfrac{1}{R_B} - \dfrac{1}{R_A}}$$

$$C = \frac{R_A R_B}{R_A - R_B} \quad \text{is the capacitance}$$

The Electroscope

An early device used to detect charge and measure electric potential is the electroscope shown in Fig. 7-19. If the metal ball on top is touched to a charged conductor, the thin metal leaves (gold or aluminum foil) acquire the same potential as the conductor. The charge on the leaves will be proportional to the potential difference between the leaves and the case. The repulsive force between the leaves due to their identical charges can be measured by observing the amount of deflection on a scale.

An electroscope can also be charged by induction in the same way as was the sphere in Fig. 7-5. A charged electroscope can be used to detect the presence of charge and to determine its sign as well. Suppose a negatively charged rod is brought near a negatively charged electroscope. The rod repels additional electrons down into the leaves and they diverge even more. A positively charged rod would attract some of the surplus electrons up into the ball and the leaves would deflect less.

Over a period of several days a charged electroscope will gradually lose its charge, because a small number of air molecules are continually being ionized due to the cosmic rays. Some of these ions can take up the surplus charge on the electroscope. The discharge rate of an electroscope is proportional to the amount of background radiation (radioactivity). A common device used to measure the radiation dose received by personnel is a small electroscope the size of a pencil. The amount of discharge of such a pocket dosimeter can be read easily by holding it up to the light. This simple device is invaluable for use in civil defense.

Fig. 7-19. A charged electroscope.

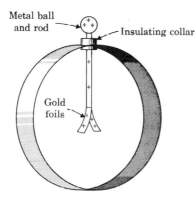

Metal ball and rod

Insulating collar

Gold foils

Summary

We are now halfway through our study of the theory of electricity. We know in principle how to calculate the forces and energies of stationary charged objects. In Chapter 8 we shall study a new kind of force that only occurs when charges are moving. This force is called the magnetic force and we shall see that it can be thought of as a relativistic correction to Coulomb's law. We shall also see that this magnetic force provides the explanation for magnetism and electric motors. Then we shall study another class of phenomena (electromagnetic induction) that occur only when electric currents are changing in value. Finally, all the laws of electricity will be combined in an attempt to prove that a changing current will radiate a traveling electromagnetic wave of velocity $v = c$. The climax of Chapter 8 will be to demonstrate the final great achievement of classical physics—the explanation of light in terms of the theory of electricity.

Problems

1. A positive charge is applied to a metal sphere. Will its mass increase, decrease, or remain the same?

2. A charge of -40 esu is placed 10 cm from a charge of $+90$ esu.
(a) What is the force between them?
(b) How many lines of force go to infinity (assuming no other charges anywhere)?

Prob. 3

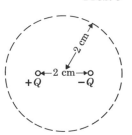

3. An electric dipole consists of two charges $+Q$ and $-Q$ of 100 statcoul each, 2 cm apart.
(a) What is the net number of lines of force leaving the dashed spherical surface?
(b) What is the electric potential at the center of the sphere?
(c) What is the electric field at the center of the sphere?

4. A negative charge of -10 statcoul sits at the center of a hollow metal sphere which contains a positive surface charge of 15 statcoul.
(a) Draw a lines of force diagram to represent the resultant electric field.
(b) What is the net number of lines of force radiating out from the sphere?

5. A charge of 6×10^3 esu is 1 m from a 3-gm pith ball that carries a charge of 300 esu. What is the initial acceleration of the pith ball?

6. On one corner of a square that has 10 cm sides is placed a charge of $+400$ esu and on the opposite corner is a charge of $+300$ esu. Find the total force on a $+10$ esu charge placed at a third corner of the square.

7. A 10 cm by 10 cm square has charges of 100 esu magnitude on all four corners. Give the magnitude and direction of E in the center of the square if the signs of Q_1, Q_2, Q_3, and Q_4 are

    (a) $+ + + +$.

    (b) $+ - + -$.

    (c) $+ + - -$.

8. Calculate the potentials in the center of the above square for Parts (a), (b), and (c). The resultant potential is the algebraic sum of the separate potentials.

9. Which of the following physical quantities have the units of electric field. B (in gauss); q/r; q/r^2; volts/cm; W/q; Cq?

10. A high energy accelerator shoots a beam of protons of kinetic energy $KE = 1.6 \times 10^{-6}$ erg into hydrogen gas.

    (a) What will be the electric potential energy in ergs between a beam proton and a hydrogen nucleus when they are at their closest possible distance of approach?

    (b) Would this potential energy be positive or negative?

    (c) What is the closest distance of approach in cm?

11. How many electrons per gram are there in hydrogen, carbon, and uranium-238?

12. A negative rod is brought near an uncharged electroscope. The leaves separate. What charge is on the leaves? The ball is then momentarily touched by hand. Then the rod is moved farther away. Now what charge is on the leaves?

13. An electroscope is charged by induction with a glass rod that has been rubbed with silk. An unknown charge is brought near the electroscope and the leaves are found to converge. What is the sign of the unknown charge?

14. Suppose that in the helium nucleus there are two protons separated by 1.5×10^{-13} cm.

    (a) What is the electrostatic force between them?

    (b) How much work must be done in order to bring two protons this close together?

15. An electron is 5.3×10^{-9} cm from a proton. How much velocity must it have to escape from the proton?

16. What is the ratio of the electrostatic to gravitational force for two protons?

17. Consider an electron of charge $-e$ and a neutron of zero charge separated by a distance R. Let m be the electron mass and M the neutron mass.

    (a) What will be the force between them in terms of the distance R and any other universal physical constants?

Prob. 7

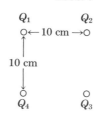

Q_1 Q_2

$\circ\leftarrow$ 10 cm $\rightarrow\circ$

10 cm

$\circ$ $\circ$

Q_4 Q_3

(b) Suppose the electron is in a circular orbit about the neutron. What is the force between them in terms of m, R, and v (the electron's circular velocity)?

(c) What is the kinetic energy of the electron in terms of G, m, M, and R?

(d) What is the electron potential energy? (Assume $U = 0$ when $R = \infty$.)

18. Two parallel plates are separated by 2 cm. The field between the plates is 20 dynes/statcoul. What is the potential difference between the plates?

19. An electron is accelerated through a potential difference of 1 volt. What is its increase in kinetic energy? Repeat for a proton.

20. Electrons are attracted to the charged sphere of a one million volt Van de Graaff. What is their kinetic energy in ergs when they reach the sphere?

21. Consider the case of two parallel infinite planes separated by 8 cm. Both have 5 esu of positive charge per cm^2.

(a) What is the electric field between the planes?

(b) What is the electric field at 3 cm to the left of the left-hand plane?

22. Suppose a 1-cm diameter carbon sphere has one extra electron per million protons.

(a) If the density is $D = 1.7$ gm/cm^3, what is the charge?

(b) What is the electric field at the surface of the sphere?

23. Plot a potential diagram for the parallel plate capacitor in Fig. 7-15.

24. The potential energy diagram of a charge q in the neighborhood of several other stationary charges looks as shown. The charge q is free to move by itself (no external forces). The total energy (not counting rest energy) of q is $W_0 = KE + U$.

Prob. 24

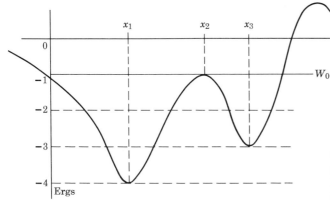

(a) What is the KE of q when at x_1?

(b) What is the KE of q when at x_2?

(c) What is U at x_1?

(d) What is W_0 at x_3?

25. An electron is in a circular orbit around a proton. What is the ratio of the potential energy to the kinetic energy of this electron? Is this ratio positive or negative? What is the ratio of the binding energy to the kinetic energy?

26. Plot a potential diagram for the spherical capacitor shown in Fig. 7-15.

27. Two metal plates of 100 cm² area are separated by 2 cm. The charge on the left-hand plate is -5 esu and the charge on the right-hand plate is -10 esu.
    (a) What is the field immediately to the left of the left plate?
    (b) What is the field between the plates?
    (c) What is the field immediately to the right of the right plate?
    (d) What is the potential difference between the two plates?

28. Assume the earth has uniform density.
    (a) If it had half its present diameter, what would be its mass in terms of M_0, the mass of the earth?
    (b) What would be the value of g on the surface of this half-sized earth?
    (c) If a hole were dug halfway to the center of the full-sized earth, what would be the value of g at the bottom of the hole?

29. Suppose the earth had an excess surface charge of 1 electron/cm².
    (a) What would be the electric field just below the surface?
    (b) What would be the electric field just above the surface?
    (c) What would be the earth's potential?

Prob. 30

Parallel

30. Two capacitors C_1 and C_2 are connected in parallel. What is the total capacity C of this combination?

Hint: $C = \dfrac{Q_1 + Q_2}{V}$

Prob. 31

Series

31. What is the capacity of two capacitors connected in series. *Hint:* Since both capacitors carry the same charge Q, we have

$$C = \frac{Q}{V_1 + V_2}$$

32. Two concentric charged cylinders have radii R_1 and R_2. Their respective charge densities are ρ_1 and ρ_2 statcoulombs per centimeter of length along the cylinder.
    (a) What is E at a distance r where $r > R_2 > R_1$?
    (b) What is E at a distance r from the common axis where r is in between the two cylinders? Give answers in terms of r, R_1, R_2, ρ_1, and ρ_2.

Prob. 33

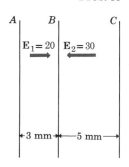

33. Consider the three charged planes as shown. The potential of plane A is zero.
    (a) What is the potential of plane B?
    (b) What is the potential of plane C?
    (c) What are the charge densities on each of the three planes?

34. What is the potential along the line of an electric dipole at a distance r from the center of the dipole? Give answer in terms of r, Q, and L.

35. We define gravitational field as the gravitational force per unit mass. If the number of gravitational lines of force per cm² is defined as equal to the gravitational field, how many gravitational lines of force will enter a body of mass M?

36. In a hydrogen-like atom a single electon of charge $-e$ is circling around a stationary nucleus of charge Ze at a distance R.

(a) What is the electrostatic force between the electron and nucleus?

(b) Let T be the time for the electron to make one complete orbit. What is the acceleration of the electron in terms of R and T?

(c) What is T in terms of e, m, Z, and R?

37. Let the symbol Mev stand for a new unit of energy (1 Mev = 1.6×10^{-6} erg). Assume that a proton as it approaches a certain atomic nucleus will "see" a potential energy curve $U(r)$ as shown. Suppose a proton is trapped in the nucleus of radius R with a total energy $W_0 = 2$ Mev as shown.

Prob. 37

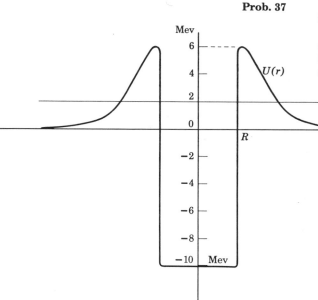

(a) What will be its kinetic energy in the nucleus?

(b) What will be its potential energy in the nucleus?

(c) According to classical physics how much additional energy does it need in order to get out of the nucleus?

(d) If external protons are shot at the nucleus, classically how much initial kinetic energy would they need in order to penetrate the nucleus? Assume the protons start from infinity.

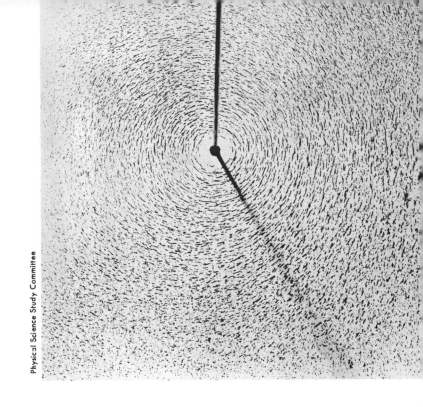

Electromagnetism

Chapter **8**

Electromagnetism

This chapter will complete our study of the theory of electricity. Chapter 9 treats more detailed applications of the theory such as electronics and circuit theory. Here we shall encounter for the first time two "new" types of electrical phenomena: a strong force between currents, (moving charges) and an electric field produced by a changing current. The new type of force is called magnetic force, and the phenomenon of an induced electric field is called electromagnetic induction. The climax of this chapter will be a demonstration showing that a changing current must radiate an electromagnetic wave propagating with velocity $v = c$. This explanation of light in terms of the theory of electricity was the last great achievement of what is called classical physics.

8-1 Electric Current

Electricity

Before we can discuss a new kind of force between currents, we must first define current. Electric current I is defined in the following equation:

Electric current

$$I = \frac{Q}{t} \tag{8-1}$$

where Q is the net amount of charge that has passed through a given area in a time t. A statampere is defined as one statcoulomb per second. The MKS unit of one coulomb per second is called an ampere. Thus

$$1 \text{ amp} = 3 \times 10^9 \text{ statcoul/sec} = 3 \times 10^9 \text{ statamp} \tag{8-2}$$

In a metal wire, the positive charges (atomic nuclei) cannot move; they are bound in a crystalline structure. However, the outer electrons or conduction electrons are completely free to move along the wire. This is contrary to all ideas of classical physics, and can only be explained using quantum mechanics as is done in Chapter 14. If N electrons per second flow through any given area in the wire, the magnitude of the current is then Ne, where e is the charge of the electron.

But what is the direction of the current? According to the convention established by Benjamin Franklin, a current flowing *into* a capacitor plate would supply positive charge to the plate. We now know, however, that a capacitor plate is made positive by conduction electrons flowing *away* from the plate. Hence the conduction electrons always flow in the direction opposite to that of the current. If the charge of the electron had been chosen as positive rather than negative, this difficulty would not have arisen.

Example 1

In the Bohr model of the hydrogen atom, what is the value of the electron current that encircles the proton?

The current will be the number of times per second the electron passes any given point in its orbit times e. Or $I = fe$, when $f = v/2\pi R$ is the orbital frequency. Then

$$I = \frac{ev}{2\pi R}$$

From page 156 we have $v = 2.18 \times 10^8$ cm/sec and $R = 5.3 \times 10^{-9}$ cm. After making the numerical substitutions, we obtain

$$I = 3.14 \times 10^6 \text{ statamp} - 1.05 \times 10^{-3} \text{ amp}$$

Current can also be produced by moving a line of charge of ρ statcoulombs per centimeter with a velocity v along its length. Then the amount of charge passing a given point in 1 sec is ρ times the length of wire which goes by in 1 sec. In this case

$$I = \rho v \qquad \qquad (8\text{-}3)$$

for a moving line charge of ρ statcoulomb/cm.

Example 2

A current of 1 amp flows through a copper wire of 1 mm² cross section. What is the average drift velocity v of the conduction electrons?

We must first calculate $\mathfrak{N}$, the number of conduction electrons per centimeter of the wire. Then $\rho = \mathfrak{N}e$ and we can get v from the equation $I = \rho v$. Assuming one conduction electron per atom, the number of conduction electrons per cm³ is the number of moles per cm³ times Avogadro's number or $\frac{D}{A} N_0$, where D is the density, A is the atomic weight, and N_0 is Avogadro's number.

$$\frac{D}{A} N_0 = \frac{8.9 \text{ gm/cm}^3}{63.6 \text{ gm/mole}} \times 6.02 \times 10^{23} \text{ atoms/mole} =$$

$$8.45 \times 10^{22} \text{ atoms/cm}^3$$

Q.1: Check the correct choice:

Current is

☐ charge per cm³ times velocity.
☐ charge per unit area times velocity.
☐ neither.

One centimeter of length of the wire has a volume of 10^{-2} cm³ and thus contains $\mathfrak{N} = 8.45 \times 10^{20}$ conduction electrons/cm. Then

$$\rho = \mathfrak{N}e = (8.45 \times 10^{20})(4.8 \times 10^{-10}) = 4.06 \times 10^{11} \text{ statcoul/cm}$$

According to Eq. 8-3 the drift velocity is

$$v = \frac{I}{\rho} = \frac{3 \times 10^9 \text{ statcoul/sec}}{4.06 \times 10^{11} \text{ statcoul/cm}} = 0.74 \times 10^{-2} \text{ cm/sec}$$

Currents can also flow in gases and liquids. Neon and fluorescent lights are examples of currents in gases. In these the current is due to moving positive ions as well as the moving electrons. However, the electrons are much faster and are the main contribution to the current. When an electron collides with a gas ion or atom, the kinetic energy of the collision can be absorbed by the atom and then radiated away in the form of electromagnetic radiation which can be seen by the eye as light.

Most liquids contain free ions and are therefore capable of conducting electricity. If two metal plates or electrodes are connected to a voltage source and are inserted in such a liquid, the current will consist of both positive and negative ions moving in opposite directions toward their respective electrodes. When the ions reach the electrodes they become neutralized. For example, if table salt (NaCl) is dissolved in water Na^+ ions flow to the negative electrode and the Cl^- ions flow to the positive electrode. When the Cl^- ions are neutralized, they chemically combine to form Cl_2 (chlorine gas) which bubbles up into the air. This process of chemical decomposition is called electrolysis and has great commerical possibilities.

Current elements

A current element is defined as the product of the current I in a wire times a short length Δl of the wire.

$$\text{Current element} = I \, \Delta l = (\rho v) \, \Delta l$$
$$= (\rho \, \Delta l)v$$

But $(\rho \, \Delta l)$ is the amount of moving charge in Δl. If we call this q, we have

$$I \, \Delta l = qv \qquad (8\text{-}4)$$

Ans. 1: The first choice has units of

$$\frac{Q}{L^3} \times \frac{L}{T} = \frac{Q}{L^2 T}.$$

The second choice has units

$$\frac{Q}{L^2} \times \frac{L}{T} = \frac{Q}{LT}.$$

Neither is correct. The units of current are $\dfrac{Q}{T}$.

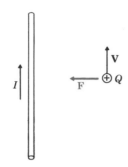

Fig. 8-1. Force between moving charge Q and current-carrying wire.

Fig. 8-2. A circuit to observe the force between parallel currents.

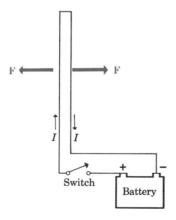

Q.2: What are the units of a current element?

We see that a moving point charge q is mathematically equivalent to a current element. We see that forces on moving charges and forces on currents must really be the same thing.

8-2 The Magnetic Force

A relativistic correction to Coulomb's law

So far in our studies of electricity only one new fundamental law of nature has been introduced (Coulomb's law), and here we strictly specified that all charges must be at rest. This was for good reason. It so happens that when we let the charges move, something new happens: a new force appears which must be added to the Coulomb force, called magnetic force. In fact, in some cases where Coulomb's law predicts zero force, there can actually be large magnetic forces. One such example is shown in Fig. 8-1. A moving charge Q in the presence of a current I will feel a strong force proportional to the current in the wire and also proportional to the velocity V of the charge Q. This is true even if the net charge on the wire is zero; that is, the electric field acting on Q is zero! Still there will be an observed force $F_m \propto \dfrac{QVI}{r}$. It is this force which gives rise to the deflection of moving electrons in a television picture tube. By changing I in wires that are wound around the tube, the force on the electron beam changes and moves the spot of light produced by electrons hitting the screen (see Fig. 8-15). Another example of this "strange" magnetic force is the attractive force that is observed between parallel currents as shown in Fig. 8-2. The essence of the electric motor is the magnetic force between two uncharged, parallel wires when currents are passed through them. All that is needed to observe this basic phenomenon is a battery, a switch, and a wire as shown in Fig. 8-2. When the switch is closed, the two wires will repel each other appreciably. Do not leave the switch closed for very long because this type of circuit is commonly known as a short circuit. In a short circuit the current will be very high, the wires will get hot, and the battery will run down quickly.

Explanation of magnetic force in terms of Coulomb's law and relativity

By all rights, this new force between neutrally charged wires should have nothing to do with the Coulomb force which only applies to charged objects. So it was given a different name—the magnetic force—to distinguish it from the electrostatic force. But are these two different kinds of forces truly independent? In addition to the gravitational interaction and the nuclear force, must we contend with two basic kinds of electrical interactions, the electrostatic and the magnetic? Fortunately the answer is no.

Nature usually turns out to be simpler than one would expect. Einstein in his theory of relativity proposed in 1905 that these two forces are one and the same. Many experiments have shown that Einstein was correct, so now we call these forces *the* electromagnetic force. Einstein proposed that the magnetic force is a mere relativistic correction to Coulomb's law. In the following paragraph we will demonstrate how the formula for magnetic force between currents can be derived from Coulomb's law and a famous effect in relativity theory—the Lorentz contraction. (The Lorentz contraction describes a contraction in length of moving objects. It is discussed in more detail in Chapter 11.)

According to Coulomb's law the net electrostatic force on Q in Fig. 8-1 should be zero, whether or not the conduction electrons in the wire happen to be moving. However, if we make use of the theory of relativity, the average spacing between the conduction electrons when they start moving will decrease by the Lorentz contraction factor $\sqrt{1 - v^2/c^2}$, where v is the drift velocity of the conduction electrons. Then according to relativity theory, the charge density ρ^- of the conduction electrons will increase by the factor $\dfrac{1}{\sqrt{1 - v^2/c^2}}$; whereas the charge density of the fixed positive charges remains the same ($\rho^+ = \rho_0$). Hence the net charge will no longer be zero. The situation in the laboratory system will look as in Fig. 8-3. We know that it would be safe to use Coulomb's law to calculate the net force on Q if Q were at rest. Then the force on Q would be $\mathbf{F} = Q\mathbf{E}$, where $\mathbf{E}$ is the net electric field produced by the wire. Although Q is not at

Ans. 2: Units of $[I \,\triangle\, l] = \dfrac{QL}{T}$.

Fig. 8-3. Moving charge Q of velocity V and current carrying wire in laboratory system. Drift velocity of conduction electrons is v. Positive charges are fixed.

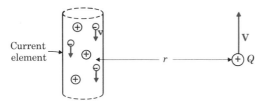

Current
element

Fig. 8-4. Same situation as above, but viewed by observer moving along with Q. Now Q is at rest, the "fixed" positive charges have velocity V pointing down, and the conduction electrons have velocity $(V + v)$.

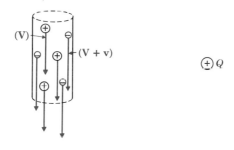

rest in the laboratory system, we can make it be at rest by moving along with it; (an observer moving up the page with velocity V will see the situation shown in Fig. 8-4). In this situation

$$\rho^+ = \rho_0\left(1 - \frac{V^2}{c^2}\right)^{-1/2} \quad \text{and} \quad \rho^- = -\rho_0\left[1 - \frac{(v + V)^2}{c^2}\right]^{-1/2}$$

The net charge is $\rho = \rho^+ + \rho^-$, and the net field at Q is given by Eq. 7-10:

$$E = \frac{2\rho}{r} = \frac{2(\rho^+ + \rho^-)}{r}$$

$$= \frac{2\rho_0}{r}\left[\left(1 - \frac{V^2}{c^2}\right)^{-1/2} - \left(1 - \frac{(v + V)^2}{c^2}\right)^{-1/2}\right] \quad (8\text{-}5)$$

Now we recall from Example 2 that the drift velocity of conduction electrons in a metal is typically about 1 mm/sec or less. Hence v/c is much smaller than one, and we can safely use the binomial theory which tells us that $(1 + a)^n \approx 1 + na$ when a is much smaller than one. Then $(1 + a)^{-1/2} \approx (1 - \frac{1}{2}a)$ or $\left(1 - \frac{V^2}{c^2}\right)^{-1/2} \approx 1 + \frac{1}{2}\frac{V^2}{c^2}$ and Eq. 8-5 then becomes

$$E \approx \frac{2\rho_0}{r}\left[\left(1 + \frac{V^2}{2c^2}\right) - \left(1 + \frac{(v + V)^2}{2c^2}\right)\right]$$

$$\approx \frac{2\rho_0}{r}\left[\frac{V^2 - (v + V)^2}{2c^2}\right]$$

$$\approx -\frac{\rho_0 v}{rc^2}[2V + v]$$

We will be interested in examples where V is greater than 1 cm/sec, or when $V \gg v$, so we can ignore the term v compared to $2V$ in the above equation. Then

$$E \approx -\frac{2\rho_0 v V}{c^2 r}$$

and

$$F = QE$$

$$\approx Q\left(-\frac{2\rho_0 v V}{c^2 r}\right)$$

$$\approx -Q\frac{V}{c} \times \left(\frac{2I}{cr}\right)$$

Q.3: The observer of Fig. 8-4 would be moving in which direction relative to the stationary wire in the laboratory system?

where $I = \rho_0 v$ is the current in the wire. The minus sign means an attractive force. Actually in relativity theory the force seen by the moving observer is not quite the same as that seen by the stationary observer. It turns out that if the above calculation is performed without any approximations using relativity theory, the exact answer in the laboratory system (the situation in Fig. 8-1) is an attractive force of strength

$$F_m = \frac{QV}{c} \times \frac{2I}{cr} \tag{8-6}$$

This is the formula for the magnetic force between a moving charge and a current carrying straight wire.

But why should magnetic forces be so strong if they are merely small relativistic corrections to Coulomb's law? In fact, we see in the above derivation that the relativistic effect is the order of $\frac{vV}{c^2}$ times the electrostatic force exerted by the conduction electrons alone, and we know that the drift velocity v is so small that $v/c \sim 10^{-12}$. The answer to this puzzle lies in the enormous amount of charge that is moving in the wire. We saw in Example 2 that there is in a typical wire over 10^{22} statcoul/cm of conduction electrons. This is typically 10^{20} times more charge than can be applied to the wire for the purposes of producing electrostatic forces.

8-3 Magnetic Field

Force per unit moving charge

The general electromagnetic force on a charge q is usually broken into two parts:

$$\mathbf{F}_{\text{electromagnetic}} = \mathbf{F}_e + \mathbf{F}_m$$

where $\mathbf{F}_e$ is what the force would be if q were at rest (the electrostatic force), and $\mathbf{F}_m$ is the velocity dependent part of the force (the magnetic force). Then electric field is defined the same as before:

$$\mathbf{E} = \frac{\mathbf{F}_e}{q}$$

It is equally useful to define a field to describe the velocity

Ans. 3: He would be moving up with respect to the wire.

dependent part of the electromagnetic force. This is called the magnetic field and is defined as follows:

$$F_m = \frac{qv}{c} \times B$$

or

$$B = \frac{F_m}{\dfrac{qv}{c}} \tag{8-7}$$

is the magnetic field acting on a test charge q when the test charge is moved in the direction which gives the largest magnetic force. The relative directions of $\mathbf{v}$, $\mathbf{B}$, and $\mathbf{F}_m$ are discussed in the next section, and may appear strange to the reader. This is only because relativity does not usually enter into his daily life. In terms of the fields, the complete electromagnetic force is then

$$F = qE + \frac{qv}{c} \times B$$

Historically, magnetic field was defined as the force on a unit magnetic pole. However, it is now known that free magnetic poles really do not exist and that all such magnetic forces are due to forces between currents or moving charges. In this book, we shall take the more modern approach of expressing magnetic phenomena as forces between currents or moving charges rather than dealing with the nonphysical (but mathematically equivalent) concept of free magnetic poles or magnetic charge. However, the theory of magnetism will be discussed in Section 8-6. From Eq. 8-7, we can see that B has the same units, dyne/statcoul, as E. In the CGS system a special name, gauss, is given for the units of B.*

Force between two parallel wires

If we substitute Eq. 8-6 into Eq. 8-7, we obtain

$$B = \frac{2I}{cr} \tag{8-8}$$

for the magnetic field produced by a straight line current I.

Q.4: How would you measure $\mathbf{F}_m$ on a charge q in the presence of an electric field?

*Strictly speaking, B is called the magnetic induction. However, the magnetic field in vacuum (or air) is equal to B, and in this book we only deal with fields in vacuum.

Note the similarity between the formula for magnetic field produced by an infinitely long current and the formula for electric field produced by an infinitely long charged wire. If we replace I/c by ρ in Eq. 8-8, we obtain $2\rho/r$, which is the formula for E produced by a line charge.

Example 3

A line of charge ρ statcoulomb/cm is moving along its length with velocity v. Find B, E, and their ratio.

According to Eq. 8-3, $I = \rho v$, so then $B = 2\rho v/rc$. According to Eq. 7-10, $E = 2\rho/r$. The ratio is $B/E = v/c$.

In order to get the force between two currents as shown in Fig. 8-5 we replace (qv) in Eq. 8-7 by its equivalent $I\,\Delta l$ (see Eq. 8-4):

$$F_m = \frac{I\,\Delta l}{c} \times B \qquad (8\text{-}9)$$

is the force on a current element ($I\,\Delta l$) sitting in a magnetic field B that is produced by other currents. In the case of Fig. 8-5, B is produced by I_1. The value of B at the position of the current element $I_2 l_2$ is $B = \dfrac{2I_1}{cr}$. So according to Eq. 8-9 the magnetic force on the length of wire l_2 is

$$F_2 = \frac{2I_1 I_2 l_2}{c^2 r}$$

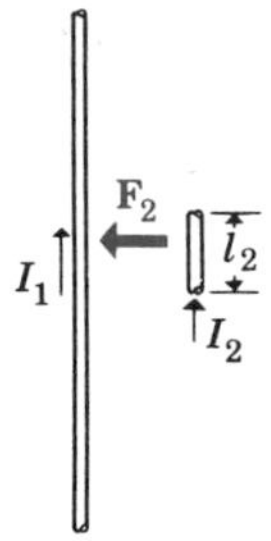

Example 4

What is the force per centimeter between two parallel currents of 1 amp that are separated by 1 cm?

In the above equation, $I_1 = I_2 = c/10$ statamp and $l_2 = r = 1$ cm. Inserting these values gives

$$F_2 = \frac{c/10}{c} \times \frac{2(c/10)}{c} = 0.02 \text{ dyne}$$

In our derivation of force between two currents using relativity, we obtained the result that the force was perpendicular to the direction of the velocity V. It can be shown using relativity theory that this must always be true. The magnetic force will always be perpendicular to the current element. In Fig. 8-6 the direction of the magnetic force is

Fig. 8-5. The force on current element $I_2 l_2$ due to the infinitely long current I_1.

Ans. 4: First determine $\mathbf{F}_e$ by holding q at rest. Then subtract the vector $\mathbf{F}_e$ from the net electromagnetic force.

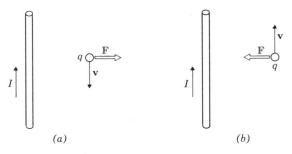

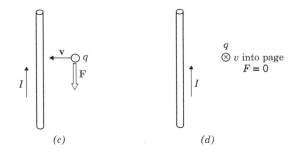

Fig. 8-6. Magnetic force **F** for four different directions of **v**.

shown for four different choices of direction of the velocity of the moving charge. It is worth noting an additional result for case (c); namely that the net force exerted by the moving charge on the wire is zero; whereas the force the wire exerts on q is $F = \dfrac{2Iqv}{c^2 r}$ pointing down. This is a gross violation of Newton's third law, and as such is an indication that magnetic forces are a purely relativistic effect. Even though at any given instant, Newton's action is not equal to reaction, after the moving charge has moved away, the change in momentum given to the moving charge will be equal and opposite to the change in momentum given to the wire. As pointed out in Chapter 3, the modern equivalent of Newton's third law is the conservation of momentum. Newton's third law may only be used when there are no effects of relativity. Because it is a relativistic effect, the behavior of magnetic force will seem peculiar to those of us who do not live in a relativistic world. We can describe these peculiar effects by what we will call right-hand Rule I and right-hand Rule II.

Right-hand Rule I

So far we have defined the magnitude of B produced by a long line of current. Since B is a vector, its direction must also be defined. The direction of B is determined by the peculiar direction in which the magnetic force acts. The direction of B is such that lines of B curl around the current, producing them as shown in Fig. 8-7. The arrowhead direction on these lines is still arbitrary. The convention used to determine this direction will be called right-hand Rule I in this book. As indicated in Fig. 8-7, if

Fig. 8-7. Right-hand rule I and lines of B produced by a long, straight current.

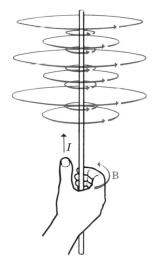

the thumb of the right hand points along the current, then the remaining four fingers when curled point along the lines of B. This rule for the direction of B agrees with the direction taken by a small magnetized needle when brought near a current. A compass needle, for example, will line up along the direction of B, and its north-seeking end will point in the positive direction of B. A compass needle that originally points toward a wire will suddenly rotate 90° when the current is turned on. Iron filings also tend to line up in the direction of magnetic field. The iron filing pattern due to the magnetic field produced by a wire is shown in Fig. 8-8.

Example 5

In Fig. 8-9 a compass needle is pointing north due to the 0.2 gauss magnetic field of the earth. A long wire running north to south is placed 4 cm above the compass needle. When the current is turned on, the compass needle reads NE instead of N. What are the magnitude and direction of the current?

Since the resultant field (along the compass needle) is 45° from the x-axis, its x- and y-components must be equal. Thus $B_x = B_y$ and $B_y = 0.2$ gauss, the field of the earth. According to Eq. 8-8, $B_x = 2I/cr$. The answer is obtained by equating this to B_y:

$$\frac{2I}{cr} = 0.2 \text{ gauss}$$

$$I = 0.1 \ cr \text{ statamp}$$

$$= 0.1 \times 3 \times 10^{10} \times 4 \text{ statamp}$$

$$= 1.2 \times 10^{10} \text{ statamp}$$

$$= 4 \text{ amp}$$

According to right-hand Rule I, this current must be flowing south.

8-4 Force on a Current

Everything is perpendicular

It can be shown using relativity theory, or by experiment, that the most general form for the force on a current element

◀ **Fig. 8-8.** Photograph of iron filings used to illustrate lines of force pattern of magnetic field produced by a long straight wire. Each iron filing behaves as a miniature magnet and lines itself up in the direction of **B**. (Courtesy Physical Science Study Committee.)

Fig. 8-9. Effect of a current placed above a compass needle. Before the current is turned on, the needle points as shown. After the current is turned on, it will point in the direction labeled **B.**

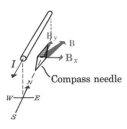

is the same as Eq. 8-9 except that the right-hand side is multiplied by the sine of the angle between l and B (or between v and B when dealing with a moving charge):

$$F_m = \frac{I\,\Delta l}{c} \times B \sin\theta$$

or   (8-10)

$$F_m = \frac{qv}{c} \times B \sin\theta$$

Right-hand Rule II

The preceding equation gives only the magnitude of F. A rule for determining the vector direction is still needed. For parallel currents we know that F must be perpendicular to B. Actually it turns out that in the general case of B produced by any configuration of currents the force on a current element must be perpendicular to the direction of B at that current element. In addition, as mentioned before, the force will also always be perpendicular to the direction of the current element. Hence magnetic force is perpendicular to both **B** and Δl at the same time. This uniquely determines the direction of $\mathbf{F}_m$ and in this book the rule for determining the direction of F_m will be called right-hand Rule II. In this rule, shown in Fig. 8-10, the outstretched right hand will be pushing in the direction of **F** if (as before) the thumb is along **I** and the fingers point along **B.**

At first thought it would seem that these electrical phenomena provide a means for the absent-minded professor to determine which is his right hand in case he should forget. One would think he could do this by measuring the force on a current and then checking to see which hand gives the right answer. However, in order to do this, he must also know the direction of B (which also requires knowledge of which is the right hand). If by mistake the professor had used his left hand for both Rules I and II, he would still end up with the correct direction for F. Thus electrical phenomena offer no way to distinguish left from right. The mirror image of any electrical experiment is also a legitimate experiment. Until 1956 most physicists firmly believed that there was no possible experiment which could distinguish right from left. This symmetry principle is called the con-

Fig. 8-10. Right-hand rule II used to obtain the direction of the force **F** on a current I in a magnetic field **B.**

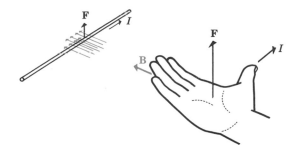

Q.5: In Fig. 8-6, what is the direction of **F** when **v** is pointing away from the wire?

servation of parity and was believed to be as firmly established as the five other conservation laws listed on page 153. As we shall learn in Chapter 16, basic physical phenomena were discovered in 1957 which overthrew the law of conservation of parity, and the absent-minded professor now has finally learned various ways to determine his right hand.

Example 6

Show that the forces in Fig. 8-6 are in agreement with the right-hand rules.

In all four cases the magnetic field acting on q is the same and is pointing into the page (this is seen by pointing one's thumb along the current I and using right-hand Rule I). Now apply right-hand Rule II to get the direction of F. We must always keep our fingers, B, pointing into the page. In case (*a*) our thumb, I, is up, so the palm then "pushes" to the left. In case (*b*), the thumb points down, so the palm "pushes" to the right. In case (*c*) the thumb points left, so the palm "pushes" down. In case (*d*) the force is zero because the charge is moving in the direction of B; (sin $\theta = 0$).

Force on a moving charge

Suppose a charge q is in a uniform magnetic field B and initially is moving perpendicular to B. Since the magnetic force will always be perpendicular to v, it will supply the centripetal force required to move the particle in a circle:

$$F_m = ma$$

$$\frac{qvB}{c} = m\left(\frac{v^2}{R}\right)$$

The radius of the circular path is then

$$R = \frac{mvc}{qB} \tag{8-11}$$

Example 7

A proton of velocity $v = 10^8$ cm/sec is moving perpendicular to a uniform magnetic field of $B = 10,000$ gauss as shown in Fig. 8-11.

(a) What will be the magnitude and direction of the force on the proton? (b) Describe the path of the proton.

According to Eq. 8-10

$$F = \frac{evB}{c} = \frac{4.8 \times 10^{-10} \times 10^8 \times 10^4}{3 \times 10^{10}} = 1.6 \times 10^{-8} \text{ dyne}$$

According to right-hand Rule II, this force must be pointed to the left in Fig. 8-11 and it will always be perpendicular to v. Thus F is a

Fig. 8-11. Path of a charged particle moving transverse to a uniform magnetic field.

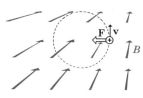

Ans. 5: Then **F** is pointing up, parallel to the current.

centripetal force which keeps the proton moving forever in the same circle of radius R. The value of R can be obtained using Eq. 8-11. For the problem in question the mass of the proton is $m = 1.67 \times 10^{-24}$ gm. Then

$$R = \frac{1.67 \times 10^{-24} \times 10^8 \times 3 \times 10^{10}}{4.8 \times 10^{-10} \times 10^4} \, \text{cm}$$

or

$$R = 1.04 \, \text{cm}$$

Figure 8-12 is a photograph of a track left by an electron traveling in a uniform magnetic field which is pointing out of the page. The radius of the circle is continuously decreasing because the electron is traveling in a liquid hydrogen bubble chamber. An electron or any other charged particle gradually loses energy while traveling through matter. Since the value of B is known, the electron momentum can be determined at any point in its path by measuring the radius of curvature.

We see from the preceding example and photograph that charged particles can be "trapped" by a magnetic field. This principle is used in those high-energy particle accelerators that use magnets. In such an accelerator the magnet supplies a magnetic field that keeps the beam of particles confined to a circle.

8-5 Ampere's Law*

How to get B

In electrostatics we were able to calculate the electric field produced by charge distributed uniformly over spheres, cylinders, and planes. Likewise, we need a method for determining the magnetic field produced by currents of shapes other than that of an infinite, straight line (Eq. 8-8). There is a more general equation for B which enables us, in principle, to calculate magnetic fields produced by currents of any given configuration. In this general equation any closed path is taken around a current. The equation states that the length L of the path times the average value of the com-

---

* The remaining 16 pages of this chapter which go more deeply into the theory of electricity are not needed for the understanding of the chapters that follow. It is recommended that a shorter or faster-paced course pass on to Chapter 9, Section 9-5 starting on page 225.

Ampere's law

Fig. 8-12. Electron track in liquid hydrogen bubble chamber. Track is curved due to uniform magnetic field pointing out of the page. Radius of curvature decreases as electron is slowed down by liquid hydrogen. (Photograph courtesy Alvarez group, Lawrence Radiation Laboratory, University of California.)

Fig. 8-13. Closed path of length $2\pi R$ around current I.

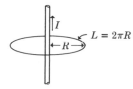

Fig. 8-14. A solenoid. The rectangular closed path is used to calculate the value of B inside the solenoid.

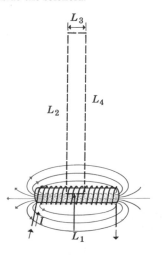

ponent of B along this path is $4\pi/c$ times the total current enclosed. The general equation is

$$(B_L \cdot L)_{\text{closed path}} = \frac{4\pi I}{c} \qquad (8\text{-}12)$$

where I is the total current enclosed.* This equation, Ampere's law, is used to calculate the magnetic field B at any point in space that is produced by a current I. We shall now use it to calculate B for three different examples: (1) the straight wire, (2) the solenoid, and (3) a surface current. In our first example we shall see that Ampere's law can be used to give the correct answer for the magnetic field produced by an infinite line of current. In Fig. 8-13 we choose a circle of radius R for the closed path around the current. Then $L = 2\pi R$ and $(B_L \cdot L) = 2\pi R B$. Now according to Ampere's law this equals $4\pi \dfrac{I}{c}$. Solving for B gives

$$B = \frac{1}{2\pi R}\left(4\pi \frac{I}{c}\right)$$
$$= \frac{2I}{Rc}$$

which checks with Eq. 8-8. In addition to Eq. 8-12 it can be shown that lines of B must be continuous. And since magnetic poles or charges do not exist in nature, the lines of B never start or stop; they just circle around in closed paths.

The solenoid

Our second application of Ampere's law will be the solenoid. A solenoid is a cylindrically wound coil or helix of n turns per centimeter. The lines of B produced by a solenoid are shown in Fig. 8-14. For the closed path take the rectangle shown in the figure. We make L_2 and L_4 so long that the far end of the rectangle is in a region of very weak B. The quan-

---

* As can be seen from Eq. 8-13 Ampere's law can also be written in the form

$$\sum_i B_i \cdot L_i = \frac{4\pi I}{c}$$

where the summation sign stands for

$$\sum_i B_i \cdot L_i = B_1 L_1 + B_2 L_2 + \cdots + B_N L_N$$

and the total path L is broken up into the pieces $L_1, L_2, \ldots, L_N$.

tity $\overline{B}_L$ in Ampere's law is the weighted average of B_L along the closed path. According to the definition of a weighted average (see page 23)

$$\overline{B}_L = \frac{B_1 L_1 + B_2 L_2 + B_3 L_3 + B_4 L_4}{L_1 + L_2 + L_3 + L_4} \tag{8-13}$$

The components (B_2 and B_4) of magnetic field along L_2 and L_4 are zero because the magnetic field is perpendicular to L_2 and L_4. Also B_3 is close to zero because it is so far from the solenoid. Hence Eq. 8-13 becomes

$$\overline{B}_L = \frac{B_1 L_1 + 0 + 0 + 0}{L}$$

and

$$\overline{B}_L \cdot L = B_1 L_1 \tag{8-14}$$

The total current enclosed by this path is the current times the enclosed number of turns nL_1. Thus the right-hand side of Eq. 8-12 is $4\pi In L_1 / c$. Equating this to the right-hand side of Eq. 8-14 gives

$$B_1 L_1 = \frac{4\pi In L_1}{c}$$

or

$$B = \frac{4\pi nI}{c} \tag{8-15}$$

inside a solenoid. Note that our derivation does not depend on where side L_1 is placed inside the solenoid. Thus B is constant everywhere inside, and the total number of lines of B produced by a solenoid must be $B_1 \cdot A$ where A is the coil area. Lines of B are defined quantitatively in the same way as lines of E. The field B equals the number of lines of B per unit area ($B = N_B / A$). The total number of lines of B produced by a solenoid is then

$$N_B = \frac{4\pi nIA}{c} \text{ lines of } B \tag{8-16}$$

The total number of lines of B is also called the total magnetic flux.

Example 7

A television picture tube has two coils wound as shown which provide the horizontal deflection of the electron beam. If the spot on

Fig. 8-15. Side view of television picture tube showing horizontal deflection coils.

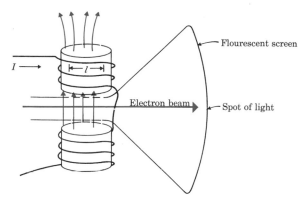

Fig. 8-16. Magnetic field produced by current flowing down a conducting surface.

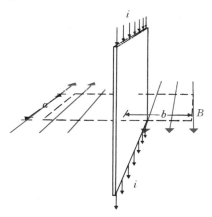

Q.6: What are the units of i in Fig. 8-16?

the screen is to sweep to the right (while facing the screen), what must be the direction of the current in the upper conductor?

This example is an exercise in the use of the right-hand rules. First, Rule II must be used to determine the direction of B. One must be careful in the use of this rule to note that in Fig. 8-15 the beam current is to the left even though the electrons are moving to the right. We want the force on this current to be into the page. In applying Rule II we point the thumb (I) left and the palm (F) into the page. This puts the fingers (B) pointing up. Now use right-hand Rule I on current element l. If the thumb (I) along l is pointed to the right, then the fingers (B) on the inside of the coil point in the desired up direction. Thus the current flows along l to the right. The answer to the problem is that the current enters the upper conductor and leaves at the lower conductor.

Magnetic field produced by current in a plane

Our final application of Ampere's law will be to determine the magnetic field produced by an infinite plane of current. Figure 8-16 shows a rectangular section of the infinite plane with the current traveling down. This situation is approximated when a current flows along a sheet of metal. Let i be the current per horizontal centimeter along the plane. We have chosen a rectangle of sides a and $2b$ which encloses a current ia. Because the plane is infinite the lines of B must run horizontally as shown in the figure. In calculating $\overline{B}_L$ around the closed path, the sides of length $2b$ do not contribute because B is perpendicular to sides $2b$. Thus

$$\overline{B}_L = \frac{B \cdot a + 0 + B \cdot a + 0}{a + 2b + a + 2b}$$

$$= \frac{2Ba}{L}$$

and

$$\overline{B}_L \cdot L = 2Ba$$

Since the enclosed current is $I = ia$, Eq. 8-12 becomes

$$2Ba = \frac{4\pi ia}{c}$$

or

$$B = 2\pi \left(\frac{i}{c}\right) \tag{8-17}$$

Again we note the similarity to the corresponding formula ($E = 2\pi\sigma$) for the electric field due to a charged plane.

8-6 Theory of Magnetism

Tiny current loops that never die

Ancient man knew of some magnetic as well as electrostatic phenomena. Some naturally occurring iron ores called lodestones are magnetized and attract other ferromagnetic samples. The ancient Greeks thought magnetic and electrostatic forces were of common origin. However, by the sixteenth century philosophers learned to base their laws of nature more on experiment than contemplation, and they concluded that magnetic and electrostatic effects were independent phenomena. No one was able to find any force between a charged object and a magnet. The discovery that there is indeed a force between *moving* charge and a magnet was made accidentally in 1820 by Hans Christian Oersted, a Danish physics teacher. At the end of a lecture on the subject he attempted to demonstrate the lack of a relationship between electricity and magnetism by turning on a current next to a magnetized needle. In the words of one of his pupils: "He was quite struck with perplexity to see the needle making a great oscillation." Thus the ancient doctrine that magnetism and electricity are related was suddenly given new life.

The purpose of this section is to explain why magnets behave the way they do purely in terms of forces between currents. How does a bar magnet behave when placed in a uniform external magnetic field? As mentioned in Section 8-4 there will be a torque tending to line it up with the field (see Fig. 8-17). The early physicists explained this effect by assuming that the bar magnet has two poles, one of strength $+m$ at the North end and one of strength $-m$ at the South end and that the force on a magnetic pole is

$$F = mB$$

The force between two magnets was explained by saying that one magnetic pole m_1 produced a magnetic field of strength $B = m_1/r^2$, which acts on a second magnetic pole m_2. Then

Fig. 8-17. Forces on a bar magnet in a uniform magnetic field.

Ans. 6: The quantity i is current per unit length which has units of $\frac{Q}{TL}$.

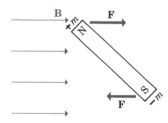

Fig. 8-18. Magnetic field pattern produced by bar magnet using iron filings. (Photograph courtesy B. Abbott, Physical Science Study Committee.) ▶

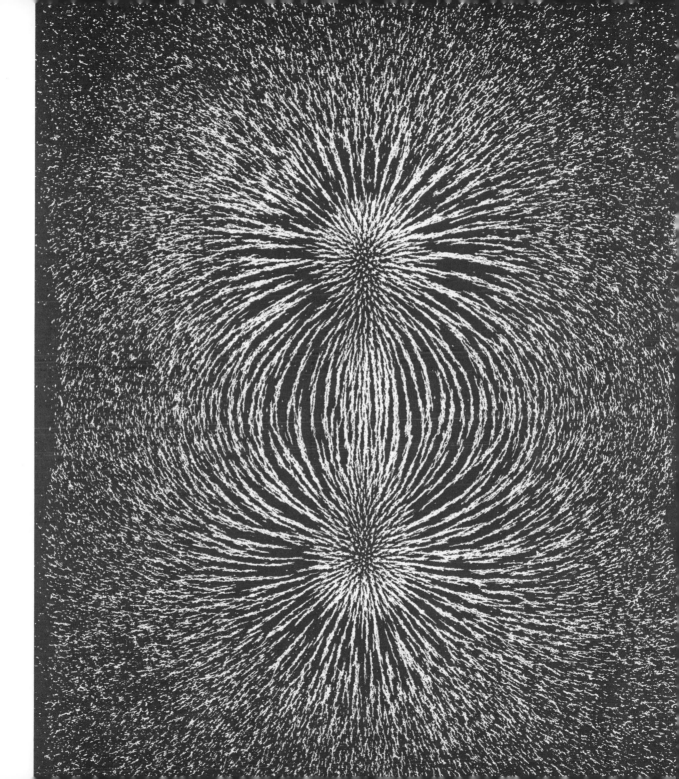

(a)

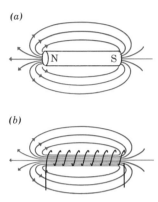

(b)

Fig. 8-20. Forces on a single turn of a solenoid in a uniform magnetic field.

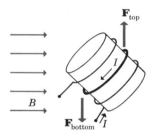

Fig. 8-21. A rectangular coil in a uniform magnetic field.

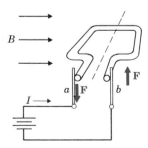

$F = m_1 m_2 / r^2$. The reader will note that this formalism of magnetostatics is mathematically exactly the same as electrostatics. Then the total number of lines of B produced by a magnetic charge (pole) of strength m is $N_B = 4\pi m$.

A plot of the lines of force produced by a bar magnet should then look the same as the lines of E on page 160 that are produced by two charges of opposite sign. Figure 8-18 is a photograph of iron filings around a bar magnet which tend to line up along the direction of B. Figure 8-19*a* is a plot of the lines of B produced by a bar magnet. Figure 8-19*b* is a plot of the lines of B produced by a solenoid. It is clear that the field pattern of a solenoid is exactly the same as that of a bar magnet. This suggests that perhaps a bar magnet is really a solenoid with some "mysterious" internal current that never dies out.

In fact, a solenoid, when placed in an external magnetic field, tends to line up with the field the same as a bar magnet. This can be seen by applying the force equation, Eq. 8-9 to the solenoid in Fig. 8-20. Right-hand Rule II tells us that the force on the bottom part of a turn is down and the force on the top part is up. The forces on the two sides can be neglected because they are equal and opposite and are also in line. Thus on any given turn there is a torque tending to rotate the coil in Fig. 8-20 counterclockwise and line it up with B.

It is this torque which makes possible the electric motor. The principle of the d-c motor is illustrated in Fig. 8-21. Application of right-hand Rule II gives forces as shown tending to rotate the coil counterclockwise about the dashed line axis. After the loop has rotated 180° there is again a current producing forces in the same direction. Once started, rotational inertia will keep this single rectangular loop rotating counterclockwise.

If a long pointer and a spring is attached to the coil of Fig. 8-21, the deflection of the pointer will increase as the applied current increases. This is the principle of the ammeter.

In 1836 Ampere proposed as an explanation of the behavior of a bar magnet that it is really a solenoid with a "built-in" current running around the outside surface.

Fig. 8-22. Bar magnet showing amperian current i'.

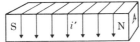

Ampere said: ". . . from the simple comparison of facts it seems to me impossible to doubt that there are really such currents about the axis of a magnet." Let the value of this Amperian current be i' statamp per centimeter of length along the bar magnet shown in Fig. 8-22. According to Eq. 8-16 the total number of lines of B is then

$$N_B = \frac{4\pi i' A}{c}$$

But when considered as a magnet of pole strength m,

$$N_B = 4\pi m$$

Thus $\quad 4\pi m = \frac{4\pi i' A}{c}$,

$$m = \frac{i' A}{c} \qquad \text{or} \quad i' = \frac{mc}{A} \tag{8-18}$$

We have just derived a formula for the pole strength of a magnet or solenoid in terms of the circulating current i'. But where does this perpetual current come from? Ampere explained this by picturing the molecules of a ferromagnetic material as having associated with them a circular current in a closed electric circuit of zero resistance. He also proposed that an external magnetic field could align these molecules parallel to each other so that their elemental magnetic fields became additive. Figure 8-23 shows how such elementary currents add up to produce a net surface current. Note that the currents inside cancel out.

This explanation of magnetism which preceded the discovery of the electron by over sixty years brilliantly anticipated our modern knowledge of atomic structure and theory of magnetism. Ampere's electric circuits of zero resistance exactly correspond to the motion of Bohr's atomic electrons. In the Bohr model, each electron presents a perpetual current similar to the single loop of a solenoid. In most atoms these electron orbits or current loops are so oriented that they cancel each other out. However, ferromagnetic substances such as iron, cobalt, and nickel have the two following properties: (1) their atoms have electron orbits and electron spin (rotating charge) which are not canceled out; and (2) the forces between neighboring atoms is such that

Fig. 8-23. End view of amperian current loops in a bar magnet. Note that the internal currents cancel each other out. The resultant surface current is shown as a dashed line. Each loop corresponds to the current in an individual molecule.

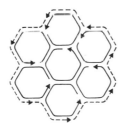

Q.7: Suppose the lines of B in Fig. 8-21 were pointing down through the coil. In which direction would the coil move?

the atoms prefer to line up so that their current loops are all pointed in the same direction. A full explanation of these points requires quantum mechanics. The quantum mechanical study of such phenomena is called solid state physics which is covered in Chapter 14. We now know that any sample of ferromagnetic material at room temperature is made up of macroscopic domains (on the order of a few thousandths of an inch) where the atoms are completely lined up. In an unmagnetized sample the domains are randomly oriented. In the process of magnetization the domains line up by movement of the domain boundaries, the domains favorably oriented with respect to the field growing at the expense of the others.

8-7 Faraday's Law of Induction

E produced by B

There is one last basic law of electricity called Faraday's law which we have not yet discussed. So far we have dealt only with steady currents fixed in space. Since Faraday's law deals only with changing currents, we have not made a mistake by ignoring it up until now. Before presenting the general equation for Faraday's law, let us first discuss a particular application—the electric generator. In Fig. 8-24 a rectangular coil in a uniform magnetic field is again shown. We recall from Fig. 8-21 that if a current is passed through the coil, it will start rotating. Suppose no current is applied, but that the coil is rotated by hand. Let the velocity of sides L_1 and L_3 be v. Then according to Eq. 8-10 there will be a force on each conduction electron in sides L_1 and L_3 of $F = evB/c$. This force would do the amount of work $W_1 = evBL_1/c$ on any electron that moves the length of L_1. The increase in energy of an electron making one complete circuit of the rectangular coil would be twice as much or $W = 2evBL_1/c$. This energy per unit charge is defined as the electromotive force (emf for short). Hence the electromotive force per turn is

$$\text{emf} = \frac{2vBL_1}{c} \qquad (8\text{-}19)$$

Note that emf has the same units as electric potential.

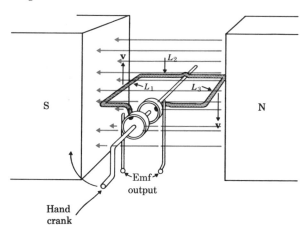

Fig. 8-24. Rectangular coil in a uniform magnetic field being rotated by hand. Such a device is actually a simple a-c generator.

Ans. 7: In no direction. Both the net force and the torque on the coil would be zero.

Fig. 8-25. Emf generated in a rotating coil as a function of time.

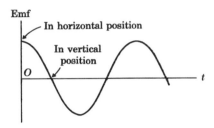

If an electric circuit were connected to the two ends of the coil, the acceleration of the electrons in L_1 and L_3 would drive them through the external circuit. The conduction electrons would give up the energy they had gained in L_1 and L_3 to the external circuit. Thus this rotating rectangular coil is a voltage source similar to a battery. If the coil consisted of N turns rather than one turn, the induced voltage or emf would be N times as great. This is the principle of the electric generator. After the coil in Fig. 8-24 has been turned through 90°, v and B are parallel and so there is zero emf at that instant of time. As the coil advances to the 180° position, the force and the emf are reversed in direction. The emf plotted against time is shown in Fig. 8-25. Such an oscillating emf is called an a-c (alternating current) voltage. Household power lines provide a-c voltage at a frequency of 60 cycles/sec which is produced by a-c generators similar in principle to Fig. 8-24.

Now we shall discuss the relation between Eq. 8-19 and Faraday's law. Let Δy be the vertical distance that side L_1 moves in time Δt starting from the 0° or hoizontal position. Then $v = \Delta y/\Delta t$ and Eq. 8-19 becomes

$$\text{emf} = \frac{2BL_1\,\Delta y}{c\,\Delta t} \tag{8-20}$$

After the time Δt the flux or number of lines of B passing through the coil is

$$N_B = B \cdot (\text{area})$$

The area is the base L_1 times the total height $(2\,\Delta y)$. Hence

$$N_B = B \cdot (L_1 \times 2\,\Delta y) = 2BL_1\,\Delta y$$

At the beginning of the time interval Δt, the flux passing through the coil was zero. Thus the change in flux during the time Δt is

$$\Delta N_B = (2BL_1\,\Delta y) - 0$$

Hence Eq. 8-20 can be written

$$\text{emf} = \frac{1}{c}\frac{\Delta N_B}{\Delta t} \tag{8-21}$$

Q.8: Which of the two terminals labelled "emf output" in Fig. 8-24 would be more positive when in the position shown?

This equation happens to be Faraday's law. So far we have

only shown that this equation applies to the situation of moving wires in a fixed magnetic field. But according to the principle of relativity, the same relation must also apply to fixed wires in a moving magnetic field; that is, Eq. 8-21 applies to a stationary coil where N_B is changing (the value of B is changing). Faraday's law says that a changing current in one fixed circuit will induce a voltage in another fixed circuit. This phenomenon of producing a voltage in a "dead" circuit just by changing a current somewhere else was discovered in 1830 by Faraday and Henry independently.

The transformer

One common application of Faraday's law is the transformer. In a transformer the primary and secondary coils are usually wound almost on top of each other, so that they both enclose the same lines of B (see Fig. 8-26). Thus both coils have the same value of $\Delta N_B/\Delta t$. Let N_1 be the number of turns of the primary coil and N_2 be the number of turns in the secondary coil. According to Eq. 8-21 the emf or induced voltage V_2 in coil 2 is

$$V_2 = (N_2)\frac{1}{c}\frac{\Delta N_B}{\Delta t}$$

Similarly the emf for coil 1 is

$$V_1 = (N_1)\frac{1}{c}\frac{\Delta N_B}{\Delta t}$$

The ratio of the two voltages is then

$$\frac{V_2}{V_1} = \frac{N_2}{N_1}$$

We see that if an a-c voltage is applied to the primary, the induced secondary voltage can be made larger or smaller by choosing the appropriate turns ratio. This convenient method for converting small voltages into large ones and vice versa is one of the advantages for using ac as compared to dc.

Fig. 8-26. The transformer.

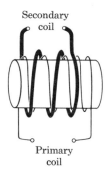

Secondary coil

Primary coil

Ans. 8: According to Right-Hand Rule II, the force on a positive charge in the left side of the coil is out of the page. This would cause positive charges to pile up on the left terminal.

8-8 Maxwell's Equations

Electricity in a nutshell

The collection of the basic laws of electricity (Coulomb's law, Ampere's law, and Faraday's law) is known as Maxwell's equations. In Faraday's law $(E_L \cdot L)_{\text{closed path}}$ is used in place of emf. As we have seen, it is the induced force on an electric charge which drives the electrons around a circuit. Since force per unit stationary charge is the definition of E, Faraday's law states that an induced electric field must always accompany a changing magnetic field. In this form Eq. 8-21 becomes

Faraday's law

$$(\overline{E}_L \cdot L)_{\text{closed path}} = \frac{1}{c}\frac{\Delta N_B}{\Delta t} \qquad (8\text{-}22)$$

Faraday's law tells us that a changing magnetic field must produce an electric field. A simple example of this is a charged particle q moving with velocity v in a uniform magnetic field B. So far the net force is $F = qvB/c$ and Faraday's law predicts zero electric field. But what about a moving observer? He sees a moving magnetic field, and, according to Faraday's law, he should then also see an electric field. Let us consider the case of an observer moving along with the charge. Since to him the velocity of the charge is zero, he sees zero magnetic force; however, the original force $F = qvB/c$ (where v is the original v) will still be there. This moving observer says the force (whatever its value) equals qE. So we have the force $qvB/c = qE$, with the result that the electric field seen by the moving observer is $E = \frac{v}{c}B$;

that is, a moving magnetic field produces an electric field.

Maxwell pondered whether the reverse might not also be true. Should not a changing electric field also produce a magnetic field? If this should be the case, then we would expect the same equation as above (Eq. 8-22), but with E and B interchanged:

$$(\overline{B}_L \cdot L)_{\text{closed path}} = \frac{1}{c}\frac{\Delta N_E}{\Delta t}$$

But we know already from Ampere's law that

Q.9: Consider a conduction electron in the rotating coil of Fig. 8-24. In the laboratory frame of reference we can calculate the net force on the electron. Will any part of this force be due to an electric field?

$$(\overline{B}_L \cdot L)_{\text{closed path}} = \frac{4\pi I}{c} \qquad (8\text{-}23)$$

So Maxwell proposed that the complete equation should be

$$(B_L \cdot L)_{\text{closed path}} = \frac{4\pi I}{c} + \frac{1}{c}\frac{\Delta N_E}{\Delta t}$$

Maxwell called this new term he added to Ampere's law the displacement current. Maxwell thought of an example which shows that something must be wrong with Eq. 8-23 in its present form. This example and the reasoning that led Maxwell to his displacement current term is given in the Appendix to this chapter.

In summary, Maxwell's equations in noncalculus form are

$$N_E = 4\pi Q \tag{8-24}$$

is the total number of lines leaving charge Q

$$(\overline{B}_L \cdot L)_{\text{closed path}} = \frac{1}{c}4\pi I + \frac{1}{c}\frac{\Delta N_E}{\Delta t} \tag{8-25}$$

$$(\overline{E}_L \cdot L)_{\text{closed path}} = \frac{1}{c}\frac{\Delta N_B}{\Delta t} \tag{8-26}$$

where the quantities in the right-hand sides of the last two equations are those enclosed by the closed path. These equations must hold for all possible closed paths.

8-9 Electromagnetic Radiation

The theory of light

Now finally armed with the correct laws of electricity, Maxwell proceeded to derive the theory of light. Specifically, he proved that a changing current will radiate electromagnetic waves with a velocity $v = c$ where c is the experimental proportionality constant appearing in the equations of electricity (Maxwell equations). The constant $1/c^2$ appearing in Eq. 8-6 was determined experimentally by measuring the force between currents. At that time nobody dreamed that this proportionality constant would have anything to do with the speed of light. Maxwell also showed that in such radiation the electromagnetic wave consists of E and B perpendicular to each other and also that both are perpendicular to the direction of the wave motion. In addition, E and B were shown to have the same magnitude. In 1864 this

Ans. 9: No. In the laboratory frame of reference $E = 0$ everywhere and hence the electrostatic force $F_e = eE$ also equals zero. The only force on the electron is the magnetic force due to its velocity. If we had been moving along with the electron, the situation would be reversed.

was indeed a crowning achievement of theoretical physics. Light waves were finally explained as high frequency electric and magnetic fields. Here the fundamental explanation of all electric, magnetic, and optical phenomena were supposedly given in one simple set of basic laws (Eqs. 8-24, 8-25, and 8-26). In addition, Maxwell's theory predicted that electromagnetic waves of any frequency, no matter how low, should be radiated with the velocity of light. This theory was first experimentally verified by Hertz in 1888, and by 1901 Marconi had succeeded in transmitting electromagnetic waves all the way across the Atlantic Ocean. These electromagnetic waves of lower frequency than light and infrared waves came to be known as radio waves.

Normally the derivation of electromagnetic radiation requires calculus and differential equations. However, now that we have worked so hard in the last two chapters to develop the complete set of Maxwell equations, it would be a shame to give up just before the grand climax. The following noncalculus explanation of electromagnetic radiation was worked out for this book with the help of Professor R. Feynman, a physicist at California Institute of Technology.

For the radiation source we shall use an infinite sheet of current. As shown on page 199, the magnetic field produced by a current running down a plane of i statcoulombs per centimeter is (see Eq. 8-17)

$$B = \frac{2\pi i}{c}$$

Now consider the case where originally there is no current and then the current i is suddenly turned on (see Fig. 8-27). The procedure will be to assume that the field $B = 2\pi i/c$ travels outward from the infinite plane with a velocity v. We also will assume that a uniform electric field E accompanies B and is perpendicular to B. In other words, we are assuming that a wave or step function of E and B travels away from the infinite plane with velocity v. Then Eqs. 8-25 and 8-26 can be used to solve for the values of E, B, and v. We will obtain the solutions $E = B = 2\pi i/c$ and $v = c$ as shown in the following paragraph. Hence such an outgoing wave traveling with $v = c$ does indeed satisfy Maxwell's equations and thus

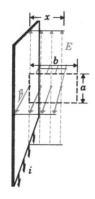

Fig. 8-27. Wave E and B traveling out from surface current of i statamp/cm. The electromagnetic wave has traveled a distance x.

Fig. 8-28. Electromagnetic pulses of 1×10^{-6} sec duration being radiated by turning on and off an infinite plane of current. A square section of the infinite plane and electromagnetic wave is shown. Lines of B are in red and lines of E in black.

$t = 0$
i just turned on

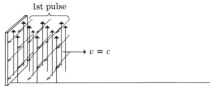

$t = 1 \times 10^{-6}$ sec
i just turned off

1st pulse

$v = c$

$t = 3 \times 10^{-6}$ sec
i just turned on

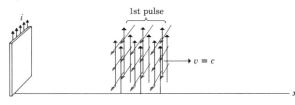

1st pulse

$v = c$

$t = 4 \times 10^{-6}$ sec
i just turned off

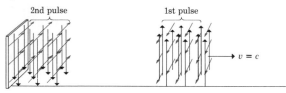

2nd pulse   1st pulse

$v = c$

is the correct solution for the situation where i is suddenly turned on.

In Figure 8-16 consider the situation shortly after the current has been turned on and the wave of $B = 2\pi i/c$ has only traveled a distance x where x is less than b. Then B is still zero at the ends of the rectangle in Fig. 8-16, and Eq. 8-25 gives

$$0 = \frac{4\pi(-ia)}{c} + \frac{1}{c}\frac{\Delta N_E}{\Delta t} \tag{8-27}$$

The minus sign is because we have chosen the up direction as positive. The electric flux N_E is the area $2ax$ times E. Thus

$$\frac{\Delta N_E}{\Delta t} = 2E\frac{a\Delta x}{\Delta t} = 2Eav$$

is the time rate of change of the electric flux through this rectangle. Substituting into Eq. 8-27 gives

$$0 = -\frac{4\pi}{c}ia + 2\frac{Eav}{c}$$

$$E = \frac{2\pi i}{c} \times \frac{c}{v}$$

Since the factor $(2\pi i/c)$ is the value of B we have

$$E = B\frac{c}{v} \tag{8-28}$$

To solve for v, another equation is needed. This is obtained using the vertical rectangle of Fig. 8-27. The electric field contributes to $(\overline{E}_L \cdot L)_{\text{closed path}}$ only along the left-hand side of the rectangle. Thus

$$(\overline{E}_L \cdot L)_{\text{closed path}} = E \cdot a$$

Equation 8-26 is then

$$Ea = \frac{1}{c}\frac{\Delta N_B}{\Delta t}$$

Since

$$N_B = Bax, \frac{\Delta N_B}{\Delta t} = Ba\frac{\Delta x}{\Delta t} = Bav$$

Then the above equation becomes

$$Ea = \frac{Bav}{c}$$

or

$$E = B\frac{v}{c}$$

The result $v = c$ is obtained by dividing this equation by Eq. 8-28.

Exactly the same derivation may be used to prove that if the current is suddenly turned on and then suddenly turned off, an electromagnetic square wave or pulse will be radiated. The radiation of such pulses is illustrated in Fig. 8-28. In fact, since E is proportional to i, it follows that if i is varied sinusoidally, then a sine wave will be radiated. Maxwell proposed that visible light consists of electromagnetic waves of the appropriate wavelength. Until this time (1864), perhaps the greatest unsolved question in physics was just what do light waves consist of? With the help of Maxwell, we now know that the light waves consist of oscillating electric and magnetic fields. This discussion is continued in Chapter 10.

Appendix 8-1

Fig. 8-29. Current flowing "through" a circular capacitor of radius R.

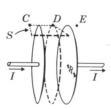

This Appendix discusses the necessity of adding the term $\frac{1}{c}\frac{\Delta N_E}{\Delta t}$ to Eq. 8-23. We will show that Eq. 8-23 when applied to a circular capacitor gives a mathematical contradiction, whereas the revised Eq. 8-25 gives the correct answer.

Consider a current I flowing into the circular capacitor of Fig. 8-29. The charge that is removed from the right-hand plate accumulates on the left-hand plate. As shown on page 197, by choosing circles of radius R around the current and by applying Eq. 8-23, the field at points C and E is found to be $B_C = B_E = 2I/cR$. However, the circle of radius R that passes through point D encloses zero current, so Eq. 8-23 predicts $B_D = 0$. We can now obtain a mathematical contradiction by also applying Eq. 8-23 to a rectangle with CD as one side and two other short sides of length s projecting out of the page. The rectangle is completed by joining these two short sides together with a second side of length CD.

Then $\quad (\overline{B}_L \cdot L)_{\text{closed path}} = B_C \cdot s - B_D \cdot s$

But since zero current flows through this rectangle, Eq. 8-23 states that

$$B_C \cdot s - B_D \cdot s = 0$$

and

$$B_D = B_C = \frac{2I}{2R} \quad \text{instead of} \quad B_D = 0$$

Maxwell discovered this contradiction and proposed the following equation which both overcomes the contradiction and agrees with experiment:

$$(\bar{B}_L \cdot L)_{\text{closed path}} = \frac{4\pi I}{c} + \frac{1}{c}\frac{\Delta N_E}{\Delta t}$$

where N_E is the flux or number of lines of E enclosed by the path. This equation can be checked for our example of the circular capacitor by again taking a circle of radius R passing through point D for the closed path. According to Eq. 7-13, the total number of lines of E between the plates is $N_E = 4\pi Q$. Then $\Delta N_E/\Delta t = 4\pi\, \Delta Q/\Delta t = 4\pi I$. Now substitute $4\pi I$ for $\Delta N_E/\Delta t$ in Eq. 8-25. Then

$$B_D \cdot 2\pi R = 0 + \frac{1}{c}(4\pi I)$$

and

$$B_D = \frac{2I}{cR}$$

which now agrees with B_C and the contradiction is resolved.

Problems

1. Does an observer moving along a charged wire measure the same current as a stationary observer?

2. Two long, straight parallel wires are 16 cm apart and carry currents of 4 amp each. Determine B at a point midway between them.
   (a) When the currents are in the same direction.
   (b) When the currents are in opposite directions.

3. A solenoid 1 m long and 8 cm in diameter is wound with 500 turns of wire.
   (a) What is B inside the solenoid when it carries a current of 6 amp?
   (b) What is the total number of lines of B?

4. Rewrite Faraday's law so that it gives emf in volts, keeping the other units the same.

5. An electron moving horizontally from east to west enters a magnetic field and is deflected downward. What is the direction of the magnetic field?

6. A 300-turn coil of 100 cm² area is rotating in a magnetic field of 5000 gauss, at 1800 rpm. What is the peak value of the emf generated?

7. An electron is shot down the axis of a solenoid. Describe the motion of the electron.

8. A thousand turn coil whose area of 100 cm² is perpendicular to the earth's magnetic field is flipped 90° in 1 sec. The average emf

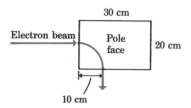

30 cm

Electron beam | Pole face | 20 cm

10 cm

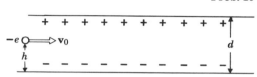

$+ + + + + + + + + +$

$-e$ ⟹ v_0

h

d

$- - - - - - - - - -$

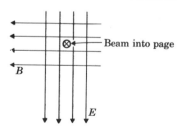

Beam into page

B

E

coming from the coil during that second was measured to be 0.6 millivolt (0.6×10^{-3} volt). What was the strength of the earth's magnetic field?

9. A beam of electrons of velocity 10^8 cm/sec is to be deflected $90°$ by a magnet as shown.

(a) What must be the direction of B to give this downward deflection?

(b) What is the radius of curvature of the electron path when between the magnet poles?

(c) What is the force in dynes on the electrons when in the magnetic field?

(d) What is the value of B in gauss?

10. An electron of mass m enters between two parallel charged plates with velocity v_0. Ignore gravity.

(a) If the value of the uniform electric field is E, the path of the electron will be part of a (circle, sinewave, parabola, ellipse).

(b) What will be the direction of the acceleration? What will be its magnitude in terms of e, E, and m?

(c) How long will it take until the electron hits one of the plates?

(d) Suppose the lines of E were magnetic rather than electric field (pointing in the same direction as E now points). Then what would be the direction of force on the electron?

11. Derive a formula for the period of revolution of a proton in a magnetic field B.

12. An 8 amp current is sent down a 2 cm diameter thin copper tube.

(a) What is B at 4 cm from the tube axis?

(b) What is B at 0.5 cm from the tube axis? (this point is inside the cylindrical surface of current). (*Hint:* does a closed path of radius 0.5 cm enclose any current?)

13. A beam of protons of velocity $v = 0.1\,c$ travels through a region of crossed magnetic field and electric field as shown. The protons are moving perpendicular into the page. The electrostatic force on the protons is 3×10^{-8} dyne.

(a) What must be the ratio E/B so that the net force on the protons would be zero?

(b) What would be the value of B in gauss?

(c) Suppose E/B was as above, what would be the direction and magnitude of the net force on a particle of charge $+e$ and velocity $v = 0.2\,c$? (This arrangement of crossed fields is used as a device to select only those particles of a given velocity.)

14. A coaxial cable consists of an inner and an outer cylinder of radii R_1 and R_2, respectively. A current of I statamp flows along each cylinder but in opposite directions. What is B at a distance r from the common axis

(a) when r is between the two cylinders?

(b) when r is outside the outer cylinder?

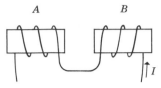

15. Consider two solenoids A and B with current flowing as indicated.

    (a) Which end of B is the N pole?

    (b) What will be the direction of the resultant force on B? *Hint:* consider solenoids as equivalent bar magnets.)

    (c) Is there a torque on B? If so, in what direction will B tend to rotate?

16. Consider a transformer which has 10^3 turns in the primary and 10^4 turns in the secondary winding. A 60-cycle per second a-c voltage is applied to the primary which produces a maximum of 10^4 lines of B passing through both coils. The flux through the windings is plotted against time in the figure.

    (a) How long does it take for the lines of B to build up from 0 to the peak of 10^4?

    (b) During this time what is the average emf induced in the secondary?

    (c) What was the average voltage across the primary during this time?

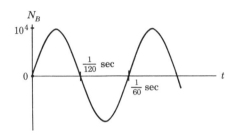

17. A jet plane with a 20-m wingspan is flying 600 mph due north at a latitude where the vertical component earth's magnetic field is 0.6 gauss.

    (a) What is the potential difference between the wing tips?

    (b) Which wing has the higher potential?

18. The mass of an electron rapidly increases as it approaches the speed of light. What is the mass of an electron moving in a circle of $R = 10$ cm when $B = 10,000$ gauss? (*Hint:* Put $v = c$ in Eq. 8-11.) What is the ratio of this mass to the electron rest mass?

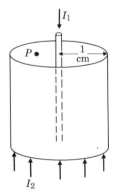

19. A coaxial cable has a current $I_1 = 3$ amp flowing down the center conductor and $I_2 = 2$ amp flowing up the outer cylindrical conductor.

    (a) What is B at a distance of 5 cm from the center?

    (b) What is B inside the cable at a distance of 0.5 cm from the center (at point P)?

20. In the Bevatron at Berkeley, California the protons reach seven times their rest mass and have velocity $v = 0.99c$. What must be the diameter of the Bevatron if it has a vertical magnetic field of 10,000 gauss over the entire proton orbit?

21. How many revolutions per second does a proton make in a cyclotron with magnetic field $B = 18,000$ gauss? Note that the answer is independent of the proton velocity or radius of the orbit.

22. A large sheet of current is oscillating sinusoidally at a frequency of 100 megacycles (10^8 cycles per sec). The peak value of the radiated electric field is 5 volts/cm.

    (a) What is the peak value of the current in statamp/cm?

    (b) What is the peak value of B in gauss?

23. A current I flows uniformly along a solid rod of radius R. What is B inside this rod at a distance r from the axis? ($r < R$).

24. Assuming Fig. 8-12 is reproduced to scale and the magnetic field is 12,000 gauss, calculate the mass of the electron as it enters the bubble chamber (assume it enters almost at the speed of light). What is the ratio of your result to the rest mass of the electron? Now use Eq. 3-6 to see how close the velocity is to the speed of light.

Cornell University

Electrical applications

Chapter

Electrical applications

9-1 Practical Units*

Volts and amps

The everyday electrical units used by electricians and electrical engineers are the coulomb for charge, the ampere for current, the volt for electrical potential, and the joule/sec, or watt, for power. As mentioned in Chapter 7, the coulomb is defined to be $c/10$ times a statcoulomb where c is expressed in cm/sec. Thus the practical units of charge and current are 3×10^9 times larger than the corresponding esu units. Presumably the practical units were established because of their more "convenient" size. However, things would have been much more convenient and practical if the conversion factor had been chosen as c rather than $c/10$. According to Eq. 7-17 the size of the practical unit of potential is determined by the choice of the coulomb for charge and joule for energy:

$$V = \frac{U}{q}$$

$$1 \text{ volt} = \frac{1 \text{ joule}}{1 \text{ coulomb}} = \frac{10^7 \text{ ergs}}{3 \times 10^9 \text{ statcoul}}$$

$$= \frac{1}{300} \frac{\text{erg}}{\text{statcoul}} = \frac{1}{300} \text{ statvolt}$$

Thus the esu unit for potential is 300 times larger than the practical unit.

9-2 Ohm's Law

A consequence of the structure of matter

If a voltage V is applied across a conductor, some value of current I will flow. Early in the nineteenth century George Ohm discovered that the magnitude of the current in metals was proportional to the applied voltage as long as the temperature was kept the same. The remarkable fact here is that this is not merely another approximation like Hooke's law, but that the relation appeared exact within the accuracy of

*Sections 9-1 to 9-4 may be omitted. Subsequent material does not depend on these sections. However, later material does depend on Sections 9-5 to 9-7.

9

measurements. *Ohm defined the resistance of a conductor as the voltage divided by the current.*

Ohm's law $R = \dfrac{V}{I}$ is independent of I for a metal

In the practical system V is in volts and I in amperes. The unit of resistance is volt/amp. This unit was given the special name ohm (by someone other than Ohm).

Just how basic is Ohm's law? Is it some new fundamental law of nature, or is it merely the consequence of the structure of matter and basic laws of interaction as was the kinetic theory? As we might expect, the latter is the case. The resistance of different substances under various conditions is explained quite well by the quantum theory of solids (see Chapter 14). In the following paragraph Ohm's law is deduced from the theory of metals.

Derivation of Ohm's law from theory of metals

The quantum theory of metals tells us that because of the wave nature of electrons, the outer atomic electrons are not bound to individual atoms in the metallic crystal. This is contrary to the ideas of classical physics and is explained in Chapter 14. These conduction electrons can travel many atomic diameters before having an atomic collision. Let L be the mean free path between collisions. The time between collisions will be $\Delta t = L/u$, where u is the average velocity of the conduction electrons (u is in all random directions; hence it does not give rise to a net current). If a voltage is applied across a piece of metal, there will be a force eE on each conduction electron, and during the time Δt each conduction electron will pick up a drift velocity $v_d = \Delta u$ given by

$$m\frac{\Delta u}{\Delta t} = eE \quad \text{(Newton's second law)}$$

$$\Delta u = \frac{eE}{m}\Delta t$$

Replacing Δt by (L/u) we obtain for the drift velocity

Q.1: Is the right-hand side of Eq. 9-1 the average drift velocity of the conduction electrons?

$$v_d = \Delta u = \frac{eL}{mu}E \tag{9-1}$$

Δu is in the same direction for all the electrons and thus gives rise to a net current. Each electron loses its drift velocity after each collision. The mean free path L is so small that v_d is always much less than u. Typically the drift velocity is less than 1 mm/sec (see p. 184). The electric current is the product of the charge per electron times $\mathcal{N}$ the total number of conduction electrons per centimeter along the wire times the average drift velocity:

$$I = e\mathcal{N}\bar{v_d}$$

The average drift velocity will be one-half the peak drift velocity. Substituting one-half the value given by Eq. 9-1 yields

$$I = e\mathcal{N}\left(\frac{1}{2}\frac{eL}{mu}E\right) = \frac{e^2\mathcal{N}L}{2mu}E$$

We see that according to Eq. 9-1 the drift velocity and hence the electric current is proportional to E, which in turn is proportional to the applied voltage. Thus as long as u (or the temperature) remains constant, the current will be proportional to the applied voltage as observed experimentally by Ohm. We can obtain a formula for the resistance of a wire of length x by noting that the voltage across it will be $V = Ex$. Substituting (V/x) for E into the above equation gives

$$I = \frac{e^2\mathcal{N}L}{2mu}\left(\frac{V}{x}\right)$$

Hence the resistance is

$$R = \frac{2mux}{e^2\mathcal{N}L}$$

9-3 Circuit Theory
Electrons prefer the path of least resistance

Many everyday electrical applications involve series and parallel combinations of resistances. The total resistance R_t of a combination of resistances is obtained by dividing the voltage applied across the combination by the total current flowing through the combination. In each of the resistance

Ans. 1: No. It is the value reached just before having another collision. The average drift velocity is halfway between this value and its initial value of zero.

Fig. 9-1. Resistances R_1, R_2, and R_3 in series combination (a), and parallel combination (b).

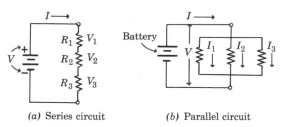

(a) Series circuit   (b) Parallel circuit

Fig. 9-2. Complex circuit of resistances. The symbol Ω stands for ohm.

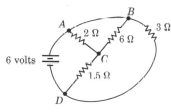

Fig. 9-3. Same circuit as Fig. 9-2 redrawn to make series and parallel combinations more apparent.

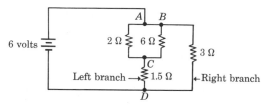

combinations shown in Fig. 9-1 the total resistance is

$$R_t = \frac{V}{I}$$

In a series circuit the same current flows through all the resistors. In a parallel circuit the total current is the sum of separate currents through each of the resistors. In the series combination the total potential difference V is

$$V = V_1 + V_2 + V_3$$

Dividing both sides by I

$$\frac{V}{I} = \frac{V_1}{I} + \frac{V_2}{I} + \frac{V_3}{I},$$

or

$$R_t = R_1 + R_2 + R_3 \quad \text{(series combination)} \qquad (9\text{-}2)$$

We see that in a series circuit the total resistance is the sum of the individual resistances.

In the parallel combination (Fig. 9-1b)

$$I = I_1 + I_2 + I_3$$

Dividing both sides by V

$$\frac{I}{V} = \frac{I_1}{V} + \frac{I_2}{V} + \frac{I_3}{V},$$

or

$$\frac{1}{R_t} = \frac{1}{R_1} + \frac{1}{R_2} + \frac{1}{R_3} \quad \text{(parallel combination)} \qquad (9\text{-}3)$$

In a parallel circuit, the reciprocal of the total resistance is obtained by adding the reciprocals of the individual resistances.

Many complicated circuits can be solved by breaking them down into combinations of simple series and parallel circuits. This procedure is illustrated in the following example.

Example 1

In the circuit shown in Fig. 9-2: (a) What is the total current supplied by the battery? (b) How much current flows through the 6-ohm resistor?

In order to find I we must first calculate R_t. This is facilitated by redrawing exactly the same circuit as shown in Fig. 9-3, so that the series and parallel combinations are made more apparent. We start with the parallel combination of the 2- and 6-ohm resistors. Let R be the resistance of this parallel combination. Then

$$\frac{1}{R} = \frac{1}{2} + \frac{1}{6} = \frac{2}{3}$$

$$R = 1.5 \text{ ohms}$$

This is in series with a 1.5-ohm resistor; hence the total resistance R' of the left-hand branch

$$R' = R + 1.5 = 3 \text{ ohms}$$

Finally this left-hand branch (R') is in parallel with the 3-ohm resistor. Thus

$$\frac{1}{R_t} = \frac{1}{R'} + \frac{1}{3} = \frac{2}{3}$$

$$R_t = 1.5 \text{ ohms}$$

and

$$I = \frac{V}{R_t} = \frac{6}{1.5} = 4 \text{ amp}$$

is the total current supplied by the battery.

To find the current through the 6-ohm resistor we must first determine the current I' through the left-hand branch, which is

$$I' = \frac{6 \text{ volts}}{R'} = 2 \text{ amp}$$

This current splits up in such a way that the voltage across the 6-ohm resistor will be the same as across the 2-ohm resistor. Thus 75% of I' goes through the 2-ohm resistor and 25% of I', or 0.5 amp, goes through the 6-ohm resistor.

The short circuit

In Figure 9-4, the bottom of the voltage source is connected to ground (the potential of the earth). In circuit theory, the potential of the earth is defined as zero potential. Note that if we connect one point of a circuit to ground, no current will flow through this connection. In Fig. 9-4 what is the potential at point A? Note that it is possible to trace a zigzag path of wires from point A to ground without passing through any resistor. In these diagrams, the connecting wires are assumed to have zero resistance. According to Ohm's law, the potential drop, or potential difference, across a wire will be

$$V = IR_{\text{wire}}$$

$$= I \times 0$$

$$= 0 \text{ volt}$$

Fig. 9-4. Illustration of a short circuit. Point A is short circuited to ground.

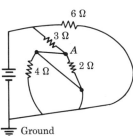

6 Ω
3 Ω
A
4 Ω
2 Ω
Ground

Thus the potential at point A is also zero. In fact, any two points connected by a resistanceless wire will be at the same potential. Since the potential difference across the 2- and 4-ohm resistors is zero, no current will flow through them. These two resistors are then said to have been short-circuited. If you were to short circuit a 115-volt home outlet, the current would be

$$I = \frac{V}{R} = \frac{115 \text{ volts}}{0 \text{ ohm}} = \infty \text{ amp}$$

In actual practice the current would not be infinite since the wires do have a small amount of resistance; however, the current would be large enough for sparks to fly and fuses to blow.

Electrical power

Electrical power is the electrical potential energy that is dissipated per unit time.

$$\text{Power} = \frac{U}{t}$$

The electrical potential energy being used up by a charge Q is $U = QV$. When a charge Q flows through a conductor, the potential energy is converted into heat energy. The electrical power or rate of heat production is

$$\text{power} = \frac{QV}{t}$$

By definition Q/t is the current I. Hence

$$\text{power} = IV \tag{9-4}$$

Example 2

A 75-watt light bulb operates at 115 volts. How much current does it draw and what is its resistance?

According to Eq. 9-4

$$I = \frac{\text{power}}{V} = \frac{75 \text{ watts}}{115 \text{ volts}} = 0.65 \text{ amp}$$

The resistance can now be determined using Ohm's law:

$$R = \frac{V}{I} = \frac{115}{0.65} \text{ ohm} = 177 \text{ ohms}$$

Fig. 9-5. The diode rectifier. An a-c voltage is applied to the input terminal A. The output voltage at B has the negative peaks "clipped" off.

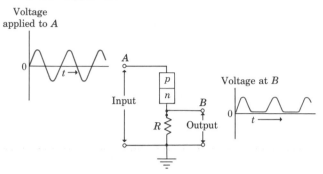

Fig. 9-6. Diode rectifier with capacitor C. Without C the output voltage would be the dashed red curve.

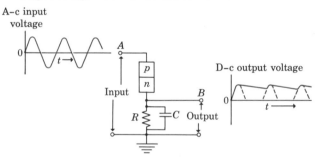

Fig. 9-7. The black curve is plot of RF (radio-frequency) voltage as a function of time. In this figure the RF or high-frequency electromagnetic wave is amplitude modulated with a pure sine wave or musical note of 1000 cycles/sec (red curve). For sake of clarity the RF frequency is greatly reduced.

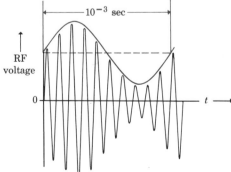

9-4 Radio and Television

Everyday electronics

Diode detector

The heart of a radio or television set is the diode detector. As shown in Fig. 9-5, the usual diode detector consists of a resistor R in series with a p-n junction diode. As will be explained in Chapter 14, a p-n junction has the property that its resistance is strongly dependent on the direction of the applied potential difference. In one direction the resistance is almost zero, and in the other direction it is very large. Any circuit element having this property is called a diode. In Fig. 9-5 we see that when a positive voltage is applied across the diode resistor combination, the diode is essentially a short circuit and the positive voltage appears across the resistor R. But when a negative voltage is applied, the diode resistance is much greater than R and most of the negative voltage appears across the diode rather than the resistor. Although the voltage on A is negative 50% of the time, the voltage on the output B is never negative. The voltage on B can be used to charge up a capacitor. This application of the diode, called a diode rectifier (see Fig. 9-6), can convert an a-c voltage into a d-c voltage. All a-c operated radio and television sets have diode rectifiers to convert the ac into dc.

Radio

An AM radio station transmits an electromagnetic wave at a fixed frequency somewhere in the broadcast band. The broadcast band includes frequencies from 0.5 to 1.6 megacycles (1 megacycle is 10^6 cps). The amplitude of the electromagnetic wave is modulated or varied with the audio signal that is being transmitted. As an example Fig. 9-7 is a plot of an electromagnetic wave modulated with a pure musical note of 1000 cps. In a radio receiver, this weak signal is picked up by an antenna, amplified by a RF (radio-frequency) amplifier, and then fed to a diode detector. The output voltage from the diode is the original audio signal of 1000 cps as shown in Fig. 9-8. This audio signal is amplified by an audio amplifier and fed into a loudspeaker. The block diagram of an AM radio receiver is shown in Fig. 9-9.

Fig. 9-8. Heavy red curve is output voltage from diode detector. Note that output voltage is original audio signal with small RF ripple. The thin black curve represents the RF signal with the negative peaks clipped off.

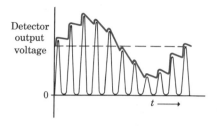

Fig. 9-9. Block diagram of AM radio receiver.

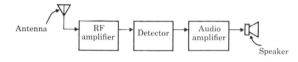

Fig. 9-10. The video signal of the letter "N" using 9 scan lines. (*a*) The corresponding image on an oscilloscope of picture tube using the video signal of (*b*) to control the electron beam intensity.

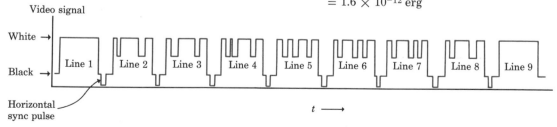

In television the video signal is also amplitude modulated onto a RF carrier. Thus a television receiver is similar to an AM receiver except that the final video signal is used to modulate the intensity of the electron beam that strikes the screen of the television picture tube. In the final video signal, the voltage is directly proportional to the brightness of the image scanned. In the time intervals between the scan lines of the video signal are voltage pulses that tell the picture tube sweep-circuit to start sweeping another line. The scanning sequence of the letter "N" is shown in Fig. 9-10*a*, and the corresponding video signal is shown in Fig. 9-10*b*. A block diagram of a television receiver, including the sweep circuits, is shown in Fig. 9-11. Commercial television in the United States uses 525 scan lines to the frame and 30 complete frames per second.

9-5 The Electron Volt

One more unit

The electron volt (ev) is just another unit of energy like the erg or joule. Its size is small enough to be convenient for measuring energies of single elementary particles.

$$1 \text{ ev} = 1.6 \times 10^{-12} \text{ erg}$$

This particular value of energy was chosen to correspond to the amount of energy given to one electron when accelerated through a potential difference of 1 volt. We can check this by using Eq. 7-17:

$$W = Q \times V$$

$$1 \text{ ev} = (e) \times (1 \text{ volt})$$

$$= 4.8 \times 10^{-10} \text{ esu} \times (\tfrac{1}{300} \text{ statvolt})$$

$$= 1.6 \times 10^{-12} \text{ erg}$$

Fig. 9-11. Block diagram of a television receiver.

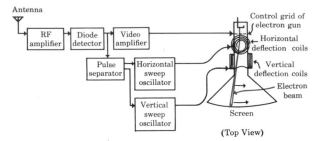

(Top View)

The kinetic energies given to particles by high-energy accelerators are usually measured in ev or Mev (million electron volts). A Van de Graaff generator (see page 158) that can produce a potential difference of 3 million volts can accelerate electrons or protons to a kinetic energy of 3 Mev. A Van de Graaff can also be used to accelerate helium nuclei of charge $Q = 2e$. In this case the energy obtained will be twice as much since $W = QV$ is proportional to the charge of the particle. A helium nucleus accelerated by a potential difference of 3 million volts acquires a kinetic energy of 6 Mev.

It is interesting to note that the mass of a 3-Mev electron has increased by a factor of seven. As mentioned in Chapter 3, the mass of a particle will increase as its velocity approaches the speed of light. According to Eq. 5-10, the rest energy of an electron is

$$W_0 = m_0 c^2$$

where $m_0 = 9.1 \times 10^{-28}$ gm is the electron rest mass. Then

$$W_0 = (9.1 \times 10^{-28}) \times (3 \times 10^{10})^2 \text{ ergs}$$
$$= 8.2 \times 10^{-7} \text{ erg}$$
$$= \frac{8.2 \times 10^{-7}}{1.6 \times 10^{-12}} \text{ ev}$$
$$= 5.1 \times 10^5 \text{ ev} = 0.51 \text{ Mev}$$

Hence a 3-Mev electron has a kinetic energy of almost six times its rest energy. Its total energy W is the sum of the rest energy plus the kinetic energy. Thus the total energy is

$$W = 7m_0 c^2$$

According to Eq. 5-10 the relativistic mass of the electron is

$$m = \frac{W}{c^2}$$

If we substitute the value $7m_0 c^2$ for W, we obtain

$$m = \frac{7m_0 c^2}{c^2}$$

or

$$m = 7m_0$$

9-6 The Cyclotron

Going around in circles can pay off

As explained on page 195, a beam of charged particles can be made to bend around in a circular path by using a uniform magnetic field. A cyclotron uses a large electromagnet to provide this uniform magnetic field. In addition, the cyclotron is so designed that after each semicircle of path the charged particles pass through an electric field and are accelerated if the electric field is pointing in the appropriate direction. As shown in Fig. 9-12 the electric field is across two hollow D-shaped electrodes. An a-c voltage is applied which switches the field direction each time the particles complete a half revolution. By this trick, the electric field is always pointing in the correct direction for accelerating the particles. The trick would not work if particles of different energies had different periods of revolution. We can show by Eq. 8-11 that the period of revolution is independent of the particle energy or radius of its circular path. According to Eq. 8-11

$$R = \frac{mc}{eB} v$$

is the radius of the circular path. For v we can substitute the quantity $2\pi R/T$ (see Eq. 2-16). Then

$$R = \frac{mc}{eB} \left(\frac{2\pi R}{T} \right)$$

or

$$T = \frac{2\pi mc}{eB} \tag{9-5}$$

which is independent of both R and v.

The synchrocyclotron

Actually the period T in Eq. 9-5 is not independent of v for velocities comparable to the speed of light. This is because the mass m must increase with velocity according to the equation

$$m = \frac{m_0}{\sqrt{1 - \dfrac{v^2}{c^2}}}$$

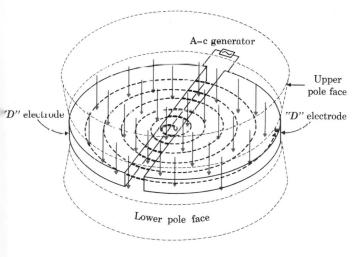

Fig. 9-12. Schematic diagram of a cyclotron. The lines of B are in red. The D-shaped electrodes are outlined in black. The pole tips of the cyclotron magnet are outlined in light broken black lines. The heavy dashed spiral is the path of a proton starting from the center. (Adapted from the *Scientific American*.)

Q.2: A cyclotron originally tuned for protons is retuned for accelerating helium nuclei. Must the a-c frequency be increased or decreased, and by how much?

For this reason conventional cyclotrons can accelerate protons only to about 20 Mev. However, if the period T of the alternating voltage is increased as the relativistic masses of the particles increase, there is no limit (other than magnet expense) on the possible energy. This type of cyclotron where the oscillator period T is varied during acceleration is called a synchrocyclotron.

Example 3

A synchrocyclotron has a magnetic field of 18,000 gauss. What will be the oscillator frequency when the protons have velocity $v = 0.01c$, and when they have velocity $v = 0.6c$?

Putting $M_p = 1.67 \times 10^{-24}$ gm and $B = 1.8 \times 10^4$ gauss into Eq. 9-5 gives

$$T = \frac{2\pi \times 1.67 \times 10^{-24} \times 3 \times 10^{10}}{4.8 \times 10^{-10} \times 1.8 \times 10^4} \text{ sec}$$

$$T = 3.64 \times 10^{-8} \text{ sec}$$

or

$$f = \frac{1}{T} = 27.5 \text{ megacycles}$$

When the velocity is as high as $0.6c$, Eq. 3-6 must be used for the proton mass:

$$M_p = \frac{1.67 \times 10^{-24}}{\sqrt{1 - (0.6)^2}} \text{ gm} = 1.25 \times 1.67 \times 10^{-24} \text{ gm}$$

We see that the mass has increased by 25%. According to Eq. 9-5 the period T will likewise increase by 25%. The kinetic energy of such a proton is 25% of its rest energy, or 25% of 938 Mev or 234 Mev.

Figure 9-13 shows the vacuum tank of the world's first cyclotron built in 1930 by E. O. Lawrence and M. S. Livingston above the world's highest energy synchrocyclotron. The giant Berkeley synchrocyclotron, also built by E. O. Lawrence, was first operated shortly after World War II. With some modifications it can now accelerate protons to kinetic energies of 730 Mev.

9-7   High-Energy Accelerators

The end is still not in sight

Recent developments in accelerator design make it possible to obtain higher energies using less massive magnets

Ans. 2: According to Eq. 9-5, the period of revolution is proportional to mass divided by charge which for helium nuclei is twice as much as for protons. Hence the frequency is $\frac{1}{2}$ the proton frequency.

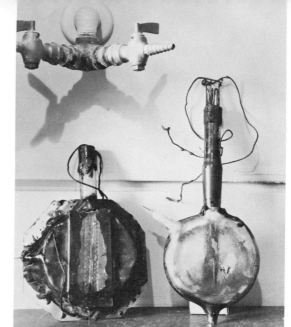

Fig. 9-13. The growth of the cyclotron. Vacuum tanks of the first cyclotron are shown at right. The world's highest energy synchrocyclotron is shown below. (Courtesy Lawrence Radiation Laboratory, University of California.)

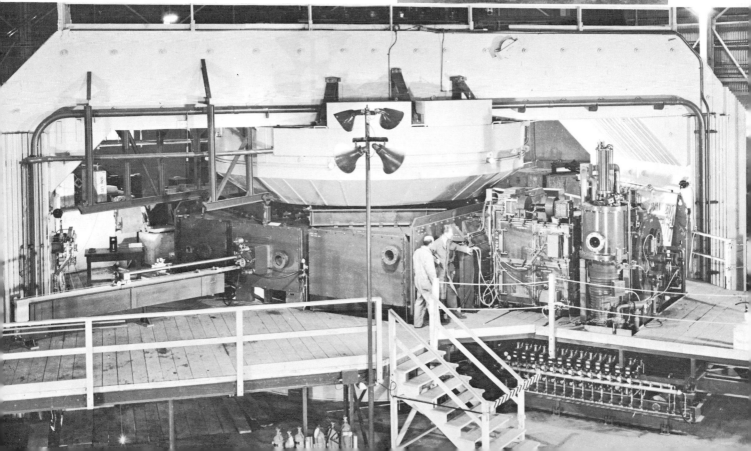

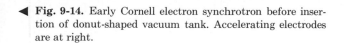

Fig. 9-14. Early Cornell electron synchrotron before insertion of donut-shaped vacuum tank. Accelerating electrodes are at right.

Fig. 9-15. Master control room of Russian 10-Bev proton synchrotron. (Courtesy Sovfoto.)

than those of the Berkeley synchrocyclotron. The main trick is to build the magnet only around what would be the outer cyclotron orbit. Instead of a cyclotron tank for the particles, an evacuated hollow donut is used. As in the cyclotron, the particles are accelerated by passing through a region of electric field. At the same time the magnetic field must be increased proportional to the increase in particle momentum; then according to Eq. 8-11 the radius of the circular orbit will stay fixed. Accelerators of this type are called synchrotrons. Figure 9-14 shows the ring-shaped magnet of one in the series of Cornell electron synchrotrons. The evacuated donut is of an elliptical cross section, 1.4×3 in., and runs through the center of the magnet. The values of R and B_{max} for this Cornell synchrotron were such that electrons could be accelerated up to 1.2 Bev (billion electron volts). Much of the pioneer work on the new elementary particles was done using a 3-Bev proton synchrotron at Brookhaven National Laboratory called the Cosmotron and using a 6.2-Bev proton synchrotron at Berkeley called the Bevatron. The Bevatron is shown in Fig. 16-3. A 10-Bev version is located near Moscow. Part of the control panel of this enormous machine is shown in Fig. 9-15.

The most recent and highest energy synchrotron is located at Brookhaven National Laboratory. As shown in Figs. 9-16 and 9-17 on the following page, it is 840 ft in diameter and is capable of accelerating protons to about 30 Bev. A similar accelerator is located at the international CERN laboratory in Geneva, Switzerland. A proton synchrotron of about twice this energy is under construction near Moscow and should be completed by 1968. In the United States plans have been made to build a 200-Bev proton synchrotron, and an even larger accelerator of about 1000 Bev is currently under study.

Example 4

What must be the diameter of a 30-Bev proton synchrotron if $B_{max} = 8000$ gauss?

According to Eq. 8-11

$$R = \frac{Mvc}{eB} \tag{9-6}$$

First, we must use the formula $W = Mc^2$ to determine the rela-

Fig. 9-16. Aerial view of the construction of Brookhaven 30-Bev proton synchrotron. The one-half mile-long magnet is contained in the ring-shaped underground tunnel. (Courtesy Brookhaven National Laboratory.)

Fig. 9-17. View inside underground tunnel of the Brookhaven 30-Bev proton synchrotron before installation of magnet. The spur to the left houses the proton injector. (Courtesy Brookhaven National Laboratory.)

tivistic mass of a 30-Bev proton. The total energy W of a proton having 30 Bev of kinetic energy is the 30 Bev plus the rest energy of 938 Mev, or a total of 30.938 Bev. Thus

$$M = \frac{30.938 \text{ Bev}}{c^2} = 5.5 \times 10^{-23} \text{ gm}$$

This is the value of M that must be used in Eq. 9-6. For v we shall use the value $c = 3 \times 10^{10}$ cm/sec because such extremely relativistic protons are almost traveling at the speed of light. Then Eq. 9-6 gives

$$R = \frac{(5.5 \times 10^{-23}) \times (3 \times 10^{10})^2}{(4.8 \times 10^{-10}) \times (8 \times 10^3)} \text{ cm} = 1.29 \times 10^4 \text{ cm} = 390 \text{ ft}$$

It is clear that there is no limit to the size or energy of synchrotrons, except perhaps the diameter of the earth. The reader will understand better the purpose of these high-energy accelerators after reading Chapters 15 and 16. Without these accelerators, it would be impossible to study the fundamental properties and interactions of the elementary particles of matter. In fact, some of the elementary particles were first discovered with the help of the high-energy accelerators. It is these fundamental properties of elementary particles upon which all of physical science is ultimately based.

Problems

1. As the mean free path of conduction electrons in a metal increases, the resistance of the metal (increases, decreases, remains the same).

2. A toaster produces more heat than a light bulb when they are connected in parallel. Which has the greater resistance?

3. Electric field can be expressed in volts/cm. A field of $E = 1$ volt/cm is how many dyne/statcoulomb?

4. The present Cornell electron synchrotron has a radius of curvature of 10 meters. The peak energy of 2.4 Bev corresponds to a magnetic field of 8000 gauss. The 10 Bev Cornell electron synchrotron now under construction is designed to reach full energy when the magnetic field is 3000 gauss. What is the diameter of the new Cornell 10 Bev synchrotron? By what factor will the electron mass be increased as it is accelerated from rest up to 10 Bev?

5. Which of the following is not a correct unit of electric potential: erg/statcoulomb, statvolt, joule/amp × sec, dyne × cm/statcoulomb, or statcoulomb/cm²?

Prob. 9

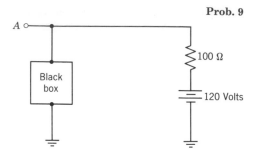

6. Two parallel plates are separated by 3 cm. The electric field between the plates is 20 dynes/statcoulomb. What is the potential difference in volts between the plates?

7. What is the horizontal sweep frequency in American TV?

8. An FM receiver picks up an RF signal of 1 microvolt (10^{-6} volt) and amplifies it to 10 volts in three identical stages of amplification. What is the amplification factor per stage?

9. The current I going into the "black box" is 0.5 amp. What is the potential at the point A? What is the resistance of the "black box"?

Prob. 10

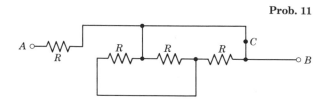

10. In the accompanying figure, what is the resistance between terminals A and B?

Prob. 11

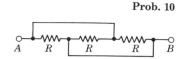

11. If $R = 12$ ohms, what is the total resistance between A and B? If the wire breaks at point C, what now is the total resistance?

Prob. 12

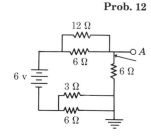

12. In the accompanying figure the 6-volt battery is connected to a total resistance of 12 ohms:
    (a) What is the potential at point A?
    (b) If the wire breaks at the point indicated by the arrow, what then will be the potential at point A?

Prob. 13

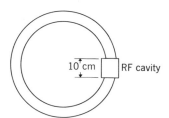

13. Electrons are accelerated by a Van de Graaff which generates a voltage of 2×10^6 volts.
    (a) What is the energy of the electron beam leaving the Van de Graaff in Mev?
    (b) The electrons are then fed into a synchrotron and circle around for 10^4 turns. If the RF cavity gives a field of 100 volts/cm and is 10 cm long, what will be the final energy of the electron beam in Mev?

Prob. 14

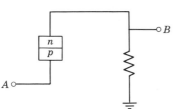

14. An a-c voltage is applied to point A in the accompanying figure. Plot a curve of the voltage appearing at B as a function of time.

Prob. 15a

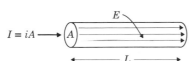

15. A cylindrical resistor has cross sectional area $A = 2$ cm^2 and length $L = 8$ cm. The current density is $i = 3$ amp/cm^2 when the electric field is $E = 6$ volts/cm. E is constant inside the rod.

    (a) What is the resistance R? in ohms?

    (b) The electrical conductivity of a metal is defined as $\sigma = i/E$ where i is the current *density* in amps/cm^2. The conductivity is then independent of the shape or current in the metal. Derive a formula for the resistance R in terms of σ, A, and L.

Prob. 16

16. What must be the resistance X in terms of R_1, R_2, and R_3 so that no current flows through the meter G?

Prob. 17

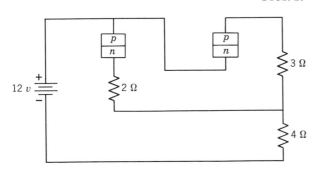

17. Consider the circuit of resistors and ideal or perfect diodes.

    (a) What is the current in the 2 ohm resistor?

    (b) What is the current in the 3 ohm resistor?

    (c) What is the current in the 4 ohm resistor?

    (d) If the battery voltage is reversed, what will be the current in the 4 ohm resistor?

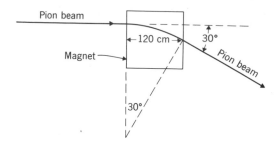

Pion beam

120 cm

Magnet

30°

Pion beam

30°

18. In an experiment at the Brookhaven AGS it is necessary to deflect a 1 Bev pion beam by an angle of 30°. This can be done by using a magnet with a pole face of area 120 cm × 120 cm. Outside the pole-face $B = 0$, and between the pole faces B is constant. For each pion in this beam $Pc = 1$ Bev $= 1.6 \times 10^{-3}$ ergs, hence $P = 1.6 \times 10^{-3}/3 \times 10^{10} = 5.33 \times 10^{-14}$ gm cm/sec.

(a) What is the radius of curvature of the pion path while in the magnetic field?

(b) What value of B in gauss is needed to achieve this?

19. If the RF frequency needed to accelerate protons in a small cyclotron of fixed B is 10^7 cps:

(a) What frequency is needed to accelerate deuterons? ($M_D = 2M_H$, $Z_D = 1$).

(b) What frequency is needed to accelerate singly ionized helium He^+? ($M_{He} = 4M_H$).

(c) What frequency is needed to accelerate doubly ionized helium He^{++}? (Doubly ionized helium is the bare helium nucleus and is also called an alpha particle.)

Prob. 20

20. In a mass spectrometer an ionized atom of charge e, mass M, and velocity v moves in a semicircle in a uniform magnetic field B. What is the mass M in terms of e, v, c, B and D, the diameter of the circle?

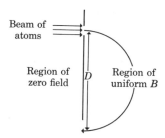

Beam of atoms

Region of zero field

D

Region of uniform B

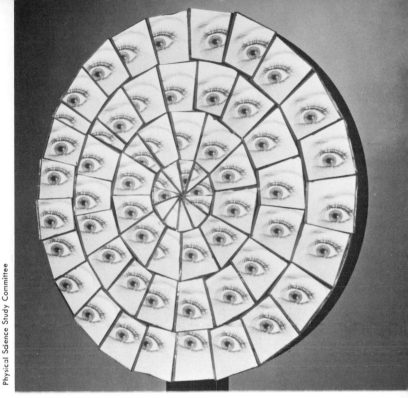

Wave motion and light

Chapter **10**

Wave motion and light

10-1 Electromagnetic Waves

Oscillation of E and B in space and time

This section is a continuation of the discussion at the end of Chapter 8. It was shown in Section 8-9 that any periodic motion of charge will radiate an electromagnetic wave of the same frequency. As discussed in Section 8-9, an electromagnetic wave consists of mutually perpendicular electric and magnetic fields and propagates in a direction perpendicular to E and B with a velocity of $c = 3 \times 10^{10}$ cm/sec. Furthermore, the magnitude of E must equal that of B.

Example 1

The transmitting antenna of a radio station is mounted vertically. At a point 10 mi due north of the transmitter the peak electric field is 10^{-3} volt/cm. What are the direction and magnitude of the radiated magnetic field in gauss?

According to the right-hand rule, the lines of B will circle around the antenna current. As shown in Fig. 10-1 the lines of B will be in the east-west direction when north of the transmitter. At a fixed point in space the direction of B will switch back and forth from east to west. The magnitude of B in gauss will be the same as that of E in esu.

$$E = 10^{-3} \text{ volt/cm} = \frac{10^{-3}}{300} \text{ statvolt/cm} = 3.33 \times 10^{-6} \text{ esu}$$

Therefore $B = 3.33 \times 10^{-6}$ gauss

Wavelength

The current in a transmitting antenna when plotted against time is usually a sine wave as shown in Fig. 10-2. At any fixed point in space the magnitude of E must also vary with time in the same way as shown in Fig. 10-2. The period of oscillation T is the time interval between successive oscillations. Not only does E oscillate as a function of time but also as a function of space: at a given instant of time the value of electric field as a function of the distance along the direction of wave motion will appear as shown in Fig. 10-3. This plot of E versus r, the distance from the antenna, is also a sine wave. The wavelength λ is defined as the distance between successive oscillations. A <u>general feature</u> of wave motion, whether electromagnetic waves, water waves, or sound waves, is that the distance covered by the wave in 1 sec

Fig. 10-1. Vertical antenna showing radiated magnetic lines of B.

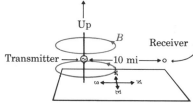

Fig. 10-2. The electrical field E at any fixed point in space as a function of time.

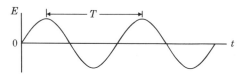

10

Fig. 10-3. The electric field E along the direction of wave propagation at a given instant of time.

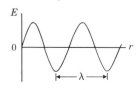

$$v = \frac{dist.}{time} = \frac{\lambda}{T} = \lambda f$$

Q.1: How many wavelengths long is the curve in Fig. 10-3?

is the distance covered in one oscillation (λ) times the number of oscillations per second (the frequency f). Thus

$$\lambda f = \text{distance covered in 1 sec}$$

But the distance covered in 1 sec is the wave velocity v. Thus

$$\boxed{\lambda f = v} \text{(the wave velocity)} \tag{10-1}$$

For electromagnetic waves

$$\lambda f = c \tag{10-2}$$

Example 2

In the United States the FM band is centered around a frequency of 100 megacycles. What will be the length of an FM antenna if each arm must be a quarter wavelength?

According to Eq. 10-2

$$\lambda = \frac{c}{f} = \frac{3 \times 10^{10}}{100 \times 10^{6}} \text{ cm} = 300 \text{ cm}$$

$$\frac{\lambda}{4} = 75 \text{ cm}$$

Thus each arm should be 75 cm or about 30 in. long.

The period of oscillation T is simply related to the frequency f by the relation

$$f = \frac{1}{T}$$

This is easily seen by writing down the number of oscillations N in a time t: $N = t/T$. Hence the number of oscillations per unit time is $N/t = 1/T$ which happens to be our definition of frequency. It has units of sec^{-1}. We use the abbreviations cps for cycles per sec and Mc for megacycle (10^6 cycles per sec). A newer abbreviation is Hz for Hertz which also stands for cycles per sec.

10-2 The Electromagnetic Spectrum

All kinds of light

Electromagnetic waves in principle could have any frequency from zero to infinity. The classification of electromagnetic waves according to frequency is called the electro-

Fig. 10-4. The electromagnetic spectrum.

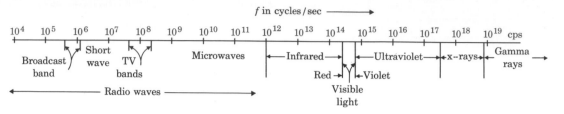

magnetic spectrum, which is shown in Fig. 10-4. Low frequency electromagnetic waves of several cycles per second have no practical application and are not generated intentionally. However, there is no way to prevent electric power lines from radiating electromagnetic waves of the a-c frequency (usually 60 cps). Such radiation is regarded as power loss, but it is very small compared to the I^2R loss due to the resistance of the power lines. Electromagnetic waves with frequencies above several thousand cycles per second are classified as radio waves. The broadcast band of radio waves is in the one megacycle region. The VHF (very high frequency) television band starts at about 50 megacycles. Then comes the UHF (ultra high frequency) band followed by SHF (super high frequency) and EHF (extremely high frequency) bands. In England there is a VHFI (very high frequency indeed) band. The highest frequency waves generated by electronic oscillators are called microwaves. The corresponding wavelengths are only a few centimeters or even millimeters long.

Even higher frequency electromagnetic waves can be generated by molecular and atomic oscillations. For example, if hydrogen gas is heated to a sufficiently high temperature, the two atoms in a hydrogen molecule will essentially vibrate back and forth in simple harmonic motion and emit electromagnetic radiation of this same frequency as discussed in Example 6 of Chapter 3. The frequency of oscillation is $f = 4.9 \times 10^{13}$ cps. According to the electromagnetic spectrum of Fig. 10-4, radiation of this frequency is classified as infrared. Electromagnetic radiation in the frequency range of 4.3×10^{14} cps to 7×10^{14} cps can be detected by eye and is called visible light. The lowest frequency appears red and

Ans. 1: Two.

the highest frequency appears violet. Frequencies higher than 7×10^{14} cps are not visible and are classified as ultraviolet radiation up to 5×10^{17} cps. Frequencies from this value to about 10^{19} cps are classified as x-rays. Electromagnetic radiation of even higher frequency is called gamma radiation. The highest energy gamma rays ever observed have been seen in the cosmic rays (see Chapter 15). The production of x-rays and gamma rays is discussed in Chapters 12, 13, and 15.

10-3　Interference

Negative meets positive

Since wave motion along a taut string is easier to visualize than an invisible electromagnetic wave, we shall first study general properties of waves propagating along a string. The velocity of a wave along a string depends only on the force or tension T on the string, and ρ, the mass of the string per unit length. The formula for the <u>wave velocity</u> is

$$v = \sqrt{\frac{T}{\rho}} \tag{10-3}$$

In the CGS system T is in dynes and ρ is in gm/cm. This formula is easiest to derive in a reference system that is moving along with the wave. To such an observer the wave will appear stationary and the string will be moving in the reverse direction with velocity v. According to such an observer the crest of a wave will appear as shown in Fig. 10-5. The portion of the string of length l will be moving to the left around a curve of radius R with the wave velocity v. The net force F on this length of string is obtained by adding the vectors $\mathbf{T}_1$ and $\mathbf{T}_2$ as shown in Fig. 10-5b. According to Newton's second law

$$F = Ma$$

where $M = \rho l$ is the mass of this length of string. The acceleration a of this mass is the centripetal acceleration $a = v^2/R$. Thus

$$F = \rho l \frac{v^2}{R} \tag{10-4}$$

Wave velocity on a string

Fig. 10-5. Crest of wave on a string. (*a*) View by observer moving along string with same velocity as that of the wave. (*b*) Net force on length of string shown in (*a*).

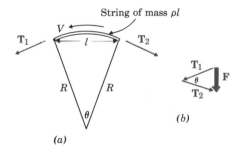

(a)

(b)

Q.2: In Fig. 10-5(*b*), is $\mathbf{F} = \mathbf{T}_2 - \mathbf{T}_1$?

Note that for small l, the force triangle in Fig. 10-5b is similar to the triangle in Fig. 10-5a. Hence

$$\frac{F}{T} = \frac{l}{R}$$

or

$$F = \frac{l}{R} T$$

Now substitute this into the left-hand side of Eq. 10-4 to obtain

$$\left(\frac{l}{R} T \right) = \rho l \frac{v^2}{R}$$

$$T = \rho v^2$$

and

$$v = \sqrt{\frac{T}{\rho}}$$

which is what we set out to prove.

If one end of the string is given a quick snap to the side (like cracking a whip), a single pulse will run down the string with velocity v. Suppose at the same time someone else generated a pulse at the other end of the string. What will happen when these two pulses meet in the center? The two pulses will cross through each other and continue to proceed in their own directions as shown in Figs. 10-6 and 10-7. This independence of <u>traveling waves</u> is a consequence of the <u>principle of superposition</u>. According to this principle, the resultant wave amplitude is the sum of the amplitudes of the separate waves. The principle of superposition of waves is merely a consequence of the fact that resultant displacements, accelerations, and forces are the sum of the separate displacements, accelerations, and forces. The principle of superposition must also apply to electromagnetic waves since the resultant electric and magnetic fields are the sum of the separate fields. We see in Figs. 10-7a and 10-7b that if the waves have opposite signs, they tend to cancel each other out when they cross. Not only pulses but continuous sine waves can be made to cancel. Figure 10-8 shows a time sequence of two sine waves starting from opposite ends of the string crossing over each other. Note that after the two waves have crossed and are overlapping there is a standing wave pattern

Ans. 2: No. $\mathbf{F} = \mathbf{T}_1 + \mathbf{T}_2$.

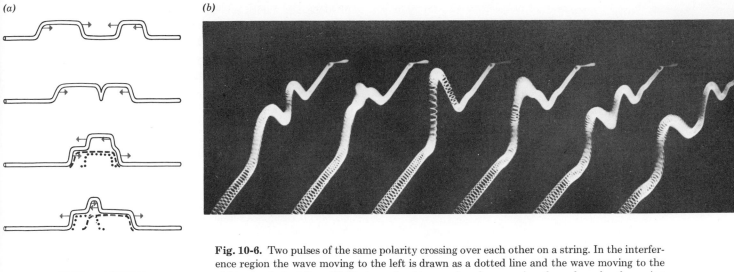

Fig. 10-6. Two pulses of the same polarity crossing over each other on a string. In the interference region the wave moving to the left is drawn as a dotted line and the wave moving to the right as a dashed line. (*b*) Photograph of two such pulses crossing through each other using a toy "slinky." (Courtesy Physical Science Study Committee.)

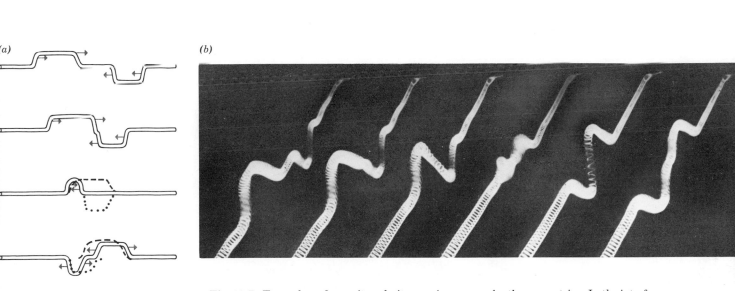

Fig. 10-7. Two pulses of opposite polarity crossing over each other on a string. In the interference region (3rd view) they tend to cancel each other out. (*b*) A photograph of two pulses of opposite polarity crossing each other on a toy "slinky." (Courtesy Physical Science Study Committee.)

Fig. 10-8. Time sequence of two sine waves starting from opposite ends of a string. If there had been only the sine wave moving to the left, the string would follow the red curve. Points A and C remain stationary after the sine waves cross and are called nodes. Note that points A and C are $\lambda/2$ apart.

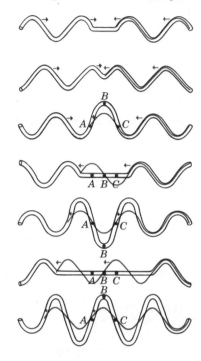

Fig. 10-9. Four possible standing waves on a string of length L fixed at both ends. The heavy lines indicate the position of the string when at its maximum displacement. The shaded areas show the regions covered by the motion of the string.

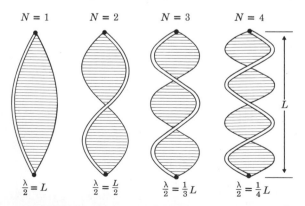

of loops and nodes on the string. Points A and C, called nodes, always remain stationary. Point B, called a loop or antinode, oscillates up and down with maximum amplitude. Note that the distance between successive nodes or successive loops is a half-wavelength.

If a taut string fastened at both ends is vibrated, standing waves such as those shown in Figs. 10-9 and 10-10 will be generated. The initial waves put on the string are reflected from both ends. The reflected waves will be of reverse polarity, thus meeting the condition that nodes be at the two ends.

Example 3

Does increasing the tension of the string in Fig. 10-10 increase or decrease the number of nodes?

According to Eq. 10-3, it increases the quantity v in the relation $\lambda = v/f$. Hence the wavelength increases and the number of nodes decreases.

Example 4

A guitar string is 30 cm long and has a mass of 100 gm. What value of tension must be put on the string in order to tune it to middle C (262 cps)?

According to Eq. 10-3 the tension is given by

$$T = \rho v^2 \qquad (10\text{-}5)$$

where ρ is 100 gm/30 cm or $\rho = 3.33$ gm/cm. We can find v from the relation $v = \lambda f$. According to Fig. 10-9, in the fundamental mode of oscillation $\lambda/2 = L$ or $\lambda = 60$ cm. Thus

$$v = 60 \text{ cm} \times 262 \text{ cps} = 1.572 \times 10^4 \text{ cm/sec}$$

Substituting this into Eq. 10-5 gives

$$T = 3.33(1.572 \times 10^4)^2 \text{ dynes}$$

$$= 8.25 \times 10^8 \text{ dynes}$$

This is a rather large force to be supported by such a light string. For this reason strings of musical instruments are usually made of strong metallic alloys.

Fig. 10-10. Photograph of standing waves on a vibrating string. In which view is the string under highest tension? Physical Science Study Committee.) ▶

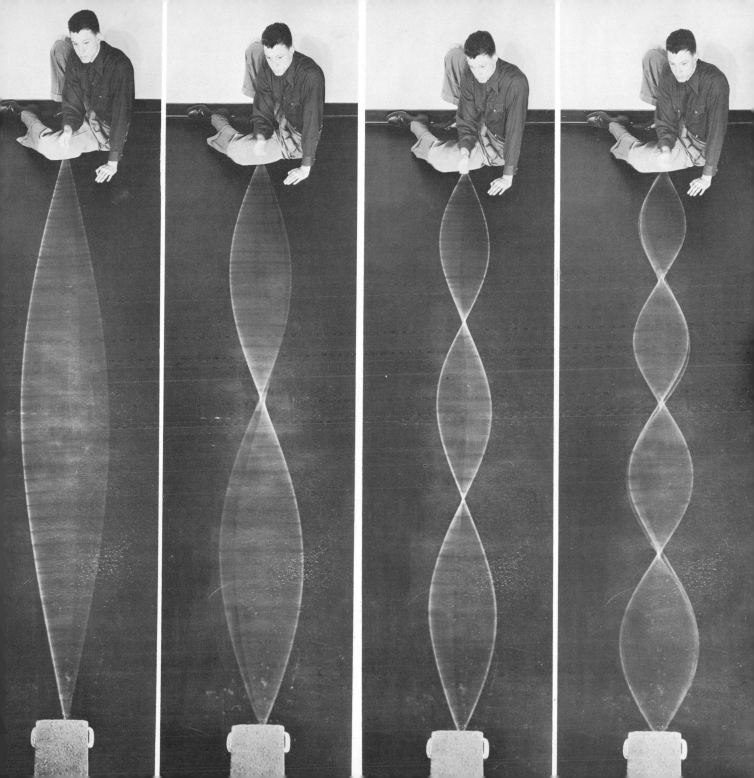

Wind instruments utilize an enclosed or partially enclosed column of air. Standing waves are induced in the confined volume of air by introducing a pressure disturbance (usually by blowing). In the simple case of a closed box of length L the same standing waves as those shown in Fig. 10-9 can be induced. Here the curves of Fig. 10-9 represent the displacement of air molecules rather than particles of a string. The general condition for a standing wave is that an integral number of half-wavelengths be contained in the distance L:

$$N\left(\frac{\lambda}{2}\right) = L \quad \text{(condition on } \lambda \text{ for a standing wave)} \quad (10\text{-}6)$$

where N is any integer greater than 0.

Example 5

The microwave oscillator in Fig. 10-11 emits plane electromagnetic waves to the right which are reflected back to the left. P_1 and P_2 are the positions of two successive intensity minima and are 5 cm apart. What is the frequency of the microwave oscillator?

Since any two successive nodes are a half-wavelength apart, $\lambda = 10$ cm. The frequency is given by

$$f = \frac{c}{\lambda} = \frac{3 \times 10^{10}}{10} = 3 \times 10^9 \text{ cps} = 3000 \text{ megacycles}$$

Surface waves

Interference effects can occur on two-dimensional media as well as on a one-dimensional string. A common example of two-dimensional wave motion is that on the surface of water. In Fig. 10-12, water waves generated by two sources whose vibrations are synchronized cross over each other forming a typical interference pattern. Note that the wave amplitude is zero along fixed curves called the lines of nodes. As with waves on a string, nodes will occur at positions where the positive crest of one wave meets the negative crest (or trough) of the other wave. Clearly, this condition is fulfilled if the path difference from the two sources to a given point is a half-wavelength. If two sine waves are displaced by a half-wavelength, their sum will be zero as seen illustrated in

Fig. 10-11. Microwave oscillator emits electromagnetic waves to the right which are reflected back on themselves. Nodes are observed at P_1 and P_2.

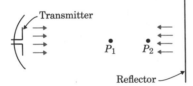

Fig. 10-12. Interference pattern of water waves generated by two synchronized vibrators tapping the water surface. (Courtesy Physical Science Study Committee.) ▶

Line of nodes

Line of nodes

Line of nodes

Line of nodes

Line of nodes

Line of nodes

Line of nodes

Line of nodes

Fig. 10-13. Two sine waves displaced by a half wavelength. At any value of x in the region of overlap, the sum of the two amplitudes will be zero.

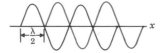

Fig. 10-14. Drawing of lines of nodes appearing in Fig. 10-12. Point P in this figure has a path difference $(D_1 - D_2) = \frac{3}{2}\lambda$.

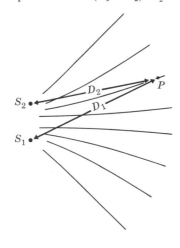

Fig. 10-15. What is the condition for a node at point P?

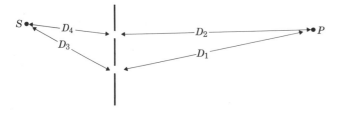

Fig. 10-13. If the path difference is $N\lambda$ where N is any integer, the two waves will reinforce each other and there will be an intensity maximum, or what is called a constructive interference. The lines of nodes of the water waves of Fig. 10-12 are drawn in Fig. 10-14. The general condition for an intensity maximum is

$$D_1 - D_2 = N\lambda \quad \text{(condition for maximum)} \quad (10\text{-}7)$$

and the general condition for a minimum or line of nodes is

$$D_1 - D_2 = (N + \tfrac{1}{2})\lambda \quad \text{(condition for minimum)} \quad (10\text{-}8)$$

Example 6

A person is at equal distances from two speakers of a stero hi-fi system and is listening to a pure musical note. He moves sideways until he hears the note fade out to a minimum. At this position he is 10 ft from the left speaker and 8 ft from the right speaker. What is the frequency of the note? The speed of sound is 1100 ft/sec.

According to Eq. 10-8 we have for the first minimum

$$D_1 - D_2 = \tfrac{1}{2}\lambda$$

where $(D_1 - D_2)$ is 2 ft. Thus $\lambda = 4$ ft. The frequency is obtained using Eq. 10-1:

$$f = \frac{v}{\lambda} = \frac{1100}{4} \text{ cps}$$

$$f = 275 \text{ cps}$$

Example 7

In Fig. 10-15 what is the condition for an intensity minimum at point P?

Now the total path difference from the source is not $(D_1 - D_2)$, but the entire path $(D_3 + D_1)$ minus the other path $(D_4 + D_2)$. This entire path difference must be $(N + \tfrac{1}{2})\lambda$ or

$$D_3 + D_1 - D_4 - D_2 = (N + \tfrac{1}{2})\lambda$$

10-4   Double-Slit Interference

Proof of the wave nature of light

In the year 1800, the prevalent theory of light was a particle theory of light that had been developed by Isaac Newton. Thomas Young was one of the few scientists of the time bold enough to challenge the teachings of Newton. Young felt that light, like sound, must consist of waves. He reasoned

Fig. 10-16. Greatly enlarged view of Young's double slit experiment. The slit separation d is usually less than a millimeter. The distance to the screen would be many times the width of this page.

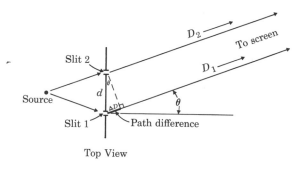

Top View

that if he were correct, he should be able to produce an interference pattern similar to the water wave pattern of Fig. 10-12. For his double source, Young used two fine slits illuminated by a single source. In 1803 he first obtained a series of interference fringes on a screen. However, according to the particle theory of light, only images of the slits on the screen should have been seen. Young's famous experiment is represented schematically in Fig. 10-16. The interference fringes one observes are shown in Fig. 10-17.

The centers of the fringes are intensity maxima corresponding to path differences $(D_1 - D_2) = N\lambda$. This path difference is the side labeled ΔD of the small right triangle in Fig. 10-16. The condition for an intensity maximum is

$$\frac{\Delta D}{d} = \frac{N\lambda}{d}$$

According to the definition of the sine of an angle, the left-hand side of the above equation is the sine of θ in the small right triangle. Thus the condition on θ for an interference maximum is

Condition for interference maximum

$$\sin \theta = \frac{N\lambda}{d} \tag{10-9}$$

Example 8

Sodium (or even table salt) when heated emits yellow light of wavelength $\lambda = 5.89 \times 10^{-5}$ cm. Suppose this sodium light is made to pass through two narrow slits that are 0.1 mm apart. If a screen is 1 m behind the slits, as shown in Fig. 10-18, what will be the spacing between fringes in the screen?

The distance a between fringes will be the distance from the central maximum to the first or $N = 1$ maximum. According to Eq. 10-9 the $N = 1$ maximum is at an angle given by

$$\sin \theta = \frac{\lambda}{d} = \frac{5.89 \times 10^{-5}}{0.1 \times 10^{-1}} = 5.89 \times 10^{-3}$$

According to Fig. 10-18,

$$\sin \theta = \frac{a}{D}$$

Thus

$$\frac{a}{D} = 5.89 \times 10^{-3}$$

and

$$a = 0.589 \text{ cm}$$

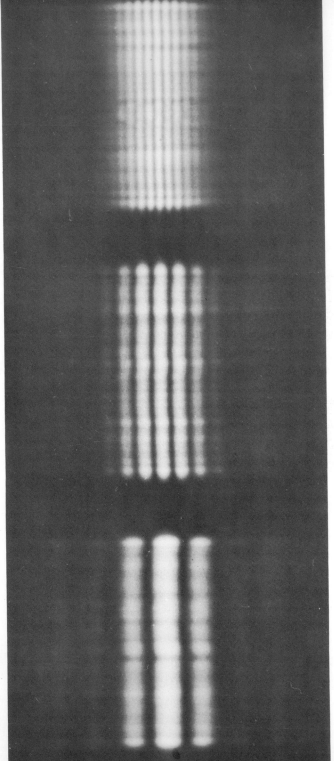

◀ **Fig. 10-17.** Young's double-slit interference pattern. This is the intensity pattern that would be obtained by placing a photographic film at the screen position of Fig. 10-16. Successive views correspond to decrease in slit separation d. (Courtesy Physical Science Study Committee.)

Fig. 10-18. Double-slit intensity pattern on screen 1 m from double slit. Intensity is plotted to the right of screen.

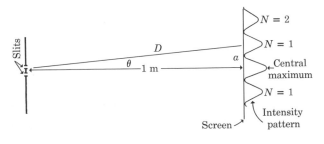

The intensity of a wave is defined to be the energy per unit volume of the wave. As we shall see in the next paragraph, this is proportional to the square of the wave amplitude. The double slit pattern of light *intensity* at a given instant of time is shown in Fig. 10-19. The waves of light are flowing to the right. Consequently, the *average* light intensity as a function of position appears as in Fig. 10-20.

We shall now demonstrate that the intensity or energy of a wave is proportional to the square of the wave amplitude. First let us consider a wave of maximum amplitude y_0 traveling along a string. If the period of the wave is T, any element of mass Δm of the string will be oscillating transverse to the string in simple harmonic motion with maximum amplitude y_0. According to Eq. 3-10 the acceleration of the mass Δm is $a = (4\pi^2/T^2)y$. The force on Δm is then $F = (4\pi^2\Delta m/T^2)y$ and according to the discussion on page 106, the energy of Δm is $\frac{1}{2}F_{max} \cdot y_0$ or $(2\pi^2\Delta m/T^2)y_0^2$. We see that the wave intensity is proportional to y_0^2, the square of the amplitude.

Now we shall demonstrate that the energy of an electromagnetic wave is proportional to the square of the amplitude E, the electric field strength. The energy in the wave will be proportional to the energy dissipated in a pickup wire or antenna of length l. The energy per unit time dissipated in the antenna is the electrical power or V^2/R, where R is the resistance of the wire. The potential V across the wire is E times the distance l. Hence the wave intensity is proportional to $(El)^2/R$ or to E^2.

10-5 The Diffraction Grating

Three or more slits are better than two

A diffraction grating can be considered as a series of double slits repeated one after the other. Figure 10-21 represents a diffraction grating with a slit spacing of d cm. First let us consider monochromatic light of wavelength λ striking the grating. In Fig. 10-21b when the path difference $a_1 = N\lambda$, then $a_2 = 2N\lambda$, $a_3 = 3N\lambda$, etc. Under this condition the waves from all the slits will reinforce each other and give an intensity maximum. The condition for intensity maximum is

---

For SHM   accel. $= 4\pi^2/T^2 \times displ.$ *

or   En. $= \frac{1}{2}(2\pi y_0/T)^2 \Delta m.$

* Proof: for SHM, restoring force $\propto$ displ.

$F = -ky = ma$

Let $y = y_0 \sin 2\pi f t$

$a = d^2y/dt^2 = -y_0(2\pi f)^2 \sin 2\pi f t$

$\therefore -k y_0 \sin 2\pi f t = -m(2\pi f)^2 y_0 \sin 2\pi f t$

$\therefore k = \frac{4\pi^2}{T^2}m$

or $a = -\frac{4\pi^2}{T^2}y$

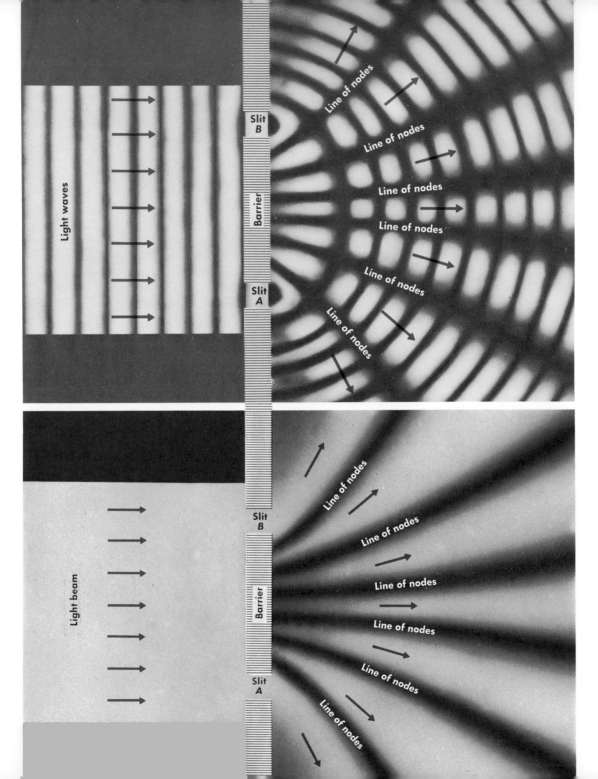

Light waves

Slit B

Barrier

Slit A

Line of nodes

Line of nodes

Line of nodes

Line of nodes

Line of nodes

Line of nodes

Line of nodes

Light beam

Slit B

Barrier

Slit A

Line of nodes

Line of nodes

Line of nodes

Line of nodes

Line of nodes

Line of nodes

Line of nodes

◀ **Fig. 10-19.** Light intensity on both sides of double slit at a given instant of time. Waves and interference pattern are propagating to the right.

then

$$\frac{a_1}{d} = \frac{N\lambda}{d}$$

or

$$\sin\theta = \frac{N\lambda}{d} \tag{10-10}$$

At any other angle each successive wave will be out of step from the preceding wave, and the resultant effect will be an almost complete cancellation.

We see from Eq. 10-10 that each value of λ has its own particular direction θ. For blue light $\sin\theta$ will be about half as large as for red light. If we shine a beam of white light on the grating, we will obtain a continuous spectrum on the screen. As we shall see in Chapter 13, when excited (by an electric arc or heat), atoms will radiate light in only certain particular wavelengths. Each kind of atom or element has its own characteristic list of wavelengths or spectrum.

Example 9

By far the most intense line in the sodium spectrum is the sodium D line. The corresponding wavelength is $\lambda = 5890$ Å (angstroms). One angstrom is defined to be 10^{-8} cm. In the first order ($N = 1$) spectrum of sodium, at what angle will the sodium D line appear using a diffraction grating of 10^4 lines/cm?

If there are 10^4 lines/cm, the spacing between lines is $d = 10^{-4}$ cm. Inserting this value into Eq. 10-10 gives

$$\sin\theta = \frac{5890 \times 10^{-8}}{10^{-4}} = 0.589$$

Using a trig table or a slide rule, we see that an angle of 36° has a sine value of 0.589. Hence the sodium D line will appear at an angle of 36° from the normal to the grating.

The grating spectrograph is of course a powerful tool in the study of structure of materials and chemical analysis. Diffraction gratings can be constructed by ruling fine lines onto a glass surface using a special precision machine called a ruling engine.

◀ **Fig. 10-20.** The double slit intensity pattern averaged over time. This is exactly the same as Fig. 10-19 except it is viewed over a time interval that is long compared to the period of oscillation.

Q.3: Does Eq. 10-10 apply to double-slit interference as well?

Fig. 10-21. (*a*) Beam of white light separated into a continuous spectrum by a diffraction grating. (*b*) Greatly enlarged view showing part of the grating.

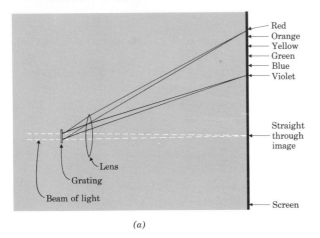

Red
Orange
Yellow
Green
Blue
Violet

Straight
through
image

Lens

Grating

Beam of light

Screen

(*a*)

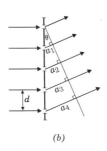

(*b*)

10-6 Geometrical Optics*

The mathematics of rays

The wavelength of light is so small compared to most optical instruments that interference effects usually do not show up. A wave train or sequence of light waves progresses forward in a straight line. Any such straight line indicating the direction of motion of light waves is called a light ray. As we shall see, light rays obey the law of reflection (from mirrors) and the law of refraction (for transparent media such as lenses). An entire mathematics or geometry can be developed using these two laws along with the usual rules of Euclidian geometry. This mathematics of rays, a subject in itself, is called geometrical optics. Since the only new physics principles involved here are the laws of reflection and refraction, we shall concentrate on these two laws and treat the rest of the subject rather lightly.

Law of reflection

The law of reflection states that when a light ray strikes a reflecting surface, the angle of incidence equals the angle of reflection. This law is illustrated by the use of water waves in Fig. 10-22. Note that the reflected wave must leave the reflecting surface at the same angle as the incident wave. As an application of the law of reflection we shall show how a concave mirror behaves as a focusing lens. It is well known that a simple lens or magnifying glass will focus parallel rays to a common point called the focus. This is also true of a concave mirror. For example, a concave shaving mirror can be used to burn a hole in a piece of paper by pointing it toward the sun and holding the paper at the focus. As shown in Fig. 10-23, the focal distance will be one-half the radius of curvature of the mirror. In this figure an arbitrary ray *AP* is selected out of a bundle of parallel rays. Let θ be the angle between this ray and the normal (*CP*) to the mirror. Note that *CP* is the radius of curvature of the mirror. According to the law of reflection, angle *APC* must equal angle *FPC* and the triangle *FPC* must then be isosceles. Hence the two

*This section may be omitted without loss of continuity.

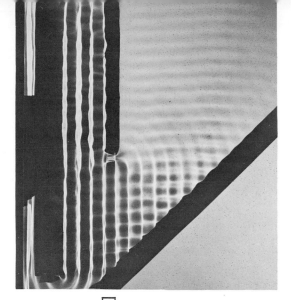

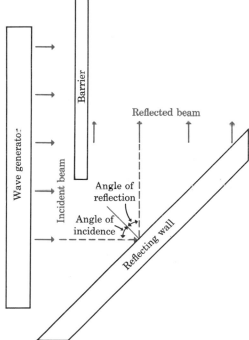

Fig. 10-22. Illustration of the law of reflection using water waves. (Photograph courtesy Physical Science Study Committee.)

sides *CF* and *FP* are equal and each is very close to half the distance from *C* to *P*, or radius of curvature.

Figure 10-24 shows how to obtain the image of an object (an arrow) graphically if the focal position *F* is known. Draw ray (1) from the arrowhead parallel to the axis of the mirror. Draw ray (2) from the arrowhead to the center of the mirror. Where the two rays intersect is the image point of the arrowhead. All other rays from the arrowhead will also pass (or almost pass) through this same image point. A concave mirror can be used to form the image of a distance object. The image can be further magnified with a magnifying glass or eyepiece. In astronomical telescopes, photographic plates are placed directly at the focus of a large concave mirror. This common type of astronomical telescope, which was developed by Isaac Newton, is called a reflecting telescope.

Refraction

The ray diagram for a converging lens or magnifying glass is almost the same. A converging lens has the property of bending parallel rays so that they are focused at a focal point *F*. In Fig. 10-25 the image position can be located graphically.* First draw ray (1) parallel to the horizontal axis. It is then bent by the lens and passes through the focus *F*. Then draw ray (2) passing directly through the center of the lens. Where the two rays cross is the image point. Again the image could be examined with an eyepiece and we would have what is called a refracting telescope. Conversely, a small object could be placed at the image position shown in Fig. 10-25 and its image would fall on top of the original arrow. This enlarged image could be examined with an eyepiece and we would have what is called a microscope.

Example 10

A projection microscope consists of a frosted glass screen 1 m behind a microscope objective of 1 mm focal length. What is the magnification factor?

The magnification factor is the ratio *AB/DE* in Fig. 10-25. Because triangles *ABC* and *EDC* are similar this ratio is the same as the image distance *BC* divided by the object distance *DC*. The ob-

*See Appendix 10-1 for an algebraic method on how to locate image positions.

Fig. 10-23. Parallel rays of light striking a concave mirror of radius CP.

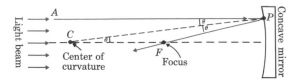

Fig. 10-24. Image formation by a concave mirror. The rays (1) and (2) show how to obtain the image position graphically.

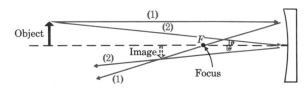

Fig. 10-25. Image formation by a converging lens.

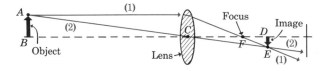

Index of refraction

ject distance DC will be very close to the 1 mm focal length. Thus

$$\frac{AB}{DE} = \frac{100 \text{ cm}}{0.1 \text{ cm}} = 1000$$

is the magnification factor.

Snell's law

The last point to be discussed here is why light rays are bent or refracted when they cross from air to glass and vice versa. We shall see that this is a direct consequence of the wave nature of light, provided that the speed of light in glass is less than that in air. At first thought this seems to violate Maxwell's proof that electromagnetic waves must travel with velocity $v = c$. However, if the velocity of a light wave in glass is measured, we obtain the experimental result $v = 0.66c$. This paradox is resolved by noting that the measured wave is actually the sum of an almost infinite number of separate waves, each propagating with velocity $v = c$ as required by the Maxwell theory. Each atomic electron in the glass is a separate wave source just as each line in a diffraction grating is a separate source. There is an oscillating force on each atomic electron due to the electric field of the incoming wave. Because of its inertia, the oscillation of each electron will lag behind the oscillation of the incoming electromagnetic wave. Using integral calculus, one can add the electric fields radiated by the lagging electrons to the electric field of the incoming wave. As might be expected, the answer is a resultant oscillating electric field propagating with a velocity which lags behind the speed of light. For glass, the resultant electromagnetic field propagates with velocity $v = 0.66c$. The ratio of c divided by the speed of light in glass is defined as n, the index of refraction of glass:

$$n = \frac{c}{v}$$

Now we can use Fig. 10-26 to see why the direction of wave propagation is altered at a glass-air interface. The figure shows a portion of two successive waves AB and $A'B'$.

Let λ' be the reduced wavelength in the glass.

$$\lambda' = \frac{v}{f} \quad \text{and} \quad \lambda = \frac{c}{f} \tag{10-11}$$

Fig. 10-26. Two successive wavefronts crossing a glass-air interface.

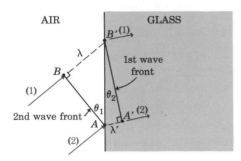

In the right triangle ABB',

$$\sin \theta_1 = \frac{\lambda}{AB'}$$

In the right triangle $A'AB'$,

$$\sin \theta_2 = \frac{\lambda'}{AB'}$$

Dividing the two above equations by each other gives

$$\frac{\sin \theta_1}{\sin \theta_2} = \frac{\lambda}{\lambda'}$$

Now substitute the values for λ and λ' given by Eq. 10-11:

$$\frac{\sin \theta_1}{\sin \theta_2} = \frac{c}{v}$$

or

Snell's law

$$\frac{\sin \theta_1}{\sin \theta_2} = n$$

From this basic equation, which is also called Snell's Law or the law of refraction, the optical properties of lenses can be calculated.

Appendix 10-1

The Thin Lens Equation

We would like to derive an algebraic relation between object distance s and image distance s' as shown in Fig. 10-27. Triangle ABO is similar to triangle $A'B'O$. From these two similar triangles, the ratio

$$\frac{A'B'}{AB} = \frac{s'}{s} \tag{10-12}$$

Triangle POF is similar to triangle $A'B'F$ and the ratio

$$\frac{A'B'}{PO} = \frac{s' - f}{f} \tag{10-13}$$

Since $PO = AB$, the right-hand sides of Eqs. 10-12 and 10-13 are equal. Equating these two right-hand sides gives

$$\frac{s'}{s} = \frac{s' - f}{f}$$

or

$$\frac{1}{f} = \frac{1}{s} + \frac{1}{s'}$$

This relationship between image and object distance is referred to as the thin lens formula. There is a sign convention to tell whether

Fig. 10-27. Object AB is distance s from lens of focal length f. The image $A'B'$ is at distance s'.

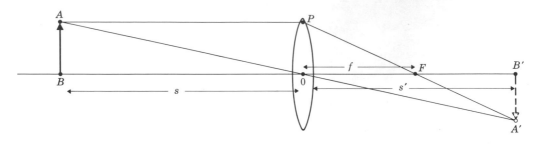

s, s', or f is positive or negative. According to this convention, the problem under study must be oriented so that the light passes through the lens from left to right. Then s' will be positive if the image is to the right of the lens, and negative if to the left of the lens. The quantity f is negative if the lens is a diverging lens. If the rays passing through the lens are converging on a virtual object (it could be the image produced by a preceding lens to the left), then s will be negative.

Problems

1. In any periodic wave motion what is the distance in wavelengths between two successive positions of zero amplitude? $\lambda/2$

⟶▷ **2.** A 150 cm string contains a standing wave as shown.
   (a) What is the total number of nodes? 6
   (b) How many wavelengths long is the string? $2\frac{1}{2}$
   (c) If waves travel with $v = 20$ m/sec on this string, what is the frequency of oscillation?

$\lambda = \dfrac{150}{2\cdot5} = 60\,cm. \quad v = \lambda f$

$\therefore f = \dfrac{20\ m/sec}{.6\ m} = 33.3\ H_3$

Prob. 2

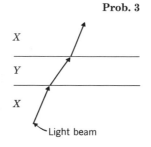

3. A light beam passes from a liquid X into glass Y and back into liquid X again. Which medium, X or Y, has the greater wave velocity for light?

4. In an electromagnetic wave what is the distance between two successive intensity maxima? $\lambda/2$

Prob. 3

X

Y

X

Light beam

5. The unit 1 A (one angstrom) is defined as 10^{-8} cm. What are the wavelength limits of visible light in terms of angstroms?

6. Sketch a standing wave made up of four nodes and three anti-nodes at a time when the antinodes are at maximum displacement. Sketch the picture both $\frac{1}{2}$ and $\frac{1}{4}$ a period later. Under each of the three sketches plot the intensity of the wave. On one of the six diagrams indicate the wavelength of the wave.

7. Does a moving charge always radiate electromagnetic waves?

8. Two musical notes are an octave apart if their frequencies are in the ratio 2:1. If middle C is 262 cps and high C is two octaves higher, what is the frequency of high C? $4 \times 262 = 1048$

9. The lowest and highest frequencies detectable by the human ear are about 20 cps and 15,000 cps, respectively. What are the corresponding wavelengths in air?

10. Consider a single sinewave of wavelength λ traveling along a string with velocity v. (There is no reflected wave.) How many times per second will the *entire* string have zero displacement?

11. If the two pulses of Fig. 10-7 had been exactly the same size, they would exactly cancel each other out in the third view (the string would be a straight horizontal line). Would the pulses stay canceled out? Explain. Draw diagrams where necessary.

Prob. 12

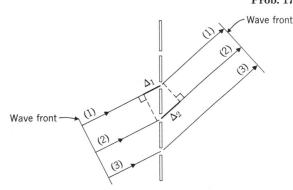

12. S_1 and S_2 are two ripple tank probes having the same frequency of vibration; however, S_1 always hits the water $\frac{1}{4}$ of the period earlier than S_2.
  (a) What is the condition for an interference maximum at P?
  (b) What is the condition for an interference minimum at P?

13. The wave velocity of a string fixed at both ends is 2 m/sec. The string contains standing waves with nodes 3.0 cm apart.
  (a) What is the frequency of vibration?
  (b) How many times per second is the string in a straight line containing no visible waves?

14. The second and third order visible spectra of a diffraction grating will partially overlap. What $N = 3$ wavelength will appear at the $\lambda = 7000$ A position of the $N = 2$ spectrum?

15. • ←——— 8 ft ———→ • ←———— 18 ft ————→ •
      S_1 S_2 P

S_1 and S_2 are two sinewave sources of sound. If they are in phase and 8 feet apart;
  (a) List 3 different wavelengths which will give a constructive interference at point P.
  (b) List 3 different wavelengths that will give a destructive interference at P.
  (c) What is the lowest frequency in cycles per second which will give a destructive interference at P? The velocity of sound is 1100 ft/sec.

16. If the two water wave sources of Fig. 10-12 are vibrating 180° out of phase (source 1 moves up when source 2 moves down), what are the conditions on $(D_1 - D_2)$ for the lines of nodes?

Prob. 17

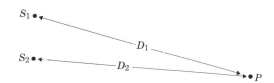

17. Parallel light strikes a diffraction grating at an angle as shown. The diffracted beam leaves as shown. Assume 10^4 rulings per cm.
  (a) What is the path difference between rays (1) and (2) in terms of Δ_1 and Δ_2?
  (b) What is the path difference between rays (1) and (3)?

(c) What is the condition on Δ_1 and Δ_2 for observing the nth maximum of wavelength λ?

18. In the accompanying figure, what are the conditions on $(D_1 - D_2)$ for an interference maximum and for an interference minimum to occur at point P?

19. In Problem 18, suppose the wave is reflected with a change in sign (180° change in phase). Now what is the condition for an intensity minimum?

20. The object is located between the focus and the center of curvature of a concave mirror.
    (a) Locate the image position by drawing at least two rays.
    (b) Will the image be inverted?
    (c) Will the image be larger than the object?

21. The above converging lens is located at the origin. Its focal length is 5 cm. In each of the following cases state whether the image is: between A and B, at B, between B and C, at C, between C and infinity, at infinity, or virtual (to the left of the lens).
    (a) The object is at $x = -5$ cm.
    (b) Object is between 0 and -5 cm.
    (c) Object is between -5 and -10 cm.
    (d) Object is at -10 cm.
    (e) Object is to the left of -10 cm.

Prob. 18

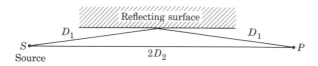

Prob. 20

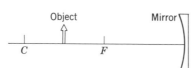

Prob. 21

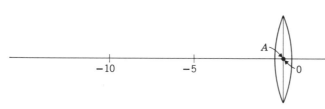

22. A point source of light is placed at the center of curvature of a mirror. Where will the image be? (The chapter opening illustrates a corresponding situation.)

23. A point source of light is placed at the focus. Where will the image be?

24. The distance of an object from a concave mirror is less than the focal length. If you looked in the mirror, would you see an image. If so, would it look larger or smaller than the original object? Would the image be inverted? (*Hint:* where is the focus of a concave shaving mirror?)

Prob. 25

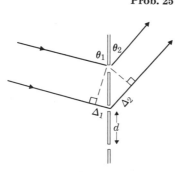

25. Consider a diffraction grating where the incident light of wavelength λ is not at right angles, but at an angle θ_1 as shown.
    (a) What is the condition for an intensity maximum in terms of the path differences Δ_1 and Δ_2?
    (b) What is the condition for a maximum in terms of θ_1, θ_2, λ, and d?

26. Light of wavelength λ strikes a single slit of width d. Let θ be the angle where the light path from the center of the slit is a half wavelength longer than the path from the edge of the slit. What is $\sin \theta$ for this condition? (At this angle an intensity minimum will be observed. This effect is called single slit diffraction.)

27. A light ray crosses an oil-glass interface. If the velocity of light is v_1 in the oil and v_2 in the glass, what is $\sin \theta_1 / \sin \theta_2$ in terms of v_1 and v_2?

28. Light of wavelength $\lambda = 5 \times 10^{-5}$ cm is diffracted by a grating of 2000 lines/cm. The screen is 3 m from the grating. What is the distance on the screen between the zeroth and the first order image?

Prob. 29

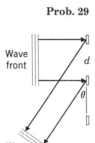

Wave front

Wave front

29. A light ray enters a flat slab of glass at an angle of 60° from the normal. If $n = 1.5$, at what angle does the ray leave the slab on the other side? The faces of the slab are parallel.

30. Consider a reflection diffraction grating (lines ruled on a mirror). The incoming beam is perpendicular to the grating. Consider the outgoing direction θ.

(a) What is the path difference in terms of θ and d?

(b) What must be the condition on θ for an intensity maximum?

The New York Times

Relativity

Chapter 11

Relativity

11-1 The Principle of Relativity

Velocity is relative

Galileo proposed as a general symmetry principle that the laws of physics should be the same to any observer moving with constant velocity no matter what the magnitude and direction of the velocity. In other words, there must be no preferred reference system or no way to determine absolute velocity. This general symmetry is called the principle of relativity. Of course, Galileo and Newton made sure that their laws of classical mechanics would obey the principle of relativity.

However, unless some changes are made in the physics we have presented thus far, the laws of electricity violate the principle of relativity. An observer at "rest" and one in motion will get two different theoretical predictions for the very same experiment. To illustrate this, we shall consider the simple example of a point charge Q at a distance R from a charged wire of ρ statcoulombs per centimeter as shown in Fig. 11-1. The force on Q is $F = QE$ where $E = 2\rho/R$ is given by Eq. 7-10. Thus

$$F = \frac{2Q\rho}{R} \tag{11-1}$$

for a stationary observer. Now consider a second observer who happens to be moving parallel to the wire with velocity v as shown in Fig. 11-2. In addition to the above electrostatic force, this observer would also measure a magnetic force. He sees an electric current $I = \rho v$ moving along the wire and the charge Q also moving parallel to the wire with velocity v. According to Eq. 8-10 the magnetic force on charge Q is $F_m = Qv/c \times B$. The magnetic field B produced by a single straight-line current is $B = 2\rho v/cR$. Thus

$$F_m = \frac{Qv}{c} \times \frac{2\rho v}{cR} = \frac{2Q\rho}{R} \frac{v^2}{c^2}$$

is the magnetic force. This attractive magnetic force must be added to the repulsive electrostatic force. Thus the moving observer calculates that the resultant force should be

$$F = \frac{2Q\rho}{R} - \frac{2Q\rho}{R} \frac{v^2}{c^2}$$

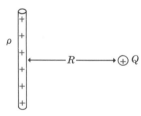

Fig. 11-1. Point charge Q at a distance R from a charged wire of ρ statcouls/cm.

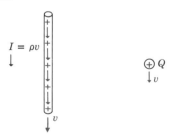

Fig. 11-2. Figure 11-1 as seen by an observer moving with velocity v in the up direction.

or

$$F = \frac{2Q\rho}{R}\left(1 - \frac{v^2}{c^2}\right) \qquad (11\text{-}2)$$

for an observer moving with velocity v. Numerically this is smaller by the factor $(1 - v^2/c^2)$ than the result (Eq. 11-1) obtained by the stationary observer. Thus it is clear that the laws of electricity give different results in different reference systems. However, according to the Galilean principle of relativity and classical mechanics, the force or acceleration of Q should be independent of the velocity of the observer.

Clearly, Galileo's principle of relativity which works for classical mechanics is inconsistent with Maxwell's equations. To resolve this contradiction, either the principle of relativity, or Maxwell's equations, or classical mechanics must be revised. We shall now consider the following three alternatives:

1. The principle of relativity exists for mechanics but not for electrodynamics. The laws of electricity hold in only one preferred reference system. Only in this system does light travel with velocity $v = c$.

2. The principle of relativity exists for both mechanics and electrodynamics, but the laws of electricity are not correct as formulated in Chapters 7 and 8. The laws of electricity can be modified in a consistent way to give the result that light always travels at velocity c with respect to its source. Such a modification of the Maxwell equations is called the emission theory.

3. The principle of relativity is correct, but the laws and concepts of mechanics require modification. This is the alternative chosen by Einstein who, by modifying the definitions of mass, energy, momentum, and the properties of space and time, was able to force the laws of mechanics along with the laws of electricity to obey the principle of relativity. The laws of electricity needed no modification. As pointed out in Chapter 8, magnetic force can be thought of as a relativistic correction to Coulomb's law. For about 50 years the theory of relativity was ready to be discovered in the equations of electricity and nobody fully realized this until Einstein in 1905.

Q.1: What is the direction of the force F in Eq. 11-2?

11-2 The Problem of the Ether

"Common sense is that layer of prejudices laid down in the mind prior to the age of eighteen."
A. Einstein

We shall first discuss possibility (1) and show how it was contradicted by experiment. In this preferred reference system light would travel with velocity c. To an observer having velocity v with respect to this preferred reference system, the velocity of light would be $(c + v)$ when moving toward the light source. This is just what one would expect if light transmission required a "physical," but massless, medium. This medium was called the ether. The preferred reference system where Maxwell's equations were assumed to be valid was interpreted as the rest system of the ether. Thus all velocities were considered absolute—they could be measured with respect to the ether.

We shall now see how this assumption was contradicted by the famous experiments of Michelson and Morley conducted in the 1880's. Since the earth is moving at a velocity $v = 18$ mi/sec around the sun, supporters of the ether theory reasoned that there must be times of the year when the earth has a velocity of at least 18 mi/sec with respect to the ether (or the ether has a velocity of 18 mi/sec with respect to the earth). Then to an observer on earth light traveling in the same direction as the ether should have a measured velocity $(c + v)$ with respect to the earth, and light traveling in the opposite direction would have velocity $(c - v)$ where v is at least 18 mi/sec. Thus the length of time it would take light to travel a distance D to a mirror and back would be

$$t = \frac{D}{c + v} + \frac{D}{c - v} = \frac{2Dc}{c^2 - v^2} = \frac{2D}{c} \times \frac{1}{1 - \frac{v^2}{c^2}} \qquad (11\text{-}3)$$

Suppose a rigid arm of length D holds a light source and a mirror as shown in Fig. 11-3. Then the time for the light to travel to the mirror and back is given by Eq. 11-3 when the arm is lined up with v, the ether velocity. If the arm is rotated $90°$ so that it is perpendicular to v, the light must travel a distance $2D'$ as viewed by an observer at rest with respect to the ether (see Fig. 11-4). Then the time to travel to the mirror

Fig. 11-3. Light path from source to mirror and back.

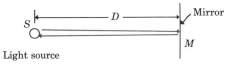

Ans. 1: It is a net repulsive force.

Fig. 11-4. Light path when source and mirror are moving with velocity v to the right as shown.

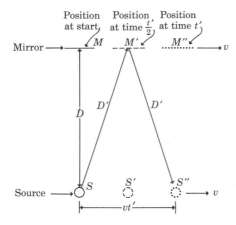

Position at start, Position at time $\frac{t'}{2}$, Position at time t'

Mirror →

Source →

and back would be

$$t' = \frac{(2D')}{c} \quad \text{or} \quad D' = \frac{ct'}{2} \tag{11-4}$$

From the right triangle in Fig. 11-4 we have

$$D'^2 = D^2 + \left(\frac{vt'}{2}\right)^2$$

Now by substituting the right-hand side of Eq. 11-4 for D', we have

$$\frac{c^2 t'^2}{4} = D^2 + \frac{v^2 t'^2}{4},$$

$$\frac{c^2}{4}\left(1 - \frac{v^2}{c^2}\right)t'^2 = D^2,$$

and

$$t' = \frac{2D}{c\sqrt{1 - \frac{v^2}{c^2}}} \tag{11-5}$$

Hence the ratio $t'/t = \sqrt{1 - v^2/c^2}$, which means that the time for the light to make the round trip is shorter when the arm is perpendicular to the ether velocity. The binomial theorem may be used to obtain the following approximation for the time difference:

$$t - t' = \frac{2D}{c} \times \frac{v^2}{2c^2}$$

Example 1

Consider two small islands 60 mi apart and a motorboat which cruises at 10 mph with respect to the water as shown in Fig. 11-5a. One island is due west of the other.

(a) How long does a round trip take when the water is still?

(b) How long does a round trip take when the current is 5 mph flowing east?

(c) How long does a round trip take when the current is 5 mph flowing south?

For part (a) the time equals the round-trip distance ($2D = 120$ mi) divided by the speed ($c = 10$ mph):

$$t = \frac{2D}{c} = \frac{120 \text{ mi}}{10 \text{ mi/hr}} = 12 \text{ hr}$$

For part (b) the net velocity heading east is 15 mph and returning is 5 mph. Then

Fig. 11-5. Boat trips between two islands.

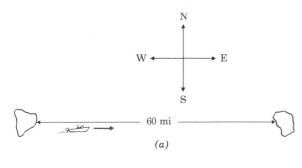

N

W ← → E

S

60 mi

(a)

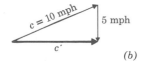

$c = 10$ mph

5 mph

c'

(b)

$$t = \frac{D}{15 \text{ mi/hr}} + \frac{D}{5 \text{ mi/hr}} = 4 \text{ hr} + 12 \text{ hr} = 16 \text{ hr}$$

This same result can be obtained by substituting $c = 10$ mph and $v = 5$ mph into Eq. 11-3. We see that the effect of the current is to slow down the trip by 4 hr.

For part (c) in order for the boat to have a resultant velocity c' pointed due east, the captain must steer a course somewhat north of east as shown in Fig. 11-5b. Then

$$c' = \sqrt{10^2 - 5^2} = \sqrt{75} = 5\sqrt{3} = 8.66 \text{ mph}$$

and

$$t = \frac{2D}{c'} = \frac{120}{8.66} = 13.86 \text{ hr}$$

Again the result can also be obtained by substituting $c = 10$ mph and $v = 5$ mph into Eq. 11-5. We see that as with the previous discussion about light paths in an ether current, the trip when the current is in the direction of travel takes longer than when the same current is perpendicular to the direction of travel.

Michelson and Morley realized that they could observe such a small time difference (less than 10^{-16} sec) by using an interferometer which already has two arms at $90°$ to each other. An interferometer is shown in Fig. 11-6. In the interferometer, light from source S is split by the half-silvered mirror M_1. The two light rays are reunited at the screen. If the two light paths take the same amount of time, there will be a constructive interference at the screen. The experiment consists of adjusting the mirror positions to give a constructive interference. Then the apparatus is rotated $90°$ by the earth's rotation and the new pattern observed. A change in the times of the light paths due to the velocity of the ether should show up as a change in the interference pattern. Even a value for v as small as 18 mi/sec should give a very noticeable effect.

But try as they might, Michelson and Morley never obtained any effect at all. One explanation given was that the ether happens by accident to have a velocity of 18 mi/sec with respect to the solar system. When the earth is at position A in Fig. 11-7, its velocity relative to the ether would be zero. However, Michelson and Morley repeated their experiment six months later when the earth was in position B. Then they should have obtained twice the expected effect, but again nothing happened.

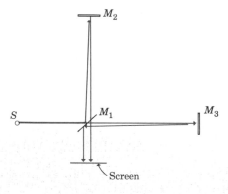

Fig. 11-6. The Michelson interferometer. Light from source S is split by half-silvered mirror M_1 and reunited at the screen.

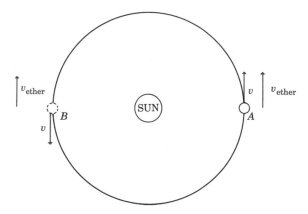

Fig. 11-7. Earth's orbit about the sun. At position A the ether drift would be zero. But at position B the effect would be doubled.

Another possible explanation was that the earth drags a local region of the ether along with it. However, this would cause the stars to appear to shift in position back and forth each year in a manner different from that which is observed. This explanation was therefore ruled out by astronomical observation.

Another attempt to explain the null result of Michelson and Morley was made by Fitzgerald and independently by Lorentz who proposed that all physical lengths are contracted in the direction of motion by the factor $\sqrt{1 - v^2/c^2}$ when moving at a velocity v with respect to the ether. Then when the interferometer is rotated 90°, its arms would change in length just by the amount necessary to cancel out the effect. To answer this objection a new interferometer with arms of unequal length was employed. Now the Fitzgerald-Lorentz contraction would lead to different results for different velocities of the interferometer relative to the ether. However, the null result was still observed for all positions of the earth in its orbit.

As a result of these extensive experiments, we conclude that the light emitted by the interferometer source always travels with velocity c with respect to the source and mirrors of the interferometer.

Another possible way to explain the Michelson and Morley result would be that of alternative (2); namely, to revise the laws of electricity so that light would always be emitted with velocity c with respect to the emitter or source of the electromagnetic waves. However, this explanation was also ruled out on the basis of astronomical observation. If this emission theory were true, the motion of a double star should appear distorted and appear to violate Kepler's laws. When one member of the binary star is moving toward the earth with velocity v, its light would travel all the way with velocity $c + v$ and arrive early, whereas the light emitted while the star was moving away would travel with velocity $c - v$ and would arrive late.

We see that every attempt to explain the null result of Michelson and Morley was defeated by additional experiments and observation.

The remaining alternative (3) was seriously pursued by

Q.2: Which time interval is larger, that given by Eq. 11-3 or by Eq. 11-5?

Albert Einstein in 1905. Actually the Michelson-Morley experiments were of little concern to Einstein. He was more bothered about the inconsistencies between Maxwell's equations and classical physics. A favorite thought problem of his was what would happen if he should chase after a light wave and catch up with it by traveling at $v = c$. So Einstein tackled the problem of what changes in the classical ideas of space and time would be necessary to make the equations of electricity consistent with the principle of relativity. Fortunately, this is a well-defined mathematical problem and has a unique solution. In fact, the purely mathematical part of this problem had essentially been solved by H. A. Lorentz a few years earlier. However, many people (including Lorentz at the time) have difficulty with the physical interpretation of the mathematical results because they seem to violate common sense. It was this physical interpretation and its extension to all of physics which was Einstein's main contribution—a contribution which greatly overshadows the mathematical curiosity obtained by Lorentz.

One of the main points in Einstein's theory is that the speed of light is always $c = 3 \times 10^{10}$ cm/sec, independent of the velocity of the observer or of the source. Thus two observers, one at rest with respect to a distant star and one traveling very fast toward the star, will measure the same value for the velocity of the light coming from the star. This result would of course be consistent with the observations of Michelson and Morley, but it seems contrary to common sense.

Einstein explained this "strange" result by attributing "strange" properties to space and time. He proposed that space as viewed by a moving observer contracts in the direction of motion, and that time as viewed by different observers will not be the same for them (events that are simultaneous for one will not necessarily be simultaneous for the other). These effects are described mathematically by the equations obtained by Lorentz and are called the Lorentz transformation.

The Lorentz transformation relates the time and distance coordinates measured by two observers moving with velocity v with respect to each other as follows:

Ans. 2: Eq. 11-3.

$$\begin{cases} x' = \dfrac{x + vt}{\sqrt{1 - \dfrac{v^2}{c^2}}} & \text{(11-6)} \\[2em] y' = y \\[0.5em] z' = z \\[1em] t' = \dfrac{t + \dfrac{v}{c^2}x}{\sqrt{1 - \dfrac{v^2}{c^2}}} & \text{(11-7)} \end{cases}$$

The Lorentz transformation

The primed coordinates refer to one observer and the unprimed to the other observer. The corresponding classical equations are

$$x' = x + vt$$

$$t' = t$$

which says that if an object is at rest in the unprimed system at $x = x_0$, in the primed system its position will be $x' = x_0 + vt$, which means it will appear to be moving with velocity v to the right. Equations 11-6 and 11-7 can be derived by assuming all observers must obtain the same value for the speed of light. Such a derivation is given in Appendix 11-1.

11-3   The Lorentz Contraction

"There was a young fellow named Fisk
Whose fencing was exceedingly brisk.
So fast was his action,
The Lorentz contraction
Reduced his rapier to a disk."

One simple consequence of the Lorentz transformation is that all moving objects will appear to be contracted by the factor $\sqrt{1 - v^2/c^2}$ in the direction of motion. This almost can be seen directly by looking at Eq. 11-6. Specifically, if two observers start out with identical meter sticks and observer B travels with velocity v with respect to observer A, then observer A will measure the meter stick of B as being $\sqrt{1 - v^2/c^2}$ meters long (see Fig. 11-8). Because of the principle of relativity the reverse must also be true: observer B

Fig. 11-8. Effect of Lorentz contraction on two identical sticks moving at a velocity $v = 0.6c$ with respect to each other.

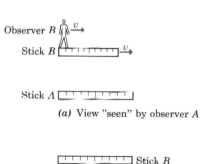

(a) View "seen" by observer A

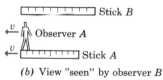

(b) View "seen" by observer B

Q.3: If an object is fixed at $x' = 0$, will its x-coordinate given by Eq. 11-6 increase or decrease with time?

Fig. 11-9. (*a*) Two identical light clocks at $t = 0$. Clock *B* is moving to the right with velocity *v*. (*b*) The light clocks τ seconds later as seen by *A*. Both light pulses have traveled a distance $c\tau$. The pulse in *A* has reached the end, but the pulse in *B* still has farther to go.

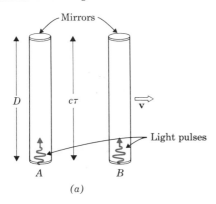

(a)

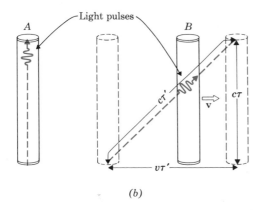

(b)

Ans. 3: It will decrease. The equation for x in terms of x' and t' is

$$x = \frac{x' - vt'}{\sqrt{1 - (v^2/c^2)}}$$

would measure *A*'s meter stick as being contracted by the same amount.

Example 2

Suppose we see a meter stick passing by with a velocity 60% of the speed of light. How long would we measure it to be?

$$L' = \sqrt{1 - 0.6^2}\, \text{m}$$
$$= \sqrt{0.64}\, \text{m}$$
$$= 80\, \text{cm}$$

In Section 11-2 we saw that the Lorentz contraction was not sufficient to explain the experimental results when an interferometer with arms of unequal length was used. This difficulty is resolved by revising our concept of time in addition to our concept of length.

11-4 Time Dilation

> *"There was a young girl named Miss Bright,*
> *Who could travel much faster than light.*
> *She departed one day,*
> *In an Einsteinian way,*
> *And came back on the previous night."*

To illustrate why Einstein found it necessary to change our concept of time, let us consider a "light clock." The construction is very simple: just two parallel mirrors a distance *D* apart.

Let τ be the time required for a pulse of light to strike the top mirror starting from the bottom. Each time the light strikes a mirror we have a "tick" of the clock. Consider two such identical light clocks ticking away together. The time between ticks is $\tau = D/c$. Now let clock *B* move to the right with velocity *v* as in Fig. 11-9*b*. Its length must still appear the same.* We as observers are "attached" to light clock *A* and we observe that there is now a longer path for the light pulse to travel from one end to the other in light clock *B*.

---

*If it were shorter than *A* as the two objects pass each other, two observers (one attached to *A* and the other attached to *B*) would both see that light clock *B* is shorter than *A*. This would give a means of detecting absolute motion which violates the principle of relativity.

As we observe light pulse B taking the diagonal path in Fig. 11-9b, we see that it must travel with the same speed $v = c$ as our light pulse (this is where the theory of relativity comes in). Hence light pulse B must take a longer time than our light pulse A. If we call this longer time τ', by applying the Pythagorean theorem to Fig. 11-9b we obtain

$$(c\tau')^2 = (v\tau')^2 + (c\tau)^2$$

$$(c^2 - v^2)\tau'^2 = c^2\tau^2$$

$$\tau'^2 = \frac{1}{1 - \dfrac{v^2}{c^2}}\tau^2$$

$$\tau' = \frac{1}{\sqrt{1 - \dfrac{v^2}{c^2}}}\tau \qquad\qquad (11\text{-}8)$$

The time interval τ' is the time we observe between ticks for the moving clock and is a longer time than τ. Any observer must find that moving light clocks tick more slowly than an identical stationary light clock.

But do light clocks behave this way because of the special nature of light? Should ordinary mechanical clocks whose parts move much slower than the speed of light also slow down by the same factor $\dfrac{1}{\sqrt{1 - v^2/c^2}}$? Einstein said yes, because it has nothing to do with the nature of the particular clock—it is due to an intrinsic property of time itself. To see this, suppose a light clock and a wristwatch are fastened together, both keeping identical time. Then they are pushed sideways with velocity v and the light clock slows down as it should, but the wristwatch does not. Now we would have a simple detector of absolute motion—whenever the two clocks agree, they are at rest; whenever the light clock runs slower, everybody knows they are in motion. This, of course, violates the principle of relativity upon which all of our discussion is based.

Since time dilation is a property of time itself, not only are all moving clocks slowed down, but so are all physical processes such as chemical reaction rates slowed down when in motion. Since life consists of a complex of chemical reactions, life also would be slowed down by the same factor.

Q.4: In Fig. 11-9(b) does the fixed observer see both light pulses moving with the same speed?

Fig. 11-10. Pion beam produced by proton beam of a synchrocyclotron striking an internal target. The pions produced by proton collisions in the target are deflected by the magnetic field and travel out into the experimental area.

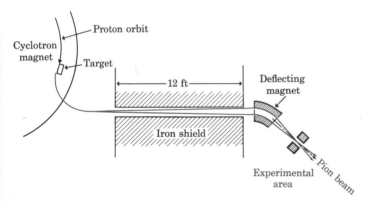

Even physical processes, such as the halflife of a radioactive sample, must be slowed down by the factor $\sqrt{1 - v^2/c^2}$. In fact, this effect on a halflife has been observed directly by using a beam of unstable particles. The pion is an unstable particle with a halflife of 1.8×10^{-8} sec. A beam of pions can be produced by using a synchrocyclotron. Figure 11-10 shows a typical synchrocyclotron-produced pion beam.

Example 3

A beam of pions have velocity $v = 0.6c$. How long will it take for half the pions to decay? How far will they travel in this time?

The observed halflife T' will be 1.8×10^{-8} sec increased by the factor

$$\frac{1}{\sqrt{1 - 0.6^2}} = \frac{1}{\sqrt{0.64}} = 1.25$$

Thus the halflife is increased by 25%, or $T' = 2.25 \times 10^{-8}$ sec. The distance traveled in this time is

$$D = vT' = 0.6 \times 3 \times 10^{10} \times 2.25 \times 10^{-8} \text{ cm} = 4.05 \text{ m}$$

Not only has time dilation been observed using microscopic "clocks" in the form of unstable particles, but in 1960 the effect was first observed using macroscopic clocks. The most stable timekeeping device that can be built is a clock utilizing what is called the Mossbauer effect. Mossbauer "clocks" utilize photons from a radioactive iron isotope imbedded in an iron crystal. Two identical Mossbauer clocks will keep the same time within one part in 10^{16}. A time shift of that amount shows up as a sudden increase in a photon counting rate. In this time-dilation experiment, an entire Mossbauer clock was rotated rapidly and found to slow down at the rate $\dfrac{1}{\sqrt{1 - v^2/c^2}}$ compared to an identical Mossbauer clock at rest. The theory was again confirmed.

11-5 The Twin Paradox

Perpetual youth?

Since the advent of space exploration, it has become common knowledge that space travelers will not age as fast as their brothers on earth. In fact, if a space traveler could travel with the speed of light, he would not age at all.

Ans. 4: Yes, however light pulse B is moving slower with respect to its light clock which is also moving.

This slowing down of time for the space traveler is seen by applying Eq. 11-8 to him. According to an observer on earth the clocks and all physical processes including life itself on a spaceship of velocity v would be slowed down by a factor $\sqrt{1 - v^2/c^2}$.

Example 4

Consider two twins A and B, age 20. Twin B takes a round trip space voyage to the star Arcturus at a velocity $v = 0.99c$. According to those of us on earth, Arcturus is 40 light years away. What will be the ages of A and B when B finishes his trip?

According to twin A, the trip would take 1% longer than the 80 years it takes light to travel to Arcturus and back. Thus A would be 20 plus 80.8 or 100.8 years old when B returns. According to twin A the clocks on the spaceship would be running slower by the factor $\sqrt{1 - 0.99^2} = \sqrt{0.02} = 0.141$. On the spaceship the elapsed time for the trip would then be 0.141 times the 80.8 years of earth time or 11.4 years. Twin B would be 20 plus the 11.4 years or 31.4 years old at the end of the trip. He would then be 69.4 years younger than his twin brother who stayed on the earth.

The space traveler does not feel that his time is running slower. In Example 4, twin B sees the distance to Arcturus shortened by the Lorentz contraction. He measures that the distance from earth to Arcturus is $\sqrt{1 - 0.99^2}$ times 40 light years or 5.64 light years. He also sees the earth moving away at the same relative velocity of $v = 0.99c$. So the twin in the spaceship calculates that the time for him to get to Arcturus is just 1% longer than the time it takes light to travel the 5.64 light years. Thus the twin in the spaceship thinks it takes 5.7 years to get to Arcturus or 11.4 years for the round trip. This result checks with the calculations of twin A on earth.

However, we run into an apparent paradox when the space traveler looks back on the earth and sees the earth clocks slowed down with respect to his clock. It seems that B should get the result that A is younger, which contradicts our previous reasoning. In fact, if velocity is truly relative, how can we arrive at an asymmetric result at all? According to symmetry should not both twins end up with the same age? At first sight it appears that Einstein's formalism gives rise to a self-contradiction. The paradox is resolved by noting that this problem is inherently asymmetric. The man on

Q.5: Should an astronaut who merely circles the earth return younger than otherwise?

earth always remains in the same reference system, whereas the spaceman changes reference systems when he turns around. A correct application of Einstein's equations also gives the result that, according to the spaceman, the over-all effect is that the earthman gets older.

There has been a long history of controversy concerning the twin paradox (also called the clock paradox). By now nearly all physicists accept the interpretation given here; however, there are still some philosophers, and even a few mathematicians, who claim that the twins actually end up the same physical age. For further discussion of the "twin paradox" the reader is referred to Dennis Sciama, *Unity of the Universe,* Anchor Books. An article by E. McMillan in the August 30, 1957 issue of *Science* shows that it is impossible for man to make practical use of time dilation in space travel. Time dilation effects are negligible except for velocities near the velocity of light. We shall see in Section 11-8 that a kinetic energy comparable to the rest mass energy of the spaceship is needed in order to reach such high velocities. Even the energy release of nuclear fission if utilized with 100% efficiency is still about 1000 times too low for this.

11-6 Einstein Addition of Velocities

$$v + c = c$$

The Lorentz contraction and time dilation are direct consequences of the Lorentz transformation. These transformation equations also determine how to transform velocity. For example, in a certain reference system an object may be traveling with velocity u (such as the jet plane in Fig. 11-11). What velocity u' would be measured by a moving observer traveling with velocity v? According to classical physics $u' = u + v$. But Einstein's result is

Einstein addition of velocities

$$u' = \frac{u + v}{1 + \dfrac{uv}{c^2}} \tag{11-9}$$

This equation is easily obtained from the Lorentz transformation equations by dividing the equation for x' by the equation for t'. If we divide Eq. 11-6 by Eq. 11-7, we obtain

Ans. 5: Yes, but $\sqrt{1 - (v^2/c^2)}$ is so close to one that it is not noticeable.

Fig. 11-11. Two jet planes traveling with speeds u and v with respect to the ground. The observer plane sees the jet on the left coming at him with a speed u' that is less than $(u + v)$.

Ground

$$\frac{x'}{t'} = \frac{x + vt}{t + \frac{vx}{c^2}}$$

Now divide numerator and denominator of the right-hand side by t.

$$\frac{x'}{t'} = \frac{\frac{x}{t} + v}{1 + \left(\frac{x}{t}\right)\left(\frac{v}{c^2}\right)}$$

The stationary observer sees an object moving with speed $u = x/t$, and the moving observer sees the same object moving with speed $u' = x'/t'$. If we make these substitutions for x/t and x'/t', we obtain the result, Eq. 11-9.

Example 5
Assume the ground speeds of the two jet planes in Fig. 11-11 are $u = 2000$ mph and $v = 1000$ mph. What is the velocity of the first plane as measured by the second plane?

$$u' = \frac{2000 + 1000}{1 + \frac{2 \times 10^6}{c^2}} = \frac{3000}{1 + 4.5 \times 10^{-12}} = 2999.999999986 \text{ mph}$$

Example 6
The neutron is an unstable particle which decays into a proton, electron, and antineutrino: $N \rightarrow P + e^- + \bar{\nu}$. Suppose the decay electron has a velocity of eight-tenths the speed of light when its parent neutron is at rest. What will be the observed electron velocity if the decay occurs while the neutron is moving in the same direction at nine-tenths the speed of light?
Our frame of reference is moving at $v = 0.9c$ to observe the electron traveling at $u = 0.8c$. Substituting these values into Eq. 11-9 gives

$$u' = \frac{0.8c + 0.9c}{1 + 0.8 \times 0.9} = \frac{1.7}{1.72}c$$

Example 7
An elementary particle called the neutrino is traveling with the speed of light $(u = c)$. An observer is moving with velocity v toward the neutrino. According to the moving observer, what is the neutrino velocity?

$$u' = \frac{c + v}{1 + \frac{cv}{c^2}} = \frac{c + v}{\frac{1}{c}(c + v)} = c$$

Q.6: In Fig. 11-8(a) will a third observer traveling at velocity $v/2$ "see" both meter sticks as the same length?

We see from Example 7 that light (or anything else) traveling with velocity c must appear also to have velocity c to all other observers, no matter how fast they are moving. As previously stated, the Lorentz transformation equations transform space and time in just such a way that a beam of light must travel with the same velocity c as seen by all observers. We have now shown this to be true because the result of Example 7 is a direct consequence of Eq. 11-9 which itself is a direct consequence of the Lorentz transformation equations.

One basic result of Einstein's equations is that no object can travel faster than the speed of light. In fact, as an object approaches the speed of light its volume approaches zero due to the Lorentz contraction. Also according to Eq. 11-9, no matter how much additional velocity is given to an object, it cannot exceed $u' = c$. On the other hand, Newton's laws permit objects to exceed the speed of light. In Section 11-8 we shall see what changes are necessary to make mechanics consistent with the Einstein transformation equations.

11-7 Simultaneity

The future can precede the past

A physical reason why one observer thinks that a given meter stick is shorter than what another observer thinks is because events which are simultaneous for one observer are not simultaneous for the other. (In order to measure the length of a meter stick, one must measure the two ends simultaneously.)

We shall now demonstrate by using a moving boxcar as an example that two events which are simultaneous to a stationary observer will not be simultaneous for an observer in the moving boxcar. Consider the boxcar of length L (when measured at rest) in Fig. 11-12. Observer B standing in the center of the boxcar measures the length to be L. We shall now devise one way for observer A to measure its length while it is in motion, and thereby verify the Lorentz contraction. Just as observer B is passing A we will let both ends of the boxcar be struck by lightning bolts which are simultaneous according to A. Now A will obtain the result that

Fig. 11-12. Observer A sees two lightning bolts hit ends of boxcar simultaneously. Observer B is "racing" toward light pulse on right and meets it first. To him, the bolt on the right must have struck first.

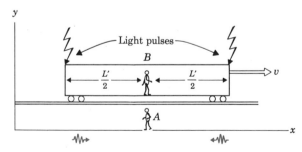

Ans. 6: No. He would "see" stick A move to the left with $v = 0.3c$, but according to the Einstein addition of velocities stick B would move to the right with $v = 0.366c$.

$L' = \sqrt{1 - v^2/c^2}\, L$ by measuring the distance between the marks left by the lightning bolts on the track.

An equally amazing result, however, is the fact that B claims that the lightning bolt on the right struck first. Certainly as viewed by A, the man in the boxcar is moving toward the light from the right-hand lightning bolt and he will reach it first. And if B reaches the right-hand pulse of light first, he reaches it first, no matter who the observer is. But according to B, the two bolts of lightning were equal distances away, and if he saw with his own eyes the one from the right occur first, then according to him it must have occurred first. An observer C starting at the same position, but moving to the left, would similarly claim that the left-hand bolt struck first. We conclude that whenever two events occur within the time required for light to travel between them, the order of occurrence is undefined—it depends on the velocity of the observer. In such cases future events can be made to precede past events by selecting an appropriate moving observer.

11-8 Relativistic Mechanics

Mass is energy, energy is mass

Einstein noticed that the classical laws of momentum and energy conservation did not agree with the Lorentz transformation equations. For example, if two bodies of equal mass M_0 have an elastic collision (by elastic we mean the relative velocity can only change direction, not magnitude), the quantity $(M_0\mathbf{v}_1 + M_0\mathbf{v}_2)$ will not be the same after the collision as it was before the collision. (This result is easily obtained by Lorentz transforming from the center-of-mass system to the laboratory system.) Hence, the total classical momentum could not be conserved. However, in this same calculation the quantity $\left(\dfrac{M_0}{\sqrt{1 - v_1^2/c^2}}\,\mathbf{v}_1 + \dfrac{M_0}{\sqrt{1 - v_2^2/c^2}}\,\mathbf{v}_2\right)$ does not change its value after an elastic collision. So Einstein proposed that if one redefined momentum as Mv where

Relativistic mass
$$M = \frac{M_0}{\sqrt{1 - v^2/c^2}} \tag{11-10}$$

then momentum would be conserved. He proposed that this new law of conservation of momentum would hold for systems of particles of unequal masses having inelastic collisions as well. Such a far-reaching prediction must be tested thoroughly by experiment. Many accurate tests have been made and Einstein's new formulations for mass, energy, momentum conservation, and energy conservation are well-accepted and believed to be correct.

One experimental test is the measurement of momentum and velocity of electrons in a high energy electron synchrotron. The measurement of velocity of the electrons in the Cambridge Electron Accelerator gives the result $v = c$ within the accuracy of measurement. The momentum can be measured within a fraction of a percent. The result for the ratio of momentum to velocity (the definition of inertial mass) is a value 12,000 times the rest mass of the electron. These Cambridge electrons when at full energy are actually six times heavier than protons! But ordinary macroscopic objects such as jet planes and space rockets have velocities so small compared to the speed of light that their true masses are for all practical purposes the same as their rest masses. The increase of mass with velocity is plotted in Fig. 11-13.

Example 8

Electrons in the Cornell University synchrotron reach a velocity of $v = 0.999999955c$. What is the mass of these electrons?

$$M = \frac{M_0}{\sqrt{1 - (0.999999955)^2}} = \frac{M_0}{\sqrt{0.00000009}} = \frac{M_0}{3 \times 10^{-4}}$$

$$M = 3300 M_0$$

Thus the Cornell electrons are 3300 times heavier than they were when at rest.*

One often encounters the reverse problem: the mass increase is given and one is asked to calculate the velocity. For this we square Eq. 11-10 and obtain

$$M^2 = \frac{M_0{}^2}{1 - \dfrac{v^2}{c^2}}$$

*We do not want to give the impression from this that MIT and Harvard are better than Cornell. By the end of 1967 Cornell should have electrons almost twice as heavy as the Cambridge electrons.

Fig. 11-13. Mass as a function of velocity. Note that at $v = 0.5c$ the mass has only increased by 15.5%.

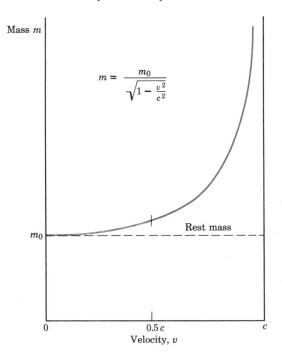

$$1 - \frac{v^2}{c^2} = \left(\frac{M_0}{M}\right)^2$$

$$\frac{v^2}{c^2} = 1 - \left(\frac{M_0}{M}\right)^2$$

$$\frac{v}{c} = \sqrt{1 - \left(\frac{M_0}{M}\right)^2} \qquad \text{(11-11)}$$

Another consequence of the Lorentz transformation equations is that the total relativistic mass and hence the quantity $(M_1c^2 + M_2c^2)$ remains unchanged when two particles interact. Einstein identified this with the conservation of energy and showed that in the limit of small velocities this approaches the classical formula for conservation of energy. Einstein proposed that

Einstein mass-energy relation

$$W = Mc^2 \qquad \text{(11-12)}$$

be the energy of a particle. We can write energy as an explicit function of velocity by substituting the right-hand side of Eq. 11-10. Then

$$W = \frac{M_0c^2}{\sqrt{1 - v^2/c^2}} \qquad \text{(11-13)}$$

So far this looks quite different from the classical formula $W = \frac{1}{2}Mv^2$. On page 283 we shall show the connection between Eq. 11-13 and the classical expression $\frac{1}{2}Mv^2$. Suppose $v = 0$, then Eq. 11-13 says the mass still has energy

$$W_0 = M_0c^2 \qquad \text{(11-14)}$$

Einstein interpreted this as an intrinsic rest energy. He said that if an amount of rest mass M_0 is ever destroyed, then an amount of energy M_0c^2 must be released. It is the direct conversion of some of this rest energy that explains the enormous energy release of nuclear bombs. Actually there is a severe constraint on how much energy can be obtained this way. As mentioned in Chapter 5 the law of conservation of heavy particles says that in any interaction the total number of protons plus neutrons must stay the same. For this reason there is no way we can ever get the 9×10^{20} ergs out of a gram of sand.

Q.7: Does the term mass mean rest mass, inertial mass, or gravitational mass?

Example 9

What is the energy contained in 1 gm of sand? How does this compare with the 7000 calories of heat delivered by burning 1 gm of coal?

The rest energy in 1 gm is

$$W = 1 \text{ gm} \times (3 \times 10^{10} \text{ cm/sec})^2 = 9 \times 10^{20} \text{ ergs}$$

The energy released by burning coal is

$$7000 \text{ calories} \times 4.18 \times 10^7 \text{ ergs/calorie} = 2.9 \times 10^{11} \text{ ergs}$$

Thus the rest energy is 3.1×10^9 as much as the chemical energy.

Example 10

If one ton of TNT gives an energy release of 10^9 calories when exploded, how much mass must be converted into energy in a one megaton bomb?

$$M = \frac{W}{c^2} = \frac{10^9 \times 4.18 \times 10^7 \text{ ergs}}{9 \times 10^{20} \text{ cm}^2/\text{sec}^2} = 4.6 \times 10^{-5} \text{ gm}$$

is the decrease in mass when one ton of TNT is exploded. The mass converted to energy in a one megaton bomb is one million times as much, or 46 gm. Actually the mass of fissionable material needed in such a bomb is about 1000 times as much. The total number of protons and neutrons in a nuclear explosion stays the same. See Chapter 15 for further discussion of this.

Our definition for kinetic energy will remain the same: kinetic energy is the energy of motion. The energy due to motion can be obtained by subtracting off the rest energy:

$$\text{KE} = W - W_0 \tag{11-15}$$

This equation can be written as an explicit function of velocity by using Eqs. 11-10 and 11-12.

$$\text{KE} = Mc^2 - M_0 c^2,$$

$$\text{KE} = \left(\frac{1}{\sqrt{1 - \dfrac{v^2}{c^2}}} - 1 \; M_0 c^2 \right)$$

or

$$\text{KE} = \left[\left(1 - \frac{v^2}{c^2} \right)^{-1/2} - 1 \right] M_0 c^2$$

We shall now show that for small velocities this function of velocity approaches $\frac{1}{2} M_0 v^2$. According to the binomial theorem,* the quantity $(1 - v^2/c^2)^{-1/2}$ can be replaced by

---

Ans. 7: In this book mass means inertial mass, and according to the principle of equivalence inertial mass and gravitational mass are the same.

*The binomial theorem states that $(1 + a)^n$ can be expanded as $(1 + a)^n = 1 + na + \dfrac{n(n-1)}{1 \cdot 2} a^2 + \cdots$.

$[1 + \frac{1}{2}(v^2/c^2)]$ when v/c is much less than 1. Then

$$KE = \left[\left(1 + \frac{1}{2}\frac{v^2}{c^2}\right) - 1\right]M_0 c^2$$

$$= \frac{1}{2}M_0 v^2$$

which is the familiar classical expression.

The first experimental confirmation of the Einstein mass-energy relationship came from the comparison of energy release in radioactive decay with the mass difference between initial nucleus and final products. As an example of how $W = Mc^2$ can be checked in the laboratory, let us consider the simplest case of beta decay, the beta decay of the free neutron. The free neutron is observed to decay into a proton, an electron, and an antineutrino (the antineutrino $\bar{\nu}$ has zero rest mass).

$$N \rightarrow P + e^- + \bar{\nu}$$

with an observed release of 1.25×10^{-6} ergs of kinetic energy. The rest mass of the neutron is measured to be greater than that of the proton plus electron by 13.9×10^{-28} gm. The energy corresponding to this amount of mass should be $W = 13.9 \times 10^{-28} \times c^2 = 1.25 \times 10^{-6}$ erg. This energy due to the mass difference checks with the observed kinetic energy of the decay products which is also 1.25×10^{-6} erg within the accuracy of measurement.

Example 11

A certain particle has a kinetic energy equal to its rest energy. What is the velocity of this particle?

$$W = KE + M_0 c^2$$

$$Mc^2 = M_0 c^2 + M_0 c^2, \quad \text{since} \quad KE = M_0 c^2$$

$$\frac{M_0}{M} = \frac{1}{2}$$

Substituting into Eq. 11-11 gives

$$\frac{v}{c} = \sqrt{1 - \left(\frac{1}{2}\right)^2} = \sqrt{\frac{3}{4}}$$

$$v = \frac{\sqrt{3}}{2}c = 0.866\,c$$

Q.8: Which is larger $KE = (M - M_0)c^2$, or the classical quantity $\frac{1}{2}M_0 v^2$?

Example 12

The Bevatron, a proton accelerator, gives protons a kinetic energy of 10^{-2} erg. By what factor is the mass of such protons increased? The proton rest mass is 1.67×10^{-24} gm.

First let us calculate the rest energy of a proton.

$$M_0 c^2 = 1.67 \times 10^{-24} \times (3 \times 10^{10})^2 = 1.5 \times 10^{-3} \text{ ergs}$$

$$\frac{M}{M_0} = \frac{M_0 c^2}{M_0 c^2} = \frac{\text{KE} + M_0 c^2}{M_0 c^2} = \frac{10^{-2} + 1.5 \times 10^{-3}}{1.5 \times 10^{-3}} = 7.68$$

We see that Bevatron protons are 7.68 times heavier than ordinary protons.

We conclude by noting that any form of energy must have a corresponding mass $M = W/c^2$. For example, light waves which have zero rest mass must have mass W/c^2 where W is the energy of the light. In principle the mass of light could be measured by trapping it in a box with totally reflecting walls. The box would then weigh more when it contains light than when it does not. However, the difference is too small to be measured even by our most accurate instruments. But there is one effect of the mass of light that has been measured. If light waves have mass, they should be attracted toward the sun due to gravity. This bending of light toward the sun gives an observable shift in the apparent positions of stars when seen near the sun (see Fig. 11-14). An eclipse is needed to see stars this close to the edge of the sun. As discussed in the next section, the inertial mass of light is an effect of special relativity, but the gravitational force on light is an effect of general relativity.

11-9   General Relativity

Relativistic gravity

What we have been calling Einstein's theory of relativity is sometimes called special relativity in order to distinguish it from the theory of general relativity, the first part of which Einstein worked out in 1911, six years after his special theory. The theory of general relativity is really a modern, relativistic theory of gravitation.

In Newton's theory of gravitation the force $F = GM_1M_2/r^2$ is one which acts instantaneously. If a force can act in-

Fig. 11-14. The bending of light in the gravitational field of the sun.

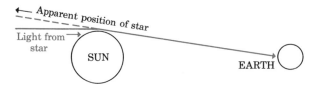

Ans. 8: The classical quantity is smaller. This is easily seen in the limit as v approaches c. Then the relativistic KE approaches infinity, but $\frac{1}{2}M_0 v^2$ approaches $\frac{1}{2}M_0 c^2$.

stantaneously, this means a signal, or energy, could be transmitted instantaneously. This violates one of the basic tenets of relativity, that is, no energy, not even a signal, can travel faster than the speed of light. Thus Einstein tackled the problem of a relativistic theory of gravitation. He was determined that his new theory should satisfy both the principle of relativity and automatically give the result that gravitational mass is always equivalent to inertial mass. Einstein's determination led him to postulate what is called the principle of equivalence. This principle states that being in a gravitational field is equivalent to being in an accelerated reference system. For example, in a rocket blasting off and accelerating upward, a passenger has the impression that gravity has suddenly been increased. In a rocket taking off with acceleration $a = 2g$ with respect to the earth, the passengers and everything else in the rocket would weigh three times their normal weight. This "pseudo-gravitational" force is exactly proportional to the inertial mass. No physics experiment performed inside the rocket could tell the occupants whether the gravitational force of the earth had suddenly increased by a factor of three or whether the rocket was accelerating with respect to the earth.

In the general theory of relativity, Einstein mathematically incorporated the principle of equivalence using a form of mathematics which is well beyond the scope of this book. In this mathematical description any mass will "distort" the region of space around it so that all freely moving objects will follow the same curved paths curving toward the mass producing the distortion. Einstein's equations relate the amount of curvature to the strength (or mass) of the source.

Classically we would say that any object moving in a curved path is being accelerated and must be under the influence of some force. It is this acceleration, which in the general theory of relativity is a property of space, that explains the phenomenon of gravitation. Since it is the space itself which is "distorted," the effect on all inertial masses will be the same and the principle of equivalence is automatically satisfied.

One of the consequences of this theory is that light when leaving a mass loses energy in overcoming the gravitational

Q.9: If a pulse of light has energy of 3 ergs, what is its mass in grams?

attraction of the mass for the inertial mass of the light. As we shall see in Chapter 12, when light loses energy, its wavelength is increased. This effect is called the gravitational red shift. Such a shift is observed in the spectral lines of the sun and of heavy stars. Thus the ticks of an atomic clock on the surface of the sun are running slower than the very same atomic clock running here on earth. As we might expect, the general theory of relativity predicts that all clocks should be slowed down when in the presence of a gravitational field. In fact, if two identical clocks on the earth are 1 m apart in height, the lower clock should run slower by about one part in 10^{16}. Frequency standards were first built in 1960 with this accuracy using the photons emitted from radioactive nuclei imbedded in a crystal (see Chapter 15). The phenomenon giving rise to such accurate frequencies is called the Mossbauer effect.

So far it has been difficult to make experimental checks of general relativity. However, with these new frequency standards, it has finally been shown in laboratory experiments that gravity slows down time. The first such experiments were performed in a 70-ft tower at Harvard University in 1960.

The only other experiments that tend to confirm general relativity were made decades ago. They are the bending of starlight around the sun (see Fig. 11-14), the red shift in the spectrum of heavy stars, and the advance in the perihelion of Mercury. The accuracy of these experimental results is such that the general theory of relativity is still considered as inadequately verified. Special relativity, on the other hand, is well verified.

11-10 Mach's Principle

Is acceleration absolute?

We have learned that there is no such thing as absolute velocity. On the other hand, acceleration is absolute, even according to modern relativity theory. The amount of absolute acceleration can be measured by an instrument called an accelerometer. A mass attached to the end of a spring can be used as a simple accelerometer. If the spring constant and

Ans. 9: The mass is $M = \dfrac{W}{c^2} = \dfrac{3}{3 \times 10^{10}} = 10^{-10}$ gm.

mass are known, the value of the acceleration can be determined from the stretch of the spring. Consider different observers accelerating relative to each other. Of the infinite number of possible relative accelerations, there will just be one particular reference system in which the measured acceleration is zero. Such a reference system is called an inertial system. Newton himself pointed out that his laws are only valid in inertial reference systems. He defined an inertial system as one which is not accelerating with respect to the "fixed" stars.

This concept of the absoluteness of an inertial system bothered some philosophers, such as Ernst Mach. Mach proposed that, in a general sense, acceleration is not absolute but that it is determined by the configuration of all the mass in the universe. According to Mach's principle, if the distribution of mass in the universe were suddenly changed, then values of local accelerations would suddenly change and what is now an inertial system might no longer be one. Mach proposed that the value of the inertial mass of any object is determined by all the other mass in the universe. One theory, consistent with Mach's principle, is that the inertial mass of a single object in a massless universe would be zero. Then all reference systems would be inertial systems.

D. Sciama in his book *Unity of the Universe* presents a rather specific theory embodying Mach's principle. He makes use of a fact obtained from general relativity that an accelerating mass produces a gravitational field proportional to $1/r$ in addition to its usual inverse square field. An approximate formula for the $1/r$ force between two accelerating masses is $F_G \approx G\dfrac{Mma}{c^2 r}$ where a is the relative acceleration. If we sit on an accelerating mass m, then all the galaxies in the universe are accelerating relative to us in the opposite direction. The net gravitational force on m would be

$$F_G \approx \sum G \left(\frac{ma}{c^2}\right)\left(\frac{M_i}{r_i}\right)$$

where the summation sign means to sum over all galaxies M_i situated at distances r_i. Assuming an average density D of matter in the universe, and a maximum radius R, we

Q.10: Do you think an accelerating charge would produce an electric field proportional to $\dfrac{1}{r}$ in addition to its usual inverse square field?

obtain*

$$F_G \simeq \frac{GDR^2}{c^2} ma$$

The proposal is that this gravitational force completely explains inertial mass. Then F_G should have the value ma. We see that it will be as large as ma if $GDR^2/c^2 = 1$. Of these three numbers, only G is know accurately. The average density of matter in the universe is estimated to be $D \simeq 10^{-28}$ gm/cm³, but it could be ten times larger or smaller. As discussed in the next section, the effective radius of the universe is $R \simeq 10$ billion light years or $R/c \simeq 3 \times 10^{-17}$ sec. Then

$$GD \left(\frac{R}{c}\right)^2 = (6.7 \times 10^{-8})(10^{-28})(3 \times 10^{17})^2 = 0.6 \simeq 1$$

We see that within experimental accuracy the theory passes the test. Not only does the theory give a physical explanation to inertial mass, but it tells us just how much inertial mass must be assigned to a given quantity of matter. Another way of saying this is that the theory requires $G = c^2/DR^2$. (Remember, G is a measure of the ratio of inertial mass to gravitational mass.) We see that in this theory G is no longer an independent physical constant, but that it is derived from other "more basic" physical constants.

The point of view taken here is that your head is the center of the universe and is always at rest. Whenever you bump your head against a wall, the distant galaxies have suddenly accelerated and exert a strong gravitational force on your head. In order to keep your head "at rest," the wall must exert an equal and opposite contact force. So the next time

---

* The volume of the ith concentric shell is $4\pi r_i^2 \, \Delta r_i$ and its mass is $M_i = 4\pi D r_i^2 \, \Delta r_i$. The summation then becomes

$$F_G \simeq G \frac{ma}{c^2} \sum \frac{(4\pi D r_i^2 \, \Delta r_i)}{r_i}$$

$$= 4\pi \frac{GDma}{c^2} \, \Sigma \, r_i \, \Delta r_i$$

Now

$$\Sigma \, \Delta r_i = R \quad \text{and} \quad \Sigma \, r_i \, \Delta r_i = \overline{r} \cdot R = \left(\frac{R}{2}\right) \cdot R$$

Hence

$$F_G \simeq 2\pi \frac{GDmaR^2}{c^2}$$

Ans. 10: Yes. It is this $\frac{1}{r}$ field which is the radiated field.

you bump your head or stub your toe, you can blame it on the distant galaxies.

We conclude this section by pointing out that as of the present it is all controversial. Mach's principle itself does not have universal acceptance among physicists.

11-10 Cosmology

Creation, sudden or continuous?

The questions raised by the principle of equivalence, general relativity, and Mach's principle are intimately related to the questions of the origin, size, and structure of the universe. Is the universe infinite or finite in size? How old is our solar system and galaxy? How were they formed? How many other galaxies are there and how are they distributed? Where did they come from? What was the universe like before these galaxies were formed? The field of physics that deals with these most basic questions is called cosmology, a very fast-moving field. For example, in the past ten years the age of our galaxy has "increased" not by ten years, but by about ten billion years. The best present estimate for the age of our Milky Way galaxy is almost twenty billion years, whereas the age of the earth is 4.5 billion years.

At present the theories of the origin and size of the universe are in a state of flux. The competing theories are not on a firm experimental basis. However, leading cosmologists feel our knowledge of these subjects is now increasing so fast that in the next ten years or so new findings might be able to give a reasonable level of confirmation to one theory and reject the competing theories.

The remainder of this section will be used mainly to review the experimental situation. The 200-in. telescope at Mt. Palomar can see about ten billion galaxies uniformly distributed in space out to a distance of about eight billion light years, at which distance the light becomes too faint to be detected. Our galaxy with a diameter of about 60,000 light years contains about 100 billion stars, one of which is our sun. Our nearest neighbor galaxy, the Andromeda Nebula, is similar in size and structure to our own galaxy (see Fig. 11-15).

Our recent detailed knowledge of thermonuclear reactions

enables us to estimate the ages of certain stars; our galaxy was formed about twenty billion years ago, although stars are still forming in it (some are only one million years old). The gravitational attractive force provides the mechanism for small particles of matter to coalesce into stars.

The processes taking place deep inside stars and the pattern of stellar evolution observed by astronomers are now fairly well understood in terms of nuclear physics.

In addition to galaxies there are newly discovered objects called quasars. Although quasars appear as stars rather than galaxies in telescopes, they are 10 to 100 times brighter than typical galaxies at the same distance. In addition, some of them emit large amounts of radio frequency energy compared to other stars or galaxies. Many have been discovered first by radio telescopes and then confirmed by optical telescope. At this writing, the physics of quasars was not understood. Perhaps some extreme effects in general relativity contribute, or perhaps there are new, yet undiscovered laws of physics at work here.

Perhaps the most significant experimental fact is the observation that all the galaxies are receding from us with velocities proportional to their distances from us. The proportionality constant is such that those barely visible galaxies at a distance of five billion light years are receding from us with about half the speed of light. The velocity of recession is determined from the shift in wavelength of the light emitted by these galaxies. When a light source is receding, the wavelength of the light is increased (called the red shift or doppler shift). The largest wavelength shift observed thus far is a wavelength which is three times its original value observed in a distant quasar. From the doppler shift formula one concludes that this quasar is speeding away from us at $v = 0.8c$.

If one calculated back in time, he would conclude that all the galaxies in the universe were crowded together in our region of space about ten to fifteen billion years ago. In fact, one of the leading cosmological theories proposes that ten to fifteen billion years ago all the matter in the universe was crowded together with the same density as that of an atomic nucleus. Then all the visible galaxies could be contained in a

Q.11: What would be the wavelength shift for a quasar receding at $v = c$?

sphere less than the diameter of Jupiter's orbit. After this "mysterious" creation, the ordinary laws of physics take over, giving rise to a gigantic explosion of the primordial nucleus. Hence, receding galaxies correspond to the bits and pieces of an exploding hand grenade. This explanation of the expanding universe is called the "big-bang" theory. A perhaps more satisfying version of this approach is the pulsating theory of the universe. In this theory, the galaxies slow down, turn around, and reverse the expansion until they all coalesce which then causes a repeated "big bang."

The major competing theory is the steady-state theory which assumes a generalized principle of uniformity. The principle of uniformity states that the universe must look the same no matter from what position in space it is viewed. This is becoming difficult to reconcile with recent observations. Also the principle of uniformity states that the universe looks the same from any position in time; that is, it has always looked the same in the past and will always look the same in the future. In the steady-state theory this principle of uniformity is satisfied by assuming matter is continuously and uniformly being created at the same rate that the average density of matter would be decreased due to the expansion of the universe. The observed rate of expansion is such that the over-all density of matter will remain fixed if one neutron (or hydrogen atom) is spontaneously produced every year in a volume of 10^{15} cm^3 or one cubic kilometer. One might object that the spontaneous production of neutrons would violate our laws of conservation of energy, conservation of momentum, and conservation of heavy particles, but it is a fantastically weak violation of these "conservation laws." We shall see in Chapter 16 that other conservation laws are violated by interactions not nearly so weak.

The steady-state theory makes several specific predictions that can be tested. For example, one prediction is that the age distribution of galaxies should be uniform not only for the near galaxies but also for the distant galaxies. Another demanding prediction is that the thermonuclear reactions in ordinary stars should be able to produce the heavy elements such as uranium out of what was originally hydrogen. So far no clear contradictions to these predictions have been ob-

Ans. 11: All wavelengths would then be infinity and we would see nothing.

served, but time will tell. In the next few years we can expect more of the predictions to be tested. If the steady-state theory manages to survive all the tests, we can perhaps expect it to receive widespread acceptability in the next ten years or so. On the other hand, if it or any other theory decisively fails just one test, that is the end of it.

APPENDIX 11-1

Derivation of Lorentz Transformation Equations

Our goal here is to derive the Lorentz transformation equations starting from the principle of relativity and the invariance of the speed of light. Our job is already half done if we make use of the derivation of the time dilation (Eq. 11-8) on page 273. We will start with the general form

$$x' = Ax + Bt \tag{11-16}$$

$$t' = Et + Fx \tag{11-17}$$

The problem is to determine the coefficients A, B, E, and F which may be functions of the relative velocity v between the primed and unprimed frames of reference. Let us arbitrarily refer to unprimed system as the stationary system, and the primed system as the moving system. For a stationary clock fixed at $x = 0$, Eq. 11-17 gives $t' = Et$. But we already have shown in Eq. 11-8 that the moving observer sees $t' = \gamma t$, where we define $\gamma = (1 - v^2/c^2)^{-1/2}$. Hence we have the solution

$$E = \gamma$$

If the moving observer is moving to the left with velocity v, he will measure the position of the clock at $x' = vt'$. Now substitute this into Eq. 11-16 also putting $x = 0$, and we obtain

$$(vt') = 0 + Bt$$

or

$$B = v\frac{t'}{t}$$

or

$$B = v\gamma$$

is the solution for B.

In order to solve for A, we place the clock at the origin of the moving observer (now $x' = 0$) and we invoke the principle of relativity by claiming that the stationary observer sees the clock moving off to the left with the same velocity v; that is, $x = -vt$. Now substitute these quantities into Eq. 11-16:

$$(0) = A(-vt) + Bt$$

or

$$A = \frac{B}{v}$$

Substituting $B = (\gamma v)$ gives the solution

$$A = \gamma$$

Equations 11-16 and 11-17 are now of the form:

$$x' = \gamma x + \gamma v t \qquad (11\text{-}18)$$

$$t' = \gamma t + Fx \qquad (11\text{-}19)$$

Now to get the final coefficient F, we must make use of the invariance of the speed of light. Let a pulse of light leave the origin at $t = 0$. Then both observers see it move away with velocity c; that is, $x = ct$ and $x' = ct'$. So we substitute these values into Eqs. 11-18 and 11-19:

$$ct' = \gamma(ct) + \gamma v t$$

$$t' = \gamma t + F(ct)$$

Now divide the upper by the lower equation to obtain

$$c = \frac{\gamma c + \gamma v}{\gamma + cF}$$

Solving this equation for F, we obtain

$$F = \frac{\gamma v}{c^2}$$

If we substitute this into Eq. 11-19 and remember that $\gamma = 1/\sqrt{1 - v^2/c^2}$, we obtain our final result.

References:

James H. Smith, "Introduction to Special Relativity," W. A. Benjamin, New York, N.Y.

D. W. Sciama, "The Unity of the Universe," Anchor Books, Garden City, New York.

Problems

1. A proton of rest energy 938 Mev is given a kinetic energy of 47 Mev. By what percent is its mass increased?

2. The fact that the galaxies are moving away from us seems inconsistent with the generalized principle of uniformity. How does the cosmological theory of Gold, Bondi, and Hoyle resolve this inconsistency?

3. By what factor is the density of an object increased when it is moving with velocity v? (Distances perpendicular to the direction of motion are not contracted.)

4. Does the bending of light near the sun make the stars appear to move away from the sun or toward the sun?

5. Will a clock on a mountain top run slower or faster than a similar clock at the base of the mountain?

6. Consider an earth satellite in orbit 100 mi above the earth's surface. Would a clock in this satellite run slower or faster than a stationary clock at the same altitude?

7. Derive a formula for velocity in terms of P, W, and c. The formula must be relativistically correct. P is momentum and W is total energy (including rest mass).

8. How many micrograms does a 100-watt light bulb radiate away in one year?

9. The velocity of an object is such that its mass increases by 10%.
(a) By what fraction does its length in the direction of motion decrease?
(b) If its rest energy is W_0, what is its kinetic energy?

10. The rest energy of a proton is 938 Mev. Consider a proton traveling at one-half the speed of light.
(a) What is its kinetic energy in Mev according to classical mechanics?
(b) What is its kinetic energy in Mev according to relativistic mechanics?

11. The energy flux from the sun at the earth is 2 calories/cm^2/min. How much of the sun's mass in grams reaches the earth per year? Would you conclude that the earth is getting heavier?

12. An electron whose rest energy is 0.5 Mev has a kinetic energy of 1 Mev. What is the ratio of its inertial mass to its rest mass?

13. Suppose a rocket motor could be built which could supply a continuous acceleration of $2g$ (as measured by the passengers) to a rocketship for one year. The relativistic formula corresponding to the classical $v = at$ is

$$v = \frac{at}{\sqrt{1 + \dfrac{a^2 t^2}{c^2}}}$$

At the end of the year what would be the time dilation factor for the passengers?

14. To an observer at rest with respect to particle A, particle A appears to decay and emit particle B to the right with a velocity $v = 0.5c$. Suppose we observe this same event when particle A is moving with a velocity $v_A = 0.4c$ to the right. What velocity would we then measure for B? (We see A moving to the right when it decays.)

15. What mass of fissionable material is needed for a 20-kiloton nuclear bomb?

16. The kinetic energy of a pion is 35 Mev. By what factor is its halflife increased? The rest energy of a pion is 140 Mev.

17. The relativistic formula for momentum is $P = Mv$. Using Eq. 11-11 show that $c^2P^2 = W^2 - (M_0c^2)^2$.

18. The rest energy of the K-meson is 495 Mev. Consider a 330 Mev K-meson beam (each K-meson has a kinetic energy of 330 Mev).
    (a) What is the total energy of each K-meson?
    (b) What is the rest mass in grams of the K-meson?
    (c) What is the velocity of these K-mesons?
    (d) What is the ratio M/M_0 for the K-mesons in this beam?
    (e) If the halflife of the K-meson (when at rest) is 1.0×10^{-8} sec, what is the observed halflife of the K-mesons in this beam?

19. Consider a beam of pions all of the same velocity. The average halflife of the pions in this beam is observed to be 67% longer than when at rest; that is, $T'/T = \frac{5}{3}$. The rest energy of a pion is 140 Mev.
    (a) What is the kinetic energy of each pion in this beam?
    (b) What is the velocity of each beam pion?
    (c) What is the ratio of the mass of a beam pion to its rest mass?
    (d) What is P/M_0C for each beam pion? (P is momentum and M_0 the rest mass.)

20. One complete fringe shift in the Michelson-Morley experiment corresponds to a path difference of 4×10^{-5} cm. For an interferometer with 10-m arms and for an ether drift of $v = 18$ mi/sec, how much of a fringe shift should be observed when the apparatus is rotated $90°$?

21. For a particle of rest mass M_0 and velocity v, which is greater: $\frac{1}{2}M_0v^2$, $\frac{1}{2}Mv^2$, or its kinetic energy?

22. Rocket ship S' moves with speed $v = 0.6c$ in the $+x$ direction relative to the x axis shown in the figure. Two squirt guns (containing ink) are located $L = 5$m apart on the x axis. The squirt guns are fired simultaneously according to clocks A and B located at x_A and x_B. The individual ink "blots" shot from guns A and B strike the rear and front windows of S', A' and B', respectively. Observers on the ground then claim that the distance between the windows A' and B' is 5 m.

Prob. 22

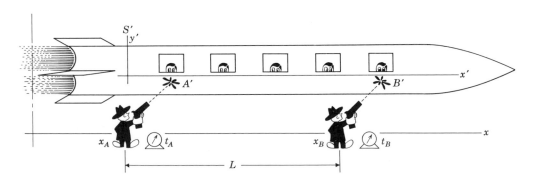

(a) What is the distance between windows A' and B' as measured on the space ship? (Using standard meter sticks.)

(b) Observers on the space ship S' carry clocks synchronized on the ship S'. Observers A' and B' will note, by their time readings:

       B fired before A,

or B and A fired simultaneously

or A fired before B. (They, of course, correct for the time it takes light to travel along the spaceship.)

(c) The observers on S' proceed to make sensible measurements of the spacing between the squirt guns A and B. How far apart are A and B as measured by observers on the rocket S'?

23. Suppose the universe is a finite sphere with radius R and average density of 10^{-29} gm/cm³. What is the velocity of escape from this sphere? (It is interesting to note that the answer to this problem turns out to be close to the observed velocity of recession of galaxies at a distance R.)

24. If a source of light of frequency f is moving with a velocity v away from a stationary observer, it can be shown that the observer will measure a lower or red-shifted frequency f' as given by the equation

$$f' = \frac{1 - \dfrac{v}{c}}{\sqrt{1 - \dfrac{v^2}{c^2}}} f$$

(a) If the observer is moving away with velocity v from a stationary source, what will be the equation for f' in this case?

(b) If the source of light is moving with velocity v toward the observer, what will be the equation for f' in this case? (Always treat v as a positive quantity.)

Quantum theory

Chapter 12

Quantum theory

12-1 Summary of Classical Physics
Modern today, classical tomorrow

So far in none of our presentation has the fundamental constant $h = 6.62 \times 10^{-27}$ erg-sec (Planck's constant) appeared. Actually it does appear in much of physics. The only reason we have so skillfully managed to avoid it is that it was not discovered until 1900 by Max Planck, and so far we have been presenting physics mainly as it was before 1900. This early physics is called classical physics.

Before we introduce the radically new concepts of quantum theory, let us stand back and look at the accomplishments and possible shortcomings of classical physics. With Newton's laws we explained falling bodies, projectiles, earth satellites, the motions of the planets, and other macroscopic motion. Also Newtonian mechanics gave us the conservation of energy, momentum, and angular momentum.

From nineteenth-century chemistry it was learned that matter is made up of molecules and atoms. This knowledge combined with Newton's laws explained away the big mystery of heat in what we call the kinetic theory of heat.

The "strange" electrical and magnetic phenomena were explained by the concept of charge and by the laws of electricity discovered a century ago that describe the "strange" interactions of both moving and stationary charges. We call these laws of electricity Maxwell's equations. The crowning achievement of classical physics was reached about 1870, when Maxwell derived the theory of light as a mathematical consequence of Maxwell's equations. This, of course, led to difficulty in explaining the ether and the problem of why the effects of the ether did not show up in the Michelson-Morley experiment. The explanation was given by Einstein in 1905 when he revised our fundamental concepts of space and time. At first encounter, relativity theory may seem shocking and a violation of "common sense." However, the physics student has an even greater shock in store for him when he meets the wave-particle duality of quantum theory.

In the 1890's the electron was discovered and so was the photoelectric effect, which is discussed in Section 12-2. By 1910 E. Rutherford had discovered that all the positive charge in an atom must be concentrated in a small, heavy nucleus. He demonstrated this conclusively by bombarding

Fig. 12-1. Head-on collision of alpha particle (mass m) with an atomic nucleus of mass M.

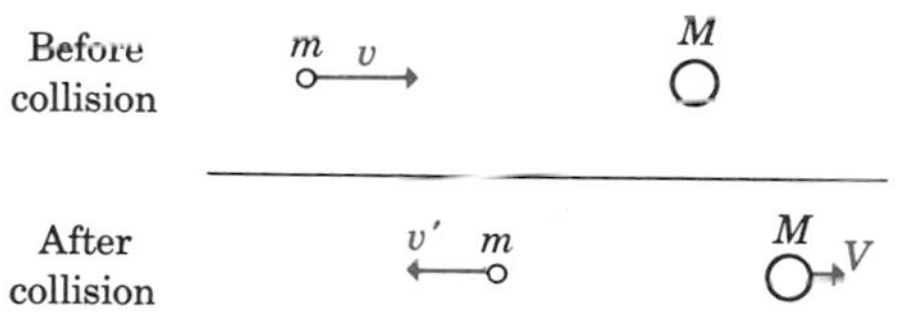

Q.1: What kind of force gives rise to the scattering of α-particles observed by Rutherford?

thin metal foils with α-particles (helium nuclei), with the result that a small, but significant number of them bounced back with very little loss of energy. The number of α-particles scattered backward determined that the charge of the scattering centers (atomic nuclei) was equal to the atomic number of the element used times the electronic charge e. As shown in Example 1, the energy of the scattered α-particles determined that almost all the mass of the target atoms was concentrated in the target nuclei.

Example 1

It is observed that when 5 Mev α-particles have head-on collisions with copper nuclei they bounce back with an energy of 3.9 Mev. Calculate the ratio of the mass of the copper nucleus to that of the α-particle.

Let m and M be the masses of the α-particle and copper nucleus, respectively. Let v be the initial velocity and v' the final velocity of m, and let V be the final velocity of M (see Fig. 12-1). We can obtain two simultaneous equations in these quantities by applying conservation of energy and conservation of momentum. According to the conservation of energy, the energy given to the copper nucleus is

$$\tfrac{1}{2}MV^2 = \tfrac{1}{2}mv^2 - \tfrac{1}{2}mv'^2$$

or

$$V^2 = \frac{m}{M}(v^2 - v'^2) \tag{12-1}$$

Since v' and V are in opposite directions, the final total momentum is $(MV - mv')$. According to the conservation of momentum, this quantity must be equal to the total initial momentum which was mv. Hence

$$(MV - mv') = mv$$

Solving this equation for V gives

$$V = \frac{m}{M}(v + v') \tag{12-2}$$

By squaring the right-hand side and equating it to the right-hand side of Eq. 12-1, we get

$$\left(\frac{m}{M}\right)^2 (v + v')^2 = \frac{m}{M}(v + v')(v - v')$$

or

$$\frac{m}{M} = \frac{v - v'}{v + v'},$$

$$\frac{m}{M} = \frac{1 - \dfrac{v'}{v}}{1 + \dfrac{v'}{v}}$$

Since the ratio v'/v is the square root of the ratio of the final to initial kinetic energies, we have

$$\frac{v'}{v} = \sqrt{\frac{3.9 \text{ Mev}}{5 \text{ Mev}}} = 0.882$$

Then

$$\frac{m}{M} = \frac{1 - 0.882}{1.882} = 0.0625$$

or

$$M = 16 \, m$$

This says that the copper nucleus mass must be sixteen times that of the α-particle or helium nucleus. This amount checks with the known atomic weights, which are 64 for copper and 4 for helium.

Physicists found it impossible to envision a stable atomic structure that was consistent both with Maxwell's equations and the experiments of Rutherford. They reasoned that the "large" size of the atom must be due to electrons in orbits which enclose the positive nucleus. These electrons would then have centripetal acceleration, and according to the Maxwell equations any accelerating charge must radiate energy. However, if this were true the electrons would continuously lose energy and quickly spiral with ever decreasing radius into the nucleus. Another difficulty was that according to classical theory, atomic radiation as viewed in a spectroscope should appear as a continuous spectrum. However, it was known that excited atoms usually radiated only certain discrete frequencies (they give a noncontinuous or line spectrum). A final problem to be explained was why are all atoms of the same element exactly the same? Why should the electrons of one carbon atom be in exactly the same orbits as the electrons of any other carbon atom? This certainly is not true of the planets of our solar system compared to another solar system.

In summary, we can list some of the phenomena that were left unexplained by classical physics at the turn of the century.

1. Specific heats of gases and solids. Also specific heat as a function of temperature (according to Section 6-8 specific heats tend to be less than those predicted by the kinetic theory).

2. The photoelectric effect.

3. A stable atomic structure.

Ans. 1: The electrostatic force.

4. Atomic radiation and absorption. Line spectra.

5. The exact similarity of all atoms of the same element.

6. The frequency spectrum of radiation emitted by a hot body (this phenomenon is called black-body radiation).

7. Radioactivity.

12-2 The Photoelectric Effect

All or nothing

Shortly after the discovery of the electron, physicists observed that when light shines on certain metal surfaces, negative charge or electrons are emitted. This is called the photoelectric effect. Figure 12-2 shows how this effect was first noticed. A metal sheet was attached to an uncharged electroscope. It was observed that negative charge would leak off only when the lights were turned on, leaving the electroscope with a net positive charge. Modern applications of this same effect are door openers, burglar alarms, television cameras, and exposure meters.

According to the classical theory of light, light is actually an oscillating electric field which would make an electron oscillate. We would expect that some of the electrons might be shaken out of the metal into the air. Since the electric field strength increases with light intensity, we would expect the maximum energy of emitted electrons to increase with intensity. If we keep the intensity constant and increase the frequency, at sufficiently high frequencies we would expect to find lower energy electrons because the electron is sluggish in response to higher frequencies due to its inertia or mass. Thus classical physics predicts that: (1) electron energy increases with intensity, and (2) electron energy decreases with frequency. By 1900 the experiments had indicated: (1) no change in electron energy with intensity, and (2) an increase in electron energy with frequency! The only effect of increasing the intensity was to increase the number of electrons emitted per second.

By 1905 the photoelectric effect was successfully explained by Einstein. His explanation also gave additional physical meaning to an earlier (1900) hypothesis of Max Planck. To explain mathematically the shape of the spectrum emitted

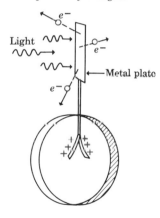

Fig. 12-2. A neutral electroscope is connected to a metal plate. When light shines on the plate, photoelectrons are ejected and the leaves then become positively charged.

Light

e^-

Metal plate

e^-

Q.2: Suppose the photoelectrons in Fig. 12-2 leave the metal plate with a kinetic energy of 3 ev. What will be the final electric potential of the metal plate?

Energy of a photon

by hot solids (black body radiation), Planck had postulated that light carried its energy in definite amounts, or "quanta." Planck assumed that the energy of a light wave had to be an integral number of quanta where the energy of each quantum is a new physical constant, h, times the frequency, f, of the light wave:

$$W = hf$$

where

$$h = 6.62 \times 10^{-27} \text{ erg-sec} \tag{12-3}$$

is Planck's constant. W is the energy of a single quantum of light. These "particles" of light are called photons. Planck was able to determine Planck's constant to within 1% of its value.

Example 2

We shall find it very useful to have a formula that tells us the wavelength of a photon in angstroms provided its energy $W = hf$ is given in electron volts. The derivation is as follows:

$$\lambda = \frac{c}{f} = \frac{hc}{hf} = \frac{6.62 \times 10^{-27} \times 3 \times 10^{10}}{hf \text{ (in ergs)}} = \frac{19.86 \times 10^{-17}}{1.6 \times 10^{-12} hf \text{ (in ev)}}$$

$$\lambda = \frac{12.39 \times 10^{-5}}{hf \text{ (in ev)}} \text{ cm}$$

$$\lambda = \frac{12,390}{hf \text{ (in ev)}} \text{ Å} \tag{12-4}$$

Warning: This formula does not apply to electrons and other particles of nonzero rest mass. One of the dangers of trying to "learn" physics by memory is that one loses sight of where the equations come from and to just what class of phenomena they apply.

Einstein proposed that in the photoelectric effect one complete photon is entirely absorbed by a single electron in an elementary act. The interaction occurs suddenly, similar to the collision of two particles. The electron in the metal now has an additional energy equal to hf. This daring proposal actually suggests that light is really made up of particles after all. The light particles or photons can only be absorbed one at a time and there is never such a thing as a fraction of a photon. This Einstein theory predicts correctly that the electron energy should increase with frequency and that it

Ans. 2: $V = +3$ volts. This means a 3 ev electron would be slowed down almost to rest by the time it got far away from the metal plate.

Fig. 12-3. Plot of KE_{max} versus frequency for a metal having a work function $\mathscr{W}$.

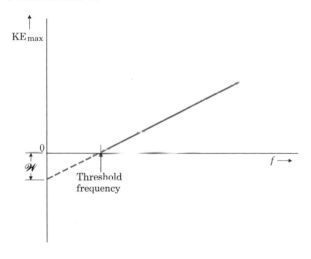

has nothing to do with the intensity. At first we might expect the kinetic energy of the emitted electrons to be equal to hf. But there are two difficulties: first, most electrons which absorb photons will lose their excess kinetic energy by collisions and not get out of the metal, and even if they do get out, most would have a remaining kinetic energy less than hf. The second difficulty is that electrons require a certain minimum amount of energy to remove them from the metal surface because of the attractive force of neighboring atomic nuclei. We shall call this energy needed to get out $\mathscr{W}$, the "work function." Pure cesium has a work function of 1.8 ev; the copper work function is 4.3 ev. In the most favorable case, the extra energy given to the electron (hf) supplies the needed energy to get out ($\mathscr{W}$) plus whatever kinetic energy is left over; that is,

$$hf = \mathscr{W} + \mathrm{KE}_{max}$$

or

$$\mathrm{KE}_{max} = hf \quad \mathscr{W} \tag{12-5}$$

We call this kinetic energy KE_{max} because other less favorable electrons are emitted with less KE.

If monochromatic light is shined on a pure metal surface, say cesium, then electrons will be emitted having energies up to KE_{max}. Now if the frequency of the light is increased, KE_{max} is observed to increase. If these experimental results are plotted as a function of f, the curve in Fig. 12-3 is obtained. Note that Planck's constant can be obtained by measuring the slope of the curve.

Example 3

A photoelectron from cesium has a kinetic energy of 2 ev. What is the maximum wavelength of light which could have ejected this electron?

$$hf = \mathrm{KE} + \mathscr{W}$$

$$hf = 2\,\mathrm{ev} + 1.8\,\mathrm{ev} = 3.8\,\mathrm{ev}$$

According to Eq. 12-4,

$$\lambda = \frac{12{,}390}{3.8}\,\text{Å}$$

$$\lambda = 3240\,\text{Å}$$

Q.3: If the frequency of light striking a metal plate is doubled, will the KE of the photoelectrons be doubled?

Fig. 12-4. Experiment to show wave properties of electrons.

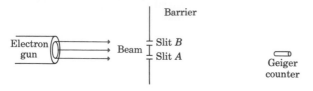

Fig. 12-5. Electron distributions according to classical physics.

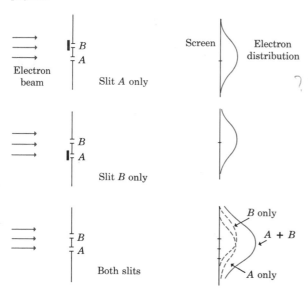

Ans. 3: No. $\dfrac{(KE)_2}{(KE)_1} = \dfrac{2hf - \mathcal{W}}{hf - \mathcal{W}}$.

12-3 Wave-Particle Duality

Particles are waves are particles

Phenomena such as the photoelectric effect indicated that light must have the properties of particles along with its known wave nature. This behavior of light becomes paradoxical when one considers Young's double-slit experiment performed with a light source so faint that only one photon at a time can enter the apparatus. Photomultiplier tubes which utilize the photoelectric effect can detect the photons one at a time. Let us use photomultiplier tubes in place of the screen in Fig. 10-18. The distance from source to screen is typically about 100 cm, so each photon "lives" for

$$t = \frac{s}{v} = \frac{100 \text{ cm}}{3 \times 10^{10} \text{ cm/sec}} = 3 \times 10^{-9} \text{ sec}$$

If the time between "clicks" from the photomultiplier tubes is less than this, only one photon at a time reaches the double slit. This experiment can be performed (see the Educational Services Inc. film on double-slit interference) and the result is that the interference pattern remains exactly the same. How can *one* photon pass through *two* slits? One way to restate the question is, how can light have both particle and wave properties in the same experiment? This question is one of the most important in all of physics and we will try to answer it in the following paragraphs.

It is now known that this wave-particle relationship or "duality" applies to all particles and waves and is the basic principle of the modern quantum theory. At first it may sound quite farfetched to claim that material particles have a wave nature similar to that of photons. Before describing just what is meant by the wave associated with any particle, let us consider the idealized experiment of Fig. 12-4.

An electron gun shoots a beam of electrons at a barrier that has two slits, A and B. On the other side there is a Geiger counter that counts each individual electron that hits it. Suppose we count 100 electrons per minute coming from slit A (slit B is closed). The counting rate from slit B alone is also 100 counts per minute. If we open only slit A and then gradually open slit B, we would expect (according to common sense and everything we have ever learned) the counting rate

Fig. 12-6. Electron distribution according to quantum theory.

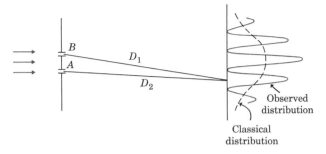

Observed distribution

Classical distribution

Fig. 12-7. Drawing from research paper by C. Jönsson in *Zeitschrift für Physik,* Vol. 161 (1961), showing his experimental arrangement for obtaining double-slit interference pattern of electrons.

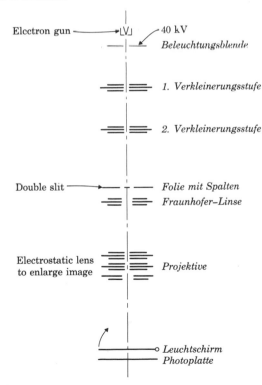

Electron gun ──────→ 40 kV
  Beleuchtungsblende

  1. Verkleinerungsstufe

  2. Verkleinerungsstufe

Double slit ──────→ *Folie mit Spalten*
  Fraunhofer–Linse

Electrostatic lens to enlarge image *Projektive*

  Leuchtschirm
  Photoplatte

to increase gradually from 100 to 200 counts per minute as slit B is opened. However, depending on the position of the counter, the true experimental result could be a gradual *decrease* from 100 to zero counts per minute! How can the act of opening slit B possibly influence those electrons that would have gone through slit A? Another violation of common sense is that a counter position can be found where the rate would increase from 100 to 400 counts per minute as slit B is opened. Then there would be twice as many electrons as obtained from the direct sum of the two separate contributions.

Figure 12-5 shows the expected classical electron distributions at the position of the counter as the counter is moved across the beam (electron intensity is plotted to the right in red). But if we actually do the experiment, we obtain a quite different pattern as shown in Fig. 12-6! Note that this experimental electron intensity pattern is of the same type as the double-slit interference pattern of light waves. If $D_1 - D_2 = N\lambda$, there would be an interference maximum; if $D_1 - D_2 = (N + \frac{1}{2})\lambda$, there would be a minimum of intensity. As we shall see in the next paragraph, electrons usually have wavelengths much smaller than that of visible light; hence it is difficult to perform Young's experiment with electrons. However, C. Jönsson in 1961 was able to obtain a genuine double-slit interference pattern of electrons on a photographic "screen." The experimental layout is shown schematically in Fig. 12-7 and the results are shown in Fig. 12-8*a*.

Each electron produces a black spot at the position where it hits the film. This photograph using a double-slit source of electrons is the result of thousands of electron impacts. For comparison, Fig. 12-8*b* shows a typical double-slit interference pattern using light.

But how can electrons that we know to be particles of definite mass and charge at the same time be waves? Actually this possibility was first proposed by a student, Louis de Broglie, in his Ph. D. thesis in 1924. He proposed that all particles must have a wave nature in the same manner that light has a wave nature. *The physical interpretation of the wave-particle duality is that the intensity of the particle wave*

Fig. 12-8. (a) A double slit interference pattern of electrons Each grain in the photographic negative is produced by a single electron. For comparison, (b) is the double slit interference pattern of light shown in Chapter 10 (Fig. 10-15). Likewise, each grain in the negative is produced by a single photon. ((a) was made by Professor C. Jönsson at the University of Tubingen.)

(a)

(b)

at any given point is proportional to the probability of finding the particle at that point. This is what is meant by the wave-particle duality. The word duality is perhaps a poor choice. What is meant is that there is the definite relationship as italicized above between the particle characteristics and the wave characteristics of any particle (or wave). De Broglie proposed a quantitative relationship between the wavelength of the particle wave and the momentum of the particle:

The de Broglie relationship $\lambda = \dfrac{h}{P}$ (12-6)

for any particle of momentum P.

Example 4

Starting with the de Broglie relationship, derive the formula $W = hf$ for particles of zero rest mass (note that this formula holds only for particles of zero rest mass). Relativistically

$$P = Mv \quad \text{and} \quad M = \frac{W}{c^2}$$

Therefore

$$P = \frac{Wv}{c^2}$$

For a particle of zero rest mass $v = c$ and then $P = W/c$. Substituting in 12-6:

$$\lambda = \frac{h}{\dfrac{W}{c}} \quad \text{or} \quad W = h\frac{c}{\lambda} = hf$$

In Eq. 12-6, the wave nature, the left-hand side, is directly and intimately related to the particle nature, the right-hand side. The proportionality constant, h, is Planck's constant, which had previously been determined by phenomena such as black body radiation, the photoelectric effect, and the hydrogen spectrum.

The wave-particle duality raises puzzling questions that require some further physical interpretation. Let us shoot only one electron at a time. Then according to this wave picture, each electron is represented by a wave train or wave packet that splits equally between the two slits. But we can put a geiger counter, cloud chamber, or other particle detector at slit A and observe that there is never in nature half of an electron. We either observe all of a particle, or else no particle at all. This is called the principle of indivisibility and is consistent with the hypothesis that the wave intensity at slit A is the probability of finding one whole electron at that position. Furthermore, if a detector is placed at slit A, the interference pattern smooths out and the classical result is then observed. To be detected, the electron must have an interaction with the detector. According to the quantum theory, we then have a new electron wave starting out from the point of interaction which produces just a single slit pattern. On the other hand, if an electron appears on the screen without being detected by the detector at slit A, we then know its wave must have gone through slit B only. Thus the presence of a detector changes the result from the interference pattern of Fig. 12-6 to the classical result of Fig. 12-5. Actually many physicists, including Einstein, have tried to contrive an experiment that would reveal the slit used by individual electrons without destroying the interference, but all such efforts have failed.

Just what is it that waves in an electron wave? We must give the same kind of answer we gave for photons. Electromagnetic waves travel freely through pure vacuum. In contrast to mechanical waves, no material of any kind is waving. Physicists use the symbol ψ for the amplitude of a particle wave. The intensity is the square of the absolute value of the amplitude or $|\psi|^2$. Hence $|\psi(x)|^2$ is proportional to the probability of finding the particle at position x. The wave amplitude $\psi(x)$ has no direct physical meaning, and in this sense nothing is waving. It is just that quantum mechanical problems are solved mathematically in the same way that water wave or other kinds of classical wave problems are solved. Classical waves and particle waves both obey the same kind of mathematical wave equation. However, in the case of classical waves, the wave amplitude is directly observable, whereas ψ is not (except in certain special cases for the photon). Another nonclassical characteristic of quantum mechanical waves is that even though the wave intensity is always a real positive number, sometimes the wave amplitude must be expressed as a complex number containing $\sqrt{-1}$. We will not deal with examples requiring complex

Q.4: Assume a thin-walled Geiger counter is placed behind slit A only. Whenever an electron goes through slit A it gives a "click" in the detector and passes through the Geiger counter tube to the screen. If the Geiger counter is turned off (but not removed), what will be the pattern on the screen?

Fig. 12-9. Set up for Example 5.

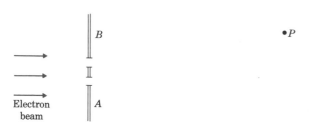

numbers in this book. Now we shall illustrate the mathematics of particle waves by a few simple examples.

Example 5

There is a Geiger counter at point P in Fig. 12-9. The part of the wave amplitude coming through slit A and reaching point P is $\psi_A = 2$ units, and through slit B is $\psi_B = 6$ units. When only slit A is open, 100 electrons per second are observed at point P.

(a) How many electrons per second are observed when only slit B is open?

(b) Assuming a constructive interference, how many electrons per second are observed when both slits are open?

(c) Assuming a destructive interference, how many electrons per second are observed when both slits are open?

Answer: We are told that the particle wave intensity $\psi_A{}^2 = 4$ corresponds to 100 electrons per second. Hence $\psi_B{}^2 = 36$ will correspond to nine times as many particles, or 900 electrons per second.

For part (b) the total wave amplitude is $\psi = \psi_A + \psi_B$, or $\psi = 8$. Since $\psi^2 = 64$ is sixteen times $\psi_A{}^2$, there will be 1600 electrons per second.

For part (c) ψ_A and ψ_B must be of opposite sign to give a destructive interference. Hence $\psi = \psi_A + \psi_B = 2 - 6 = -4$. Now $\psi^2 = 16$ which is four times $\psi_A{}^2$. This corresponds to 400 electrons per second.

Fig. 12-10. Intensity pattern of electrons hitting screen in Example 6.

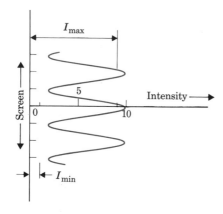

Example 6

What will be the intensity pattern of a double-slit interference experiment if slit B is four times as wide as slit A?

Answer: Four times as many electrons can get through slit B, hence the intensity $\psi_B{}^2 = 4\psi_A{}^2$, or $\psi_B = 2\psi_A$. The total intensity observed on a screen at a maximum is proportional to $(\psi_A + \psi_B)^2$, or

$$I_{\max} = (\psi_A + 2\psi_A)^2 = 9\psi_A{}^2$$

At an intensity minimum, we use a minus sign giving

$$I_{\min} = (\psi_A - 2\psi_A)^2 = \psi_A{}^2$$

The ratio $\quad \dfrac{I_{\max}}{I_{\min}} = \dfrac{9}{1}$

The intensity pattern would appear as in Fig. 12-10.

Example 7

Now we go back again to two slits of equal sizes; however, we put a small, thin detector behind slit A in an attempt to determine which slit each electron goes through. Suppose, however, that the detector is not perfect; that is, it is so thin that an electron has one chance out of four of passing through it without interacting. What now is the interference pattern?

Ans. 4: One still observes the classical pattern on the screen. The electrons are still interacting in the Geiger tube whether or not we bother to record those interactions.

Answer: Let us call ψ_A the part of the beam going through slit A that is undetected and can therefore interfere with the electron waves going through slit B. Since only $\frac{1}{4}$ get through the detector $\psi_A{}^2 = \frac{1}{4}\psi_B{}^2$. Let us denote the intensity of the detected electrons at slit A by $\psi_{A'}{}^2$ where $\psi_{A'}{}^2 = \frac{3}{4}\psi_B{}^2$ and cannot interfere. We must treat $\psi_{A'}$ as a new localized source of electron waves which have no fixed phase relation with respect to ψ_B. In such cases of independent (also called incoherent) sources of particles one must add intensities, not amplitudes. The grand total intensity is then the sum of the two intensities:

$$I = (\psi_A + \psi_B)^2 + \psi_{A'}{}^2$$

Now substitute $\frac{1}{2}\psi_B$ for ψ_A and $\frac{3}{4}\psi_B{}^2$ for $\psi_{A'}{}^2$

The result is $\dfrac{I_{\max}}{I_{\min}} = \dfrac{(\frac{1}{2}\psi_B + \psi_R)^2 + \frac{3}{4}\psi_B{}^2}{(\frac{1}{2}\psi_B - \psi_B)^2 + \frac{3}{4}\psi_B{}^2} = \dfrac{\frac{9}{4}\psi_B{}^2 + \frac{3}{4}\psi_B{}^2}{\frac{1}{4}\psi_B{}^2 + \frac{3}{4}\psi_B{}^2} = \dfrac{3}{1}$

The resulting interference pattern is shown in Fig. 12-11. If the detector were removed the intensity minima would be zero and one would then have a pure double-slit interference pattern.

12-4 Electron Diffraction

An accident

De Broglie's hypothesis was first verified by the experimental observation of electron diffraction in 1927 by two American physicists, C. J. Davisson and L. H. Germer. It is interesting that in this experiment, as in others that were of extreme importance to physics, the great discovery was "accidental." Davisson and Germer were not looking for electron diffraction. In fact, in the early stages of their experiment, they had never even heard of electron diffraction. In 1926, Davisson took some of his preliminary data to an international conference in Oxford, England. European physicists suggested to him that his results might be interpreted as electron diffraction rather than the classical electron scattering that he had been studying. Just a few months later, Davisson and Germer obtained data that conclusively demonstrated the wave nature of electrons and gave Planck's constant to an accuracy of about 1%. They scattered low-energy electrons off the surface of a single metallic crystal. The regular rows of atoms at the surface act as the lines of a very fine diffraction grating. The electron wavelength is determined by knowing the atomic spacing.

Fig. 12-11. Intensity pattern of electrons hitting screen in Example 7.

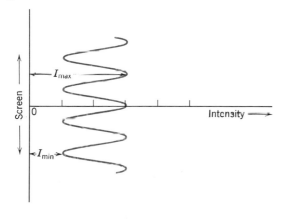

Q.5: Two light beams strike a screen. There are 3 times as many photons per sec in beam 1 as in beam 2. What is the ratio of the electric fields in the two beams?

Example 8

A beam of electrons is accelerated through a potential of 100 volts. What is the electron wavelength?

$$\text{KE} = \frac{P^2}{2m} = eV, \quad \text{where} \quad eV = 100 \text{ ev} = 1.6 \times 10^{-10} \text{ erg}$$

$$P = \sqrt{2meV} = 5.4 \times 10^{-19} \text{ gm cm/sec}$$

$$\lambda = \frac{h}{P} = \frac{6.62 \times 10^{-27}}{5.4 \times 10^{-19}} = 1.23 \times 10^{-8} \text{ cm} = 1.23 \text{ A}$$

An arrangement for observing electron diffraction from a crystal surface is shown in Fig. 12-12. The detector could be a fluorescent screen, as in the movie film *Matter Waves* produced by Educational Services, Inc. Planck's constant can be determined by observing the direction θ at which an intensity maximum appears. As can be seen in Fig. 12-11b, the path difference $\Delta D = d \sin \theta$ and this will be equal to the wavelength h/P at the first intensity maximum. Hence

$$\frac{h}{P} = d \sin \theta$$

and

$$h = P d \sin \theta.$$

Shortly after de Broglie's proposal in 1924, the English physicist, G. P. Thompson, set out in a systematic way to observe electron diffraction. His method of approach was to make higher energy electrons pass through a thin metal foil. Since x-rays have about the same wavelength, Thompson hoped to get an electron diffraction picture similar in appearance to previously obtained x-ray diffraction pictures. By 1928 Thompson obtained electron diffraction pictures looking almost the same as x-ray diffraction pictures. It is interesting that in this case the "accidental" approach was quicker than the carefully studied, deliberate approach. This may not fit the reader's conception of the scientific method, but it is true, living science. The experiment of Davisson and Germer is a good example of a valid scientific method. If an experimenter observes a strange effect, even by accident, that he does not understand, he should carefully pursue it until it is understood.

By now interference patterns of neutrons, protons, and even whole atoms as well as electrons have been observed.

Fig. 12-12. (*a*) Apparatus to observe electron diffraction from a crystal surface. (*b*) Surface of crystal same as in (*a*) but greatly enlarged.

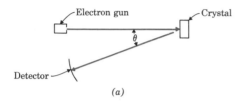

(*a*)

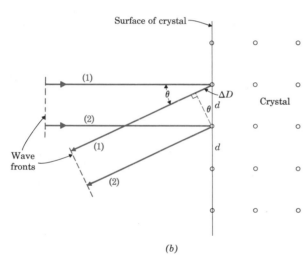

(*b*)

Ans. 5: Since the ratio of intensities is $\frac{3}{1}$, the ratio of amplitudes will be $\dfrac{E_1}{E_2} = \dfrac{\sqrt{3}}{1}$.

In many different ways, the wave nature of matter has been very well established. No violations of the theory have been found.

12-5 The Uncertainty Principle

Not even Mother Nature can know everything

An interesting consequence of quantum theory is that we cannot simultaneously specify the exact position and velocity of any particle. This "peculiar" consequence of the theory is known as the uncertainty principle. As an extreme example, let us assume the exact momentum of a particle is known; then it has a unique wavelength given by $\lambda = h/P$ and is a continuous plane wave. Since the intensity of this wave is uniform, it is equally probable to find such a particle anywhere in space. Conversely, if we know that the particle is located in a small region of space, its wave function is a short wave packet that does not have any unique λ (or unique P). Thus it is impossible to know simultaneously both the exact position and velocity of any particle. W. Heisenberg showed that the uncertainty in momentum, ΔP, and the region of localization of the particle, Δx, must obey the relation

Heisenberg uncertainty principle

$$(\Delta x)(\Delta P) \simeq h \tag{12-7}$$

where the symbol $\simeq$ means approximately equal.

Equation 12-7 can be obtained by noting that a localized particle or wave packet is really the sum of many pure (infinitely long) sine waves that have almost the same wavelength. The heavy curve of Fig. 12-13a is a typical wave packet of length L. Note that it consists of only two pure sine waves of wavelengths λ_1 and λ_2. Over the length L wave 1 contains one more cycle than wave 2. Then

$$\frac{L}{\lambda_1} = N + 1 \quad \text{and} \quad \frac{L}{\lambda_2} = N \tag{12-8}$$

where N is some integer. If the two waves are out of phase at $x = 0$, they will be in phase at $x = L/2$, and out of phase again at $x = L$. The sum of two such waves, shown in Fig. 12-13a, is a typical wave packet of length L. By subtracting the second equation from the first in Eq. 12-8, we

Q.6: As the voltage on the electron gun in Fig. 12-12 is increased, will the pattern on the screen expand or contract?

Fig. 12-13. (a) A typical wave packet of length L. (b) Two pure sine waves which when added together give the above wave packet.

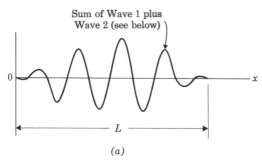

Sum of Wave 1 plus Wave 2 (see below)

L

(a)

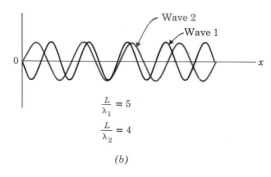

Wave 2
Wave 1

$\dfrac{L}{\lambda_1} = 5$

$\dfrac{L}{\lambda_2} = 4$

(b)

Ans. 6: As the voltage increases, the wavelength decreases and so does θ. Hence the spots move in closer to the electron gun.

have

$$\frac{L}{\lambda_1} - \frac{L}{\lambda_2} = 1$$

$$L\left(\frac{1}{\lambda_1} - \frac{1}{\lambda_2}\right) = 1$$

Now for $1/\lambda$ substitute P/h (the deBroglie relation):

$$L\left(\frac{P_1}{h} - \frac{P_2}{h}\right) = 1$$

$$L(P_1 - P_2) = h$$

or

$$(\Delta x)(\Delta P) = h$$

where $\Delta x = L$ is the region where the particle may be found. If $\psi = \psi_1 + \psi_2$ where ψ_1 and ψ_2 are two pure sine waves of wavelengths λ_1 and λ_2, then the probability of finding the particle with momentum P_1 is $\dfrac{\overline{\psi_1^2}}{(\psi_1 + \psi_2)^2}$, and with momentum P_2 is $\dfrac{\overline{\psi_2^2}}{(\psi_1 + \psi_2)^2}$.

The thoughtful reader may object to this "derivation" because pure sine waves are of infinite length and there will be other constructive interferences not shown in Fig. 12-13 running to the left and to the right. However, these outside packets can be "killed off" by adding in more waves of wavelengths that are close to the average of λ_1 and λ_2. Such waves cannot get out of phase very much with our central packet, but will get out of phase with the more distant packets and, if appropriately adjusted, can cancel all of them out. This mathematical procedure, called Fourier analysis, requires an infinite number of pure sine waves using integral calculus.

Let us now check up on the uncertainty principle by trying to localize the position of a particle with an optical microscope. With a microscope, the position of a particle can at best be located to within about one wavelength of the light being used. Hence $\Delta x \approx \lambda$. Now if the particle is at rest, $\Delta P = 0$ and we have a violation of Eq. 12-7! Or do we? Let us look for some quantum mechanical effect. We used light, and the quantum theory tells us that light is quantized into photons of momentum $P = h/\lambda$. In order to detect the particle, at least one of the photons in the converging beam of light must either be scattered or absorbed by the particle. Hence the momentum given to the particle might be as much as h/λ. The particle then has an uncertainty in its momentum $\Delta P \approx h/\lambda$ at the time it is observed at a position $\Delta x \approx \lambda$. If we multiply these two uncertainties together, we get

$$(\Delta P) \cdot (\Delta x) \approx \left(\frac{h}{\lambda}\right) \cdot (\lambda) = h$$

which checks with Eq. 12-7. In this example we see that quantum mechanics is self-consistent. Physicists have searched hard for inconsistencies, but none has been found.

Example 9

Suppose at a time $t = 0$ a free electron is observed to be localized in a region $\Delta x_0 \approx 10^{-8}$ cm. How will the wave packet spread out with time? What will be Δx after 0.1 sec? After 1 sec?

Since the momentum spread in the wave packet is $\Delta p \approx h/\Delta x_0$, the velocity spread will be $\Delta v = \Delta p/m \approx \dfrac{h}{m \, \Delta x_0}$. After a time t, the corresponding spread in distance will be

$$\Delta x = (\Delta v)t = \left(\frac{h}{m \, \Delta x_0}\right)t$$

For $t = 0.1$ sec

$$\Delta x \approx \frac{6.6 \times 10^{-27}}{9.1 \times 10^{-28} \times 10^{-8}} \times 0.1 \approx 700 \text{ km}$$

We see that after 1 sec the electron will have spread over the entire United States. It would be equally probable to find it anywhere in the United States. It rapidly becomes a hopelessly lost electron.

In the above example we see that even if we have as good a knowledge as possible of the position and momentum of a single particle, the amount of knowledge rapidly decreases with time. In principle the present quantum theory permits an exact determination of the wave function for all future time once the initial Δx's and Δp's of all the particles are known. But this is of little help in predicting the future since the wave function rapidly spreads uniformly over the world. We rapidly reach the point of zero information on the locations of particles.

We see that modern quantum theory offers a possible way out of a philosophical "dilemma" posed by classical physics.

Q.7: In Fig. 12-13, is Δx equal to $\Delta \lambda$?

In the days of classical physics it was pointed out that if one knew the exact positions and velocities of all the particles in the universe at a time t_0, one would in principle be able to calculate the future (and past) course of the universe from the exact laws of physics. The universe was thought of as one giant machine. Using such reasoning, philosophers could conclude that all human actions (even human beings are made up of protons, neutrons, and electrons) would be completely predetermined. Of course, it was realized that such calculations of the future or past would forever be impossible because of the enormous number of particles in the universe. But still, such reasoning was bothersome to believers in free will.

As we see from the uncertainty principle, there is a more fundamental obstacle to the carrying out of such calculations, thus classical determinism is no longer "forced" upon the physicist. This does not mean that we can invoke quantum mechanics as a proof of free will.

We have encountered examples other than the uncertainty principle that equally well refute the necessity of classical determinism. For example, according to the generally accepted interpretation of quantum theory, there is no way of determining which electron will absorb a photon in the photoelectric effect. All we can do is calculate the probability that the photon will be absorbed by a given electron. The same holds for the position on the screen in Fig. 12-6 where a single electron will show up. The wave interference pattern only tells us the probability of finding a given electron for each point on the screen. The same type of phenomenon holds for the decay of a radioactive nucleus such as uranium. There is no way to tell when an individual uranium nucleus will decay. According to the quantum theory all we can ever know is the probability that it may decay in a given interval of time. The predicted probabilities can then be compared with averages of a large number of observations.

We see that in the realm of interactions and structure of small particles, the quantum theory is radically different from the classical theory. If the quantum theory is correct, as we believe, there is no hope in studying elementary phenomena and the elementary structure of matter using

Ans. 7: No, Δx is the uncertainty in position which is L.

classical physics. Therefore in the following chapters, which deal with atomic structure and associated phenomena, classical physics is abandoned and the new concepts introduced in this chapter are used.

References:
R. Feynman, *Lectures on Physics,* Vol. I, Chs. 37, 38, Addison-Wesley, Reading, Mass.
A. Kompaneyets, *Basic Concepts in Quantum Mechanics,* Reinhold Publishing Corporation, New York.

Problems

1. What is the total energy of a photon in terms of λ, h, and c?

2. What is the wavelength of a 1-Mev photon in angstroms?

3. A photon of energy 2 Mev gets converted into a positron-electron pair. If both the positron and electron have equal energies, what will be the kinetic energy of each?

4. What is the kinetic energy of a photon? Its rest mass is zero.

5. What is the inertial or relativistic mass of a photon in terms of h, λ, and c?

6. If the relativistic mass of a photon is 10^{-15} gm, what will be the numerical value of its momentum in CGS units? What is the wavelength?

7. What is the momentum of a 1-ev electron? What is its wavelength in angstroms?

8. A photon and an electron both have a kinetic energy of 1 ev. Which has the longer wavelength?

9. Write a formula for the kinetic energy of a nonrelativistic electron in terms of its mass, wavelength, and Planck's constant.

10. For each metal there is a photoelectric threshold, λ_0. Radiation of wavelength greater than λ_0 will never eject electrons. What is λ_0 for copper ($\mathcal{W} = 4.3$ ev)?

11. A photon ejects a 2-ev electron from a metal which has a 2-ev work function. What is the minimum energy of this photon?

12. The human eye can barely detect a yellow light (6000 A) that delivers 1.7×10^{-18} watts to the retina. How many photons per second does the retina receive?

13. The intensity of a wave is the square of the wave amplitude. This is also true for the matter-waves or probability-waves of the quantum theory. Suppose in the double slit experiment, the wave amplitudes at a point on the screen coming from slit A and slit B are $+3$ units and $+5$ units, respectively and the counting rate from slit A alone is 60 counts/sec.

    (a) What is the counting rate from slit B alone?

    (b) What is the counting rate when both slits are open?

14. Repeat the above problem for the case where the amplitudes at the screen from slit A and slit B are $+3$ and -5 units respectively.

15. Suppose three identical slits are used for the electron diffraction experiment and that the electron counter is at a position where all waves are in phase.

    (a) If the single slit counting rate is 100 counts per sec for each of the slits by themselves, what is the rate when all three slits are opened up together?

    (b) If the intensity of the beam from the electron gun is doubled, by what factor would the above answer be increased?

16. Thermal neutrons are in temperature equilibrium with objects at room temperature. At room temperature $kT = 1/40$ ev. The neutron mass is 1.67×10^{-24} gm. What is the average kinetic energy of a thermal neutron? What is the wavelength of such an energy neutron?

17. Two very thin slits are separated by 0.01 mm. A 1-ev electron beam impinges on these slits. A screen is 10 m behind the slits. What is the separation between successive minima at the screen?

Prob. 18

18. The figure for Prob. 18 is a plot of the wave amplitude of an electron. (Non-relativistic velocity.)

    (a) What is the momentum of the electron?

    (b) What is the kinetic energy of the electron?

    (c) What is the uncertainty in position of the electron?

    (d) What is the uncertainty in momentum of the electron?

    (e) What is the wavelength of the electron?

19. When the author first learned about neutrons, he was taught that the neutron is made up of an electron and proton held together by electrostatic attraction. Assume the neutron radius is 10^{-13} cm.

    (a) According to the uncertainty principle, what would be the ΔP of such an electron? The electron must be localized within the neutron.

    (b) The lowest average momentum such an electron could have would be $\frac{1}{2}\Delta P$. What would be the energy of such an electron in Mev? (Use the relativistic relation $W = \sqrt{(m_0 c^2)^2 + P^2 c^2}$.)

    (c) How much energy in electron volts is required to overcome the electrostatic force and move the electron from 10^{-13} cm to infinity?

    (d) From your answers to (b) and (c), what do you think of this theory of the neutron?

20. A photon can have a billard-ball-type collision with a single free electron. This phenomenon was first observed by A. H. Compton in

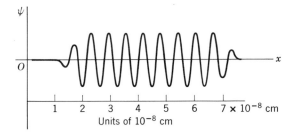

Units of 10^{-8} cm

1923 and is called the Compton effect. Since the recoil electron takes away some energy, $\frac{1}{2}mv^2$, the photon is reduced in energy from hf to hf' where $f' < f$. Assume a head-on collision (the photon "bounces" off the electron and reverses its direction by 180°).

(a) Write an equation in f, f', and v based on the conservation of momentum.

(b) Write a second equation in f, f', and v based on the conservation of energy.

Atomic theory

Chapter 13

Atomic theory

13-1 The Electron Waves in a Box

A free electron in captivity

Very crudely, an atom or molecule may be considered as a tiny box that has electrons trapped inside it. All atoms and molecules have the essential feature of holding the electron in a fixed region of space. Independent of the exact nature of such a box, it will possess certain general quantum mechanical properties that explain the classical paradoxes of why atoms radiate only discrete frequencies, and why atoms do not collapse. First we shall consider the simplest case of a single electron in an idealized box. Then we shall go on to the case of a single electron trapped in the attractive electrostatic field of a proton. This second case is that of the hydrogen atom.

Because of the wave nature of particles, a free electron confined to a box of length L must behave similarly to a sound wave bouncing back and forth in a room with 100% reflecting walls. From Chapter 10 we are familiar with the situation of a continuous wave being reflected at the end of a string or a wall. We recall that when a pure sine wave is reflected from the end of a string the result is a pattern of standing waves with a node at the reflecting surface (see Fig. 10-9). Since the probability of finding an electron just outside the box is zero, the electron wave function must go to zero at the walls of the box. The wave amplitude is determined by the requirement that the total probability for finding the electron somewhere in the box is one.

Figure 13-1 shows what the electron wave amplitudes look like for electrons traveling in the x-direction. Physicists usually use the Greek symbol ψ (psi) for the amplitude of the particle wave function. Because of the condition that the electron wave must go to zero at the walls, the only permissible electron waves are those which have an integral number of half wavelengths in the distance L. Thus $N(\lambda_N/2) = L$ or $\lambda_N = 2L/N$ where N is an integer greater than zero. We see that only certain wave functions or electron states are permitted. This is the same condition as Eq. 10-6 for standing waves on a string. Because of the de Broglie relation, Eq. 12-6, the electron can only have the corresponding

Fig. 13-1. The three lowest energy electron standing waves in a "box" of length L. Note similarity to standing waves on a string (Fig. 10-9).

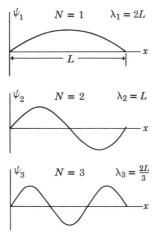

$$\psi_1 \qquad N = 1 \qquad \lambda_1 = 2L$$

$$\psi_2 \qquad N = 2 \qquad \lambda_2 = L$$

$$\psi_3 \qquad N = 3 \qquad \lambda_3 = \frac{2L}{3}$$

13

momentum values $P_N = h/\lambda_N$. Now substitute $(2L/N)$ for λ_N in the deBroglie relation and obtain $P_N = Nh/2L$ as the only possible values for the momentum. The corresponding energy values are

$$W_N = \tfrac{1}{2}mV_N^2 = \tfrac{1}{2}\frac{P_N{}^2}{m}$$

or

Energy levels in a box

$$W_N = \frac{h^2}{8mL^2}N^2 \qquad (13\text{-}1)$$

N is called the quantum number and can be any integer except zero. Thus the lowest energy the electron can ever have is $W_1 = h^2/8mL^2$. This is called the zero point energy. The electron is just not permitted to have less than this zero point energy because no state of lower energy exists. (The only lower energy state in Fig. 13-1 which goes to zero at the walls also goes to zero everywhere else in the box—corresponding to no electron being present.) In the classical theory, the electron would radiate electromagnetic waves everytime it bounced (accelerated) against the walls until it reached zero kinetic energy. Thus we begin to understand the feature of quantum mechanics that permits a hydrogen atom in its lowest energy state to be stable (it is forbidden to collapse to a state of lower energy because no such state exists).

When the laws of electrodynamics are applied to quantum mechanical phenomena, it is found that a charged particle can radiate one photon at a time. However, since our electron in the box can have only certain definite or discrete energies, the possible photon energies it can emit must also be discrete. According to the conservation of energy the frequencies of this radiation will be given by

$$hf = W_{N'} - W_N \qquad \text{where} \quad N' > N \qquad (13\text{-}2)$$

An electron of energy $W_{N'}$ can suddenly drop down to a lower energy W_N by emitting a photon whose energy is $(W_{N'} - W_N)$. For example, an electron in the $N = 4$ state would emit a photon of one of the three following energies: $W_4 - W_3$, $W_4 - W_2$, or $W_4 - W_1$. The first possibility would be followed by either $W_3 - W_1$ or $W_3 - W_2$ (which would be followed by $W_2 - W_1$). The second possibility

Q.1: What is the wavelength of the $N = 4$ standing wave for Fig. 13-1? (Fill in the box.)

$\lambda_4 = \square\, L.$

Fig. 13-2. Energy level diagram for electron in a one-dimensional box showing four lowest energies and the six possible transitions.

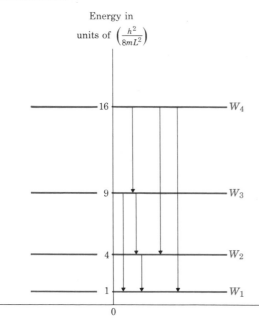

Energy in
units of $\left(\dfrac{h^2}{8mL^2}\right)$

would be followed by $W_2 - W_1$. We see that altogether there are six different possibilities or values of photon energy corresponding to six different frequencies. Hence a box containing electrons initially in the $N = 4$ state would emit a spectrum containing lines of six different frequencies. These six possible transitions are illustrated as vertical arrows in Fig. 13-2.

Now we have an inkling of the physical arguments which lead in the case of atoms to a spectrum of discrete frequencies rather than a continuous spectrum. It is all because of the wave nature of matter which requires that electrons can be put only into certain definite standing waves whether confined to a box or confined to the region around an atomic nucleus.

Example 1

An electron is trapped in a box of length 10^{-8} cm in the x-direction. For motion in the x-direction only:

(a) What is its zero point energy in ev?

(b) What wavelength "light" is emitted in the transition from the $N' = 2$ state to the $N = 1$ state?

According to Eq. 13-1 the zero point energy is

$$W_1 = \frac{h^2}{8mL^2} = 6.02 \times 10^{-11} \text{ erg} = 37.5 \text{ ev}$$

The photon energy is

$$hf = W_2 - W_1 = \frac{h^2}{8mL^2} (2^2 - 1^2)$$

$$hf = 3 \times 37.5 = 112.5 \text{ ev}$$

Using Eq. 12-4 which relates photon energy in ev to wavelength in Å: $\lambda = 12{,}390/112.5 = 110$ Å. This radiation would not be visible. It is ultraviolet radiation.

13-2 The Hydrogen Atom

A fuzzy-walled box

So far we have treated only the special case in which the kinetic energy of the electron is the same everywhere in the box. Now we shall discuss the more general case where the kinetic energy can vary as a function of the position in the "box."

As an electron "falls" toward a proton, its kinetic energy

Ans. 1: $\lambda_4 = \frac{1}{2}L$.

increases. What would be the behavior of the electron wave function? According to de Broglie, if its momentum increases, its wavelength must decrease. In general

$$\lambda = \frac{h}{P}$$

or

$$\lambda = \frac{h}{\sqrt{2m(\mathrm{KE})}}$$

For an electron-proton system, W, the sum of the kinetic plus potential energy, is constant. Then $\mathrm{KE} = W - U$ and

$$\lambda = \frac{h}{\sqrt{2m(W - U)}} \tag{13-3}$$

For the hydrogen atom

$$U = \frac{-e^2}{r}$$

The electron wavelength would then be

$$\lambda = \frac{h}{\sqrt{2m(W + e^2/r)}}$$

Thus in the general situation of a particle acted on by a force one obtains a wave function of continuously changing wavelength. In 1925, E. Schroedinger proposed an equation which permits us to construct quantitatively such a wave.* In a region of long wavelength (small KE) Schroedinger's equation gives a wave that curves gradually toward the x-axis. For larger KE the wave would curve more strongly toward the x-axis. For a fixed value of KE, such a wave would be a pure sine wave. However, Schroedinger's equation even treats the mathematical case of a negative value for $\mathrm{KE} = W - U$ (this is impossible classically). In this case the wave function curves away from the x-axis.

Figure 13-3 is a plot of the potential energy curve $U(x)$ that corresponds to the walls of a box of length L. For a perfect box with inpenetrable walls, $U(x)$ would rise to infinity at the walls. This is called a square well with infinitely high sides which we studied in Section 13-1. Now we shall con-

Fig. 13-3. (a) Plot of potential energy curve $U(x)$ corresponding to box of length L. (b) Wave function of lowest standing wave.

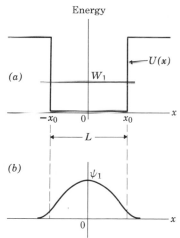

Q.2: If L were decreased in Fig. 13-3 keeping the depth of the well the same, would W_1 increase or decrease?

* For a "derivation" of Schroedinger's equation see Appendix 13-1.

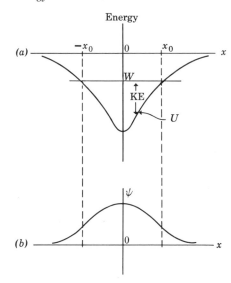

Fig. 13-4. (*a*) A plot of a potential energy somewhat similar to that of the electron in the hydrogen atom. (*b*) The corresponding lowest energy electron wave.

Ans. 2: W_1 would increase.

sider a square well with finite sides which is a more realistic model of an atom. The energy of the lowest standing wave is shown as the horizontal red line in Fig. 13-3*a*. Note that now the wave function ψ_1 does not quite go to zero at the edges, but that at $x = x_0$ there is an abrupt change in curvature. In the region outside the well ($x > x_0$), the quantity $(W - U)$ is negative and ψ curves away from the x-axis. Inside the potential well, $(W - U)$ is a constant and positive corresponding to a constant kinetic energy, hence ψ must be a pure sine wave in this region.

Next in Fig. 13-4 we go one step closer to the hydrogen atom and consider a $U(x)$ that is similar to the electron-proton potential energy, or for that matter the potential energy of an electron in a large atom containing other electrons.

We will now follow the convention that the potential energy U is zero when x is very large; that is, $U = 0$ when $x = \infty$. Hence if the total energy W is negative, then the electron will be bound to the proton. In this potential diagram the negative energy W is shown as a horizontal red line as explained on page 107. The vertical distance from this line to the curve is then $W - U$, or the kinetic energy. Note that at $x = x_0$ the KE is zero. As the electron travels from x_0 toward $x = 0$, its KE increases. Classically the region $x > x_0$ is forbidden. In this region $(W - U)$ or the KE would be negative. Figure 13-4*b* shows the corresponding wave function. Note that at $x = x_0$ the curvature changes and that in the classically forbidden region, $x > x_0$, there is still some probability of finding the electron.

If the value of W (the horizontal line in Fig. 13-4*a*) were somewhat higher, the corresponding wave function would have greater curvature. Certain discrete values of W will then give the required standing waves which must drop off to zero for large x. Figure 13-5 shows the three lowest possible standing waves. Each corresponds to a different energy, W_1, W_2, and W_3. Note that as N gets larger, so does x_0. This is a general feature of atoms—the larger the quantum number N, the larger the extent of the electron wave.

Now let us consider finally the precise potential well of an electron in the hydrogen atom. The potential is $U(r) =$

Hydrogen energy levels

Fig. 13-5. The three lowest energy wave functions corresponding to the potential energy shown in Fig. 13-4a. The dashed line shows ψ for an energy slightly higher than W_1.

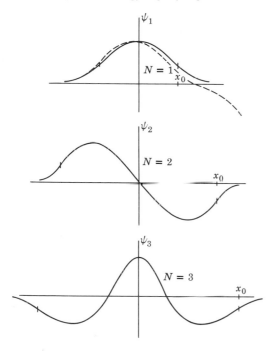

Q.3: Suppose the dashed line curve of Fig. 13-5 had been drawn for an energy slightly lower than W_1. How would it look if it started from zero at $x = -\infty$?

$-e^2/r$ and is plotted in Fig. 13-6. The energies W_1, W_2, and W_3 corresponding to the first three standing waves are shown along with their wave functions ψ. In order to get the exact values for the energy levels, we must solve Schroedinger's equation with a Coulomb potential. This is a problem in advanced calculus.* We present the exact solution:

$$W_N = - \left(\frac{2\pi^2 me^4}{h^2}\right)\frac{1}{N^2} \tag{13-4}$$

where N is any integer greater than zero. The permitted energies are all negative because of the convention that the potential energy is zero when the two charges are an infinite distance apart. The hydrogen atom electron wave functions are somewhat different than those of Fig. 13-5 for two reasons. They must be three dimensional and the shape of $U(r)$ is somewhat different from that of Fig. 13-4. Figure 13-7 is an attempt to show how three-dimensional hydrogen atoms would "look" if they could be seen.† In this figure the electron charge density or square of the wave amplitude is shown. Remember, the probability of finding a particle is the square of its wave amplitude. The views in Fig. 13-7 show electrons as if they were clouds with density proportional to the intensity of the particle wave. It is like viewing puffs of smoke. As we shall see in Section 13-6, the $N = 2$ drawings and the $N = 3$ drawings of Fig. 13-7 correspond closely to the outer electrons in the atoms from lithium to sodium.

Orbital angular momentum

Depending on its eccentricity, a classical planetary orbit of a given energy can have any value of angular momentum from 0 to mvr, where v is the circular velocity and r is the radius of the circular orbit having the specified energy (see

---

* An approximate derivation is given in Appendix 13-2.

†Occasionally we hear it said that it is impossible to draw a picture of an atom, or, that these electron clouds are not real in the sense that they can never be observed. It is true that whenever an electron is observed, it is observed at a single point rather than as an extended object. However, from another point of view, these electron clouds can be observed and hence are real. In principle the charge distribution of a certain type of atom could be measured in the same way as the charge distributions inside the proton and neutron have been measured. As explained in Chapter 15 this is done by scattering a beam of electrons on the object in question and carefully measuring the pattern of scattered electrons.

Fig. 13-6. (*a*) Plot of the electron potential energy in the hydrogen atom showing the first three energy levels. (*b*) The corresponding hydrogen wave functions (for $l = 0$).

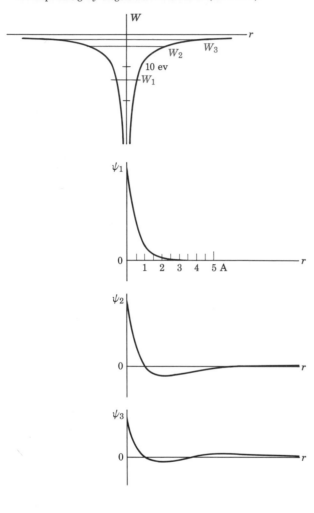

Eq. 5-1). However, a particle in a standing wave can have only certain discrete values of angular momentum in the same sort of way that the energy can have only certain discrete values. A rigorous derivation of the quantized values of the angular momentum would require the use of Schroedinger's equation and higher mathematics. However, we can get some idea of what is going on by the following plausibility argument. At a fixed value of r, let us examine the standing wave of the electron as we make one complete trip around the circumference $2\pi r$. Along this path ψ can increase and decrease in a sine wave pattern, but if it is to have the same value when we get back to the starting point, then it must have an integral number of wavelengths. Let l stand for this integer. Hence

$$2\pi r = l\lambda_l$$

But

$$\lambda_l = \frac{h}{P_l}$$

or

$$2\pi r = l\frac{h}{P_l}$$

and

$$rP_l = l\frac{h}{2\pi}$$

which is the angular momentum. This is a very general result. *Orbital angular momentum can only have the values $lh/2\pi$ where l is any positive integer including zero.* The maximum possible value of l is determined by the relation $l = 2\pi r/\lambda_l$ or by the minimum possible value of λ_l. This in turn is determined by the maximum possible curvature the wave can have, which is determined by how large the kinetic energy can be as shown in Figs. 13-5, and 6. As we might expect the result is that these "circular waves" of wavelength λ_l can have no more oscillations than the "radial waves" shown in Fig. 13-6. The result is that the maximum possible value for l is $(N - 1)$.

So far we have only discussed the magnitude of the angular momentum vector. As we might expect, not only is the magnitude quantized but also the direction. The direction is specified by another quantum number, m_l, which can be any

Ans. 3: It would never cross the x-axis and it would go to $+\infty$ as x approached $+\infty$.

Fig. 13-7. The lowest energy electron wave intensities (or charge densities) for the hydrogen atom. ▶
These drawings are projections on a plane of the three-dimensional Schroedinger wave intensities.
Side and top views are shown for each value of N, l, and m_l up to N-3.

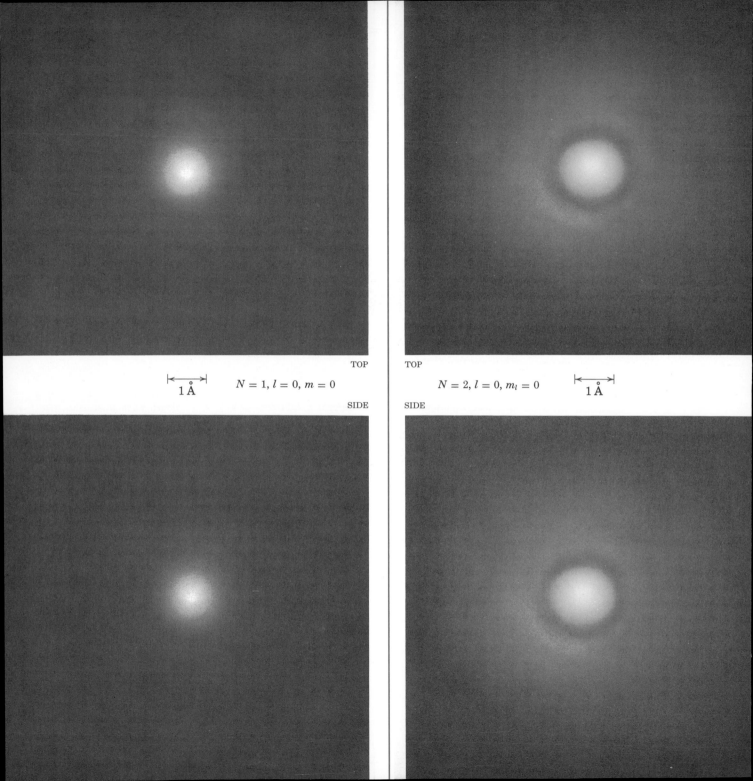

TOP

$\overset{\longleftrightarrow}{1\,\text{Å}}$ $N = 1, l = 0, m = 0$

SIDE

TOP

$N = 2, l = 0, m_l = 0$ $\overset{\longleftrightarrow}{1\,\text{Å}}$

SIDE

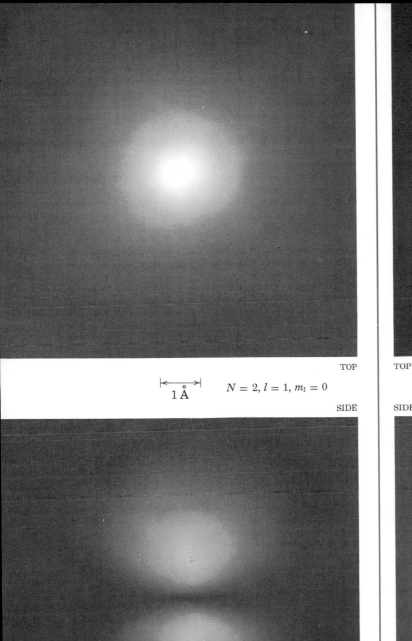
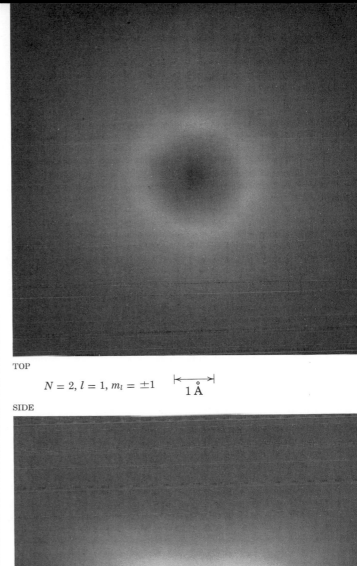

TOP

$N = 2, l = 1, m_l = 0$

TOP

$N = 2, l = 1, m_l = \pm 1$

SIDE

SIDE

1 Å

1 Å

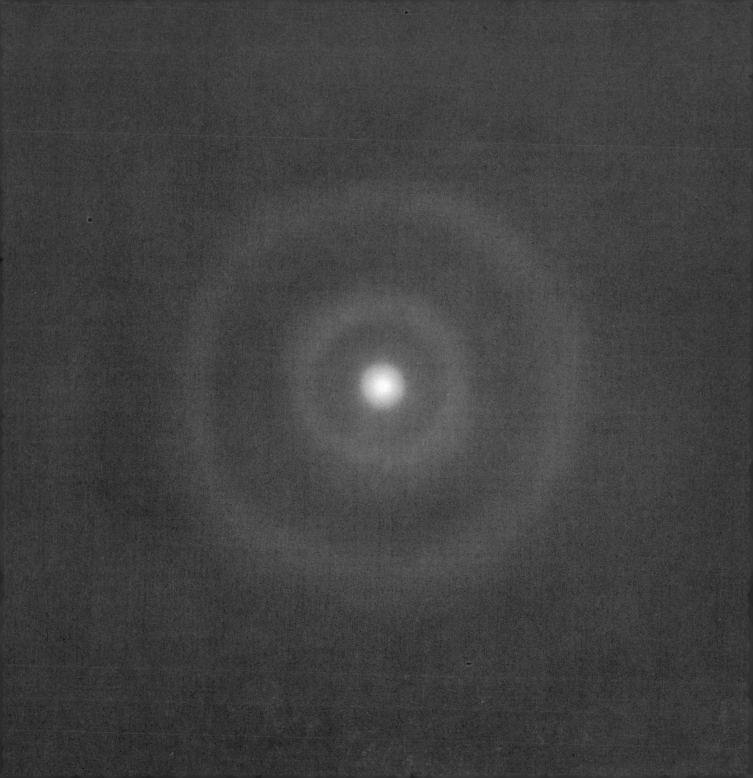

$N = 3, l = 0, m_l = 0$

|←— 1 Å —→|

Fig. 13-8. Direction of quantized angular momentum vector in terms of the quantum numbers l and m_L.

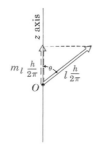

Q.4: How many different electron wave functions are there for $N = 3$?

integer from $-l$ to $+l$. The physical meaning of m_l is that the quantity $m_l h/2\pi$ is the magnitude of the component of the angular momentum vector along a specified direction (usually called the z-axis). In Fig. 13-7 this specified direction is the up-down direction. Classically the angular momentum vector would look as in Fig. 13-8 where θ is the angle between the angular momentum vector and the z-axis. The quantized angle θ is determined by the relation

$$\cos \theta = \frac{m_l \dfrac{h}{2\pi}}{l \dfrac{h}{2\pi}} = \frac{m_l}{l}$$

Altogether three quantum numbers, N, l, and m_l, are needed to specify completely a three dimensional electron wave function (or "orbit") in the hydrogen atom. Figure 13-7 shows the electron clouds corresponding to the various possibilities for l and m_l up to $N = 3$.

We see that the $(N = 2, l = 0)$ and $(N = 2, l = 1)$ standing waves look quite different, and in general we would expect them to correspond to different energies. However, the Coulomb potential $U = -e^2/r$ has the special feature that the energy depends only on N and not on l. Hence Eq. 13-4 works for all values of l having the same N. As we shall see when we get to multielectron atoms, the potential seen by the outer electron is not just that of a pure point charge at the nucleus, and then the energy does depend on l as well as N.

Example 2

How many different electron wave functions are there for $N = 2$?

For $N = 2$ there are only two possible values of l ($l = 0$ and $l = 1$). For $l = 0$, the only possible value of m_l is 0. For $l = 1$ there are three possible values of m_l (-1, 0, and $+1$). Thus there are three ($l = 1$) wave functions plus one ($l = 0$) wave function, or a total of four possible "orbits" for $N = 2$.

13-3 The Hydrogen Spectrum

A trial and error formula

In this section the atomic spectrum of hydrogen will be studied. For a review on the theory and construction of a spectrograph, see page 254.

According to the conservation of energy (see Eq. 13-2) all the lines in the hydrogen spectrum must obey the relation

$$hf = W_{N'} - W_N \qquad \text{where} \quad N' > N$$

According to Eq. 13-4

$$W_{N'} - W_N = \frac{2\pi^2 m e^4}{h^2}\left(\frac{1}{N^2} - \frac{1}{N'^2}\right)$$

Thus

$$hf = 13.6\left(\frac{1}{N^2} - \frac{1}{N'^2}\right)\text{ev} \qquad (13\text{-}5)$$

The numerical coefficient of 13.6 ev is obtained from the known values of m, e, and h.

Hydrogen gas at room temperature has virtually all of its atoms in the ground state and emits no light. However if the gas is heated to very high temperatures, some of the atoms are excited to higher energy levels. An electron in a higher energy level will make a transition to a lower level by emitting a photon of the corresponding energy difference. An atom or electron having more kinetic energy than 10.2 ev can in a collision with a hydrogen atom give up exactly 10.2 ev of its kinetic energy in exciting the hydrogen atom from W_1 to W_2. Note that $W_2 - W_1 = 10.2$ ev.

Example 3

For transitions to the ground state, what is the longest wavelength light that can be emitted? Is it visible by eye?

According to Eq. 13-4

$$hf = 13.6\left(\frac{1}{1^2} - \frac{1}{N'^2}\right)$$

The choice $N' = 2$ will give the smallest f or largest λ.

$hf = 13.6(1 - \tfrac{1}{4}) = 10.2$ ev

$\lambda = \dfrac{12{,}390}{10.2} = 1210$ Å

Since the visible spectrum is 4000 to 7500 A, this line could not be seen by eye.

The entire series of lines for $N = 1$ would form the spectrum shown in Fig. 13-9. This is called the Lyman series and was discovered in 1906 using ultraviolet spectroscopy. Notice that the series is an infinite series of lines converging on the

Fig. 13-9. The Lyman series of lines in the hydrogen emission spectrum.

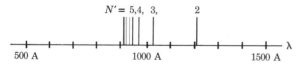

Ans. 4: The increase over $N = 2$ comes from $l = 2$ which contributes 5 possible values of m_l. Hence the total is $5 + 4 = 9$.

wavelength 908 Å which corresponds to $N' = \infty$. According to Eq. 13-4 an infinite quantum number corresponds to zero energy or an ionized hydrogen atom (a free electron and proton at large distance). Hence, if 13.6 ev of energy is given to a hydrogen atom in its ground state, the electron will have enough energy to escape from the proton. This particular value of 13.6 volts is called the ionization potential of hydrogen.

For $N = 2$, the lowest four values of $N'(N' = 3, 4, 5, 6)$ give lines in the visible region (see Figs. 13-10 and 11). In 1885 only these four lines of the hydrogen spectrum were

Fig. 13-10. All possible lines in the hydrogen spectrum up to $\lambda = 7000$A.

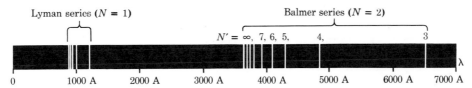

known. That year a Swiss schoolteacher, Johann Balmer, using trial-and-error discovered the following numerical relation among these four lines:

$$f = C\left(\frac{1}{2^2} - \frac{1}{N'^2}\right)$$

where C is a constant equal to 3.29×10^{15} cps. Balmer predicted there should be a fifth line corresponding to $N' = 7$ of wavelength 3970 Å, which should be at the extreme low end of the visible spectrum. This and other lines for $N' > 7$ were then promptly discovered. Balmer also speculated that the term $1/2^2$ in his formula might be replaced by $1/1^2$ or $1/3^2$ to predict even more lines in the ultraviolet and infrared regions, respectively. These two series of lines, discovered in 1906 and 1908, are called the Lyman and Paschen series, respectively. Thus in the early 1900's the numerical form of Eq. 13-5 was sitting and waiting for somebody to invent a theory that would produce this equation out of fundamental physical constants such as the charge and mass of the electron. The first man to do this was Niels Bohr in 1913.

Fig. 13-11. Photograph of hydrogen spectrum obtained using an ultraviolet spectrometer. From *Concepts of Modern Physics,* by Arthus Beiser, Copyright 1963, McGraw-Hill, Inc. McGraw-Hill Book Company. Used by permission.

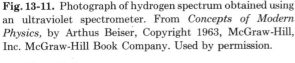

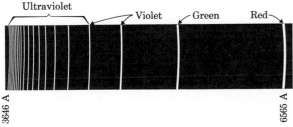

Example 4

Three lines having frequencies $f_1 = 1.2 \times 10^{15}$, $f_2 = 2.0 \times 10^{15}$, and $f_3 = 2.5 \times 10^{15}$ cps are observed in the absorption spectrum of an unknown element which we shall call Element X. The problem is to find six lines which should appear in the emission spectrum of this hypothetical element.

An absorption spectrum is obtained by passing a continuous spectrum through the gas of Element X in the ground state and appears as dark lines in a continuous spectrum. Only those frequencies are absorbed which correspond to transitions from the ground state to higher energy levels; that is, each of the above three lines is a transition from the ground state to a higher energy level. Let us call W_A the energy of the ground state, and W_B, W_C, W_D the three higher energy levels giving rise to the observed frequencies f_1, f_2, and f_3 as shown in Fig. 13-12. The transitions W_D to W_A, W_C to W_A, and W_B to W_A will appear in the emission spectrum as well as in the absorption spectrum. The three transitions W_D to W_C, W_D to W_B, and W_C to W_B do not involve the ground state and can only appear in the emission spectrum. Because of the relation $W = hf$, the energies are directly proportional to their frequencies. We can draw a vertical frequency scale and read off that the frequency for the transition $(W_D - W_C)$ is just $(f_3 - f_2) = 0.5 \times 10^{15}$ cps. The other two frequencies are

$$f_{DB} = (f_3 - f_1) = 1.3 \times 10^{15} \text{ cps}$$

and

$$f_{CB} = (f_2 - f_1) = 0.8 \times 10^{15} \text{ cps}$$

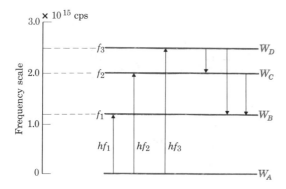

Fig. 13-12. Figure for Example 4 showing the six possible transitions between four energy levels.

13-4 The Bohr Model

The step between classical and modern

The big breakthrough in physics came about 1926 with the swift development of quantum mechanics. However, 13 years earlier Niels Bohr proposed a theory that very accurately explained the entire hydrogen spectrum and also provided a physical model for a stable atomic structure. Bohr tied together Balmer's formula and the photon concept of Einstein and Planck. He reasoned that there must be certain definite energy levels in the hydrogen atom and that according to the photon concept the differences in these energies would give rise to photons of energies hf. If Balmer's formula is multiplied by Planck's constant, we obtain

$$hf = 13.6 \left(\frac{1}{N^2} - \frac{1}{N'^2} \right) \text{ev}$$

This suggested to Bohr that the hydrogen energy levels must be $W_N = -13.6(1/N^2)$ in ev. These energy levels and the

Fig. 13-13. The five lowest energy levels of the hydrogen atom are shown in red. The corresponding quantum jumps or spectral lines have frequencies proportional to the lengths of the arrows. The first three red lines are the same as those of Fig. 13-6a.

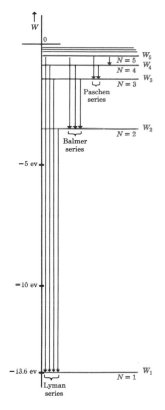

corresponding quantum jumps or spectral lines are shown in Fig. 13-13. Now Bohr was faced with the problem of how to derive these energy levels from theory alone. Bohr's theory, although now known to be wrong, is so simple and of such great historical importance that we present it here. Bohr thought of the possible electron orbits as classical circular planetary orbits and he looked for some rule that would allow only certain energies or orbit radii. The rule he invented was that the angular momentum

$$mvR = N\frac{h}{2\pi} \tag{13-6}$$

Note that the Bohr postulate differs from our present knowledge of the hydrogen atom in two ways. First, we now know that classical orbits do not apply and that the electron must be expressed as a wave. Second, we know that the angular momentum is not $N(h/2\pi)$, but it is $l(h/2\pi)$, which is always less than Bohr's value. Thus we must consider it as somewhat of a lucky accident that the Bohr theory leads to the correct hydrogen energy levels. As most of us have already experienced, it is not uncommon to get the right answer for the wrong reason. This can even happen to great men.

We shall now go through Bohr's derivation of the energy levels of an electron in the field of a nucleus of charge Ze. The Bohr postulate (Eq. 13-6) gives for the radius of the Nth orbit

$$R_N = N\frac{h}{2\pi mv} \tag{13-7}$$

Since the centripetal force is supplied by electrostatic attraction, we have

$$\frac{mv^2}{R_N} = \frac{Ze^2}{R_N^2}$$

or

$$mv^2 = \frac{Ze^2}{R_N} = -U \quad \text{(the potential energy)} \tag{13-8}$$

and

$$v^2 = \frac{Ze^2}{mR_N}$$

When we substitute the right-hand side of Eq. 13-7 into the

Q.5: In a circular Bohr orbit the magnitude of the potential energy is equal to the KE. True or false?

above equation, we have

$$v^2 = \frac{Ze^2}{m\left(\dfrac{Nh}{2\pi mv}\right)}$$

and

$$v = \frac{2\pi Ze^2}{Nh} \tag{13-9}$$

The energy level W_N is defined as

$$W_N = \text{KE} + U$$

According to Eq. 13-8, $U = -mv^2$, so the above equation becomes

$$W_N = \tfrac{1}{2}mv^2 - mv^2 = -\tfrac{1}{2}mv^2 \tag{13-10}$$

The final result is obtained by squaring the right-hand side of Eq. 13-9 and substituting it into the right-hand side of Eq. 13-10.

Energy levels of single electron and nucleus of charge Ze

$$W_N = -\frac{2\pi^2 mZ^2 e^4}{h^2}\frac{1}{N^2} = -13.6\left(\frac{Z^2}{N^2}\right)\text{ev} \tag{13-11}$$

This is the same answer as is given by the present-day theory of quantum mechanics (see Eq. 13-4). Bohr's model also gives a simple answer for the size of atoms. The formula for R_N is obtained by substituting Eq. 13-9 into Eq. 13-7. Then

Radius of Bohr atom

$$R_N = N^2\frac{h^2}{4\pi^2 Zme^2} \tag{13-12}$$

For the ground state of hydrogen ($N = 1$, $Z = 1$), $R_1 = h^2/4\pi^2 me^2 = 0.53$ Å. This is in close agreement with the size of the electron cloud predicted by quantum mechanics (see Fig. 13-7). For the $N = 2$ state, the Bohr formula predicts an orbit four times larger in diameter. This is also in good agreement with Fig. 13-7. An approximate derivation using the modern wave theory is given in Appendix 13-2.

Example 5

In the ground state Bohr atom what is the electron velocity in terms of the speed of light?

Dividing Eq. 13-9 by c and setting $N = 1$ gives

Ans. 5: False. The correct relation is $\tfrac{1}{2}mv^2 = -\tfrac{1}{2}U$.

Fig. 13-14. Scale drawing of the five lowest Bohr orbits in hydrogen showing possible electron transitions.

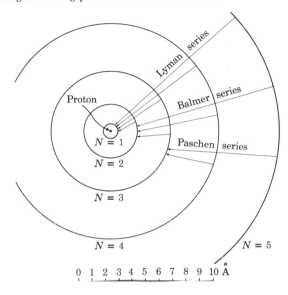

$$\frac{v}{c} = \frac{2\pi Z e^2}{hc}$$

For the hydrogen atom $Z = 1$ and $v/c = 2\pi e^2/hc = 1/137$.

Example 6

What is the relationship between the spectrum of He$^+$ and the hydrogen spectrum?

Singly ionized helium (He$^+$) consists of a helium nucleus ($Z = 2$) with just one orbital electron. Putting $Z^2 = 4$ into Eq. 13-11 gives

$$W_N = -4 \times 13.6 \left(\frac{1}{N^2}\right) \text{ev}$$

Thus the photons emitted by He$^+$ will have the energies

$$hf = 4 \times 13.6 \left(\frac{1}{N^2} - \frac{1}{N'^2}\right) \text{ev}$$

which is exactly a factor 4 greater than that for hydrogen. Similarly the Li^{++} spectrum can be obtained from the hydrogen spectrum by just multiplying all the hydrogen frequencies by the factor $Z^2 = 9$.

Bohr visualized atomic radiation as an electron suddenly "jumping" from an outer circular orbit to an inner orbit. Figure 13-14 shows the five lowest Bohr orbits for hydrogen drawn to scale. The arrows indicate the jumps corresponding to lines in the Lyman, Balmer, and Paschen series in the hydrogen spectrum. Note that the lengths of these arrows are not proportional to the photon energies as in Fig. 13-13.

The Bohr model was quite successful. It predicts the size of the atom, and the prediction checks fairly well with the experimental measurements. It yields an expression (the Balmer formula) for the wavelengths of all lines of the hydrogen spectrum, and this expression checks very closely with the measurements. It also gives the numerical value of the constant in Balmer's formula in terms of m, e, c, and h only, and it provides us with a visualizable model (which we now know is incorrect). The model explains atomic radiation as due to orbit jumps, regardless of whether the corresponding spectral lines are known or as yet beyond the detection region of existing instruments. Another success of the Bohr model is that it correctly predicts that the spectra of He$^+$, Li^{++}, Be^{+++}, etc., would have the same form as the hydrogen spectrum. According to Eq. 13-11 such frequencies would

Q.6: A nucleus of charge Z has $(Z - 1)$ singly charged negative particles attached to it. What would be the spectrum emitted by a single orbital electron?

Fig. 13-15. Niels Bohr skiing at Los Alamos. (Courtesy Mrs. Laura Fermi and University of Chicago Press.)

Ans. 6: Since the net charge of the shielded nucleus is $Z = 1$, the spectrum would be that of hydrogen.

be Z^2 times the hydrogen frequencies. This checks with experiment and the modern theory. However, a serious failure of the Bohr model is that it is unable to explain the spectra of un-ionized atoms starting with helium (a nucleus of $Z = 2$ surrounded by two electrons). Although complicated by three mutually interacting particles, the helium energy levels can be calculated using quantum mechanics. Using the modern theory and electronic computers, the helium spectrum has now been calculated to great accuracy and found to check with experiment. Physicists and chemists are confident that present-day quantum mechanics can in principle explain all atomic spectra and chemical properties.

13-5 Atomic Structure

"The underlying physical laws necessary for the mathematical theory of a large part of physics and the whole of chemistry are [now] completely known."
P. M. Dirac (1929)

In an atom with many electrons Bohr postulated that only 2 electrons can be in the $N = 1$ shell, 8 in the $N = 2$ shell, 8 in the $N = 3$ shell, 18 in the $N = 4$ shell, 18 in the $N = 5$, and 32 in the $N = 6$. This was to explain the periodicity 2, 8, 8, 18, 18, 32 observed in the chemical and physical properties of the elements (see Figs. 13-18 and 13-21 and also Table IV) at the end of the book.

Figure 13-16 shows pictures of such atoms constructed by Bohr's enlarged theory. These rules of how many electrons go in each shell were "arbitrarily" made up by Bohr to help explain the chemical properties and ionization potentials of the different elements. As we shall soon see, Bohr's prediction for the number of electrons in $N = 3$ and $N = 4$ shells was incorrect. Physicists begin to suspect a theory when it depends on too many "arbitrary" postulates such as Bohr's theory of the atoms. Most physicists and philosophers feel that nature must be simple and that the smaller the number of postulates or fundamental principles, the closer we are to physical reality.

To the best of our present knowledge modern quantum theory, which is based on very few postulates, explains all of atomic structure and all of chemistry. Since atoms consist of many electrons, we must first discuss what happens in quantum mechanics when there is more than one identical particle trying to occupy the same region in space at the same time. According to classical physics, no two bodies can occupy the same space at the same time. However, this classical concept is completely foreign to quantum mechanics. If electromagnetic waves can pass through solids as they do, then certainly photons can occupy the same space as other particles. In fact, as many photons can be crowded into the same quantum state as is desired. But then why are not all the electrons of all atoms normally in the $N = 1$ state? This would surely be the lowest possible energy state for an atom.

The exclusion principle

In 1925 W. Pauli observed that if one postulated that no more than two electrons can occupy the same state or electron orbit, then the electron structure of atoms would automatically be explained. For example, in the $N = 2$ shell, the angular momentum quantum number l can be 0 or 1. But for $l = 1$ there are three possible states corresponding to the quantum number $m_l = -1$, 0, and $+1$. Thus there are a total of four states which, if they can hold two electrons each, make up a "shell" of 8 electrons total. Back in 1925 the Pauli exclusion principle was just another "arbitrary" postulate.

Electron spin

However, in 1926 it was discovered that every electron has an intrinsic angular momentum of magnitude $\frac{1}{2}(h/2\pi)$. This can be visualized by thinking of an electron as a "spherical" mass spinning around an axis of rotation with a fixed angular momentum of $\frac{1}{2}(h/2\pi)$. This intrinsic spin can never be increased or decreased. It is unique for each type of elementary particle. Shortly after the discovery of electron spin P. M. Dirac, Pauli, and others found it possible to work out a satisfactory theory of what we call spin-$\frac{1}{2}$ particles. One of the requirements of such a wave theory of spin-$\frac{1}{2}$ particles is that the quantum mechanical equations should predict the same

Fig. 13-16.

Scale drawings of the electron orbits of some atoms according
to the Bohr theory in 1923.

Bohr's picture of the radium atom.
▼

10^{-8} cm

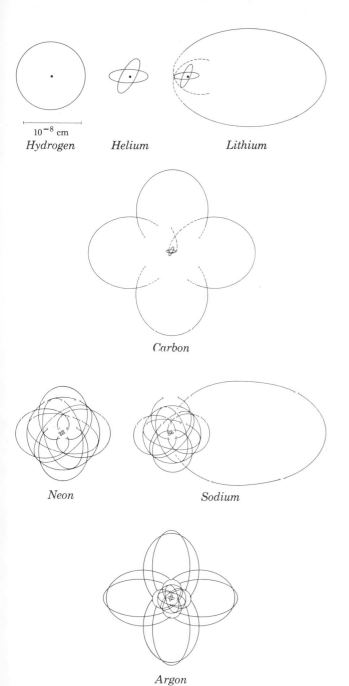

10^{-8} cm

Hydrogen　　　　*Helium*　　　　*Lithium*

Carbon

Neon　　　　*Sodium*

Argon

physical results independent of the velocity of the observer. When the relativistic equations were found which met these requirements, they also automatically obeyed the Pauli exclusion principle. So it turns out that the exclusion principle is not another arbitrary postulate pulled out of the air, but it seems to be a direct consequence of the intrinsic spin of the electron. Just why the electron has the mass, spin, and charge that it does are problems still to be solved.

We shall now go through part of the periodic table element by element.

13-6   The Periodic Table of the Elements

The physics of chemistry

The purpose of this section is to illustrate how we can deduce chemical properties of the elements without resorting to chemistry experiments. In principle, all of chemistry can be deduced from the theory of the spin-$\frac{1}{2}$ electron. In actual practice, however, the chemical properties of all the elements are determined by experiment; the calculations are too difficult to bother with.

Now that we are armed with the exclusion principle, we can specify where each electron goes in an atom. For example, consider a bare neon nucleus ($Z = 10$). If it is given just one electron, the electron will quickly drop down to the $N = 1$ orbit. The same is true for a second electron. These 2 electrons completely fill up the $N = 1$ orbit. If the 8 remaining electrons are given to the neon nucleus containing two $N = 1$ electrons, these 8 electrons will completely fill up the four possible $N = 2$ orbits. These 4 orbits are the ($l = 0$), the ($l = 1$, $m_l = -1$), the ($l = 1$, $m_l = 0$), and the ($l = 1$, $m_l = +1$). We shall now systematically describe the atomic structures predicted by the quantum theory, starting with hydrogen. We shall see that without resorting to detailed calculations, it is possible to give numerical estimates of the valences and ionization potentials of each element.

$Z = 1$ (*hydrogen*)

We have completed our discussion of this case. The single electron is in the $N = 1$ state which has an energy of

Fig. 13-17. Wolfgang Pauli. (Courtesy American Institute of Physics.)

−13.6 ev. Thus the binding energy or ionization energy is 13.6 ev. Since ionization potential is the minimum voltage required to ionize an atom, the ionization potential of hydrogen is 13.6 volts. The ionization potentials of the elements are plotted versus Z in Fig. 13-18. Note the periodicity pattern 2, 8, 8, 18, 18, 32.

$Z = 2$ (helium)

We have already noted that the helium ion He$^+$ which has a single electron must have the same energy levels as hydrogen, only four times as great. This is due to the factor Z^2 in Eq. 13-11. The ionization potential of He$^+$ would be 4×13.6, or 54.4 volts. This checks with experiment.

If a second electron is brought near He$^+$ it will first see an object that appears to have a charge of $(Z - 1)$. But when this second electron gets down to the $N = 1$ shell, half the time it is closer to the nucleus than the original electron and then it sees a nuclear charge of Z. The straight average of these two values is $(Z - \frac{1}{2})$. Thus we predict that the effective nuclear charge will be $Z_{\text{eff}} = 1.5e$ for an electron in helium. We can generalize Eq. 13-4 to read

$$W_{N,l} = -13.6 \frac{Z_{\text{eff}}^2}{N^2} \text{ ev}$$

where Z_{eff} depends both on N and l. From this estimate of Z_{eff}, we would expect the ionization potential of helium to be about $(1.5)^2 \times 13.6$, or 30 volts. Actually, because of the positive potential energy of repulsion of the two electrons we would expect the binding to be not quite so large. Experimentally the ionization potential of helium is 24.6 volts. This is the highest ionization potential of any element. Helium is very inert chemically because of its large ionization potential and because there is no more room for a third electron in the $N = 1$ shell. No chemical forces are strong enough to supply 24.6 ev in order to form to positive ion He$^+$. If we tried to form the negative ion He$^-$, the extra electron must be in a $N = 2$ standing wave which is well outside both the nucleus of charge $+2e$ and the two $N = 1$ electrons of negative charge. Hence the net charge at the center of the $N = 2$ wave is zero and there is no attractive potential available to hold

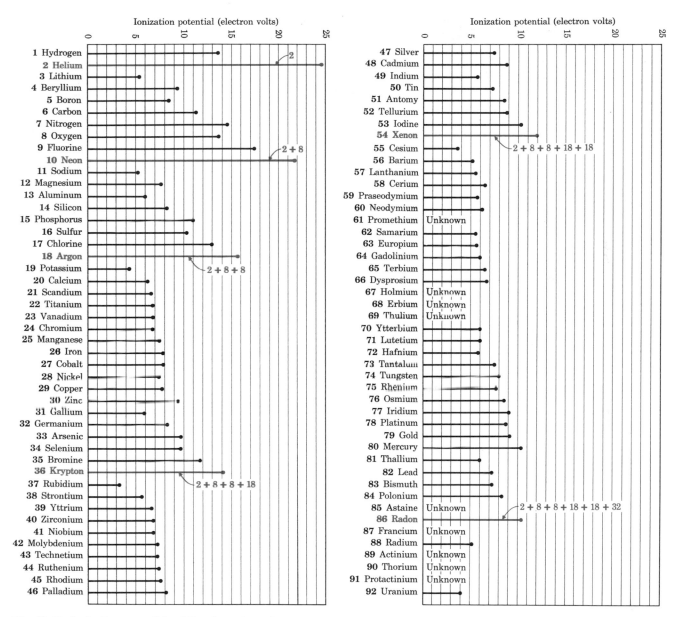

Fig. 13-18. Ionization potentials of the elements up to uranium are presented in this chart; the noble gases are shown in red.

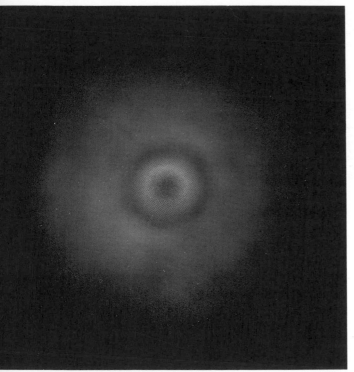

Fig. 13-19. Modern picture of a lithium atom. The $N = 1$ electron cloud is in red and the outer electron ($N = 2$) is in gray.

$\longleftarrow$ 1 Å $\longrightarrow$

an $N = 2$ wave; that is, for $N = 2$, $Z_{\text{eff}} \approx 0$. Consequently helium does not form molecules with any element. It and the other closed shell atoms are called the noble gases. Several of the heavier noble gases can form certain special compounds.

$Z = 3$ (*lithium*)

Doubly ionized lithium Li^{++} will have a hydrogen-like spectrum with the energy levels $(3)^2$ or 9 times that of hydrogen. Singly ionized lithium has a helium-type spectrum with a Z_{eff} of about $(3 - \frac{1}{2})$ instead of the $(2 - \frac{1}{2})$ for helium. Because of the exclusion principle, neutral lithium must have its third electron in the $N = 2$ shell. For this electron Z_{eff} should be somewhat larger than one. Thus we expect the ionization potential of lithium to be somewhat more than $13.6/N^2 = 13.6/2^2 = 3.4$ volts. The experimental value is 5.4 volts which corresponds to a Z_{eff} of 1.25. The second ionization potential (for removing the second electron) is 75.6 volts. Thus lithium should always appear in compounds with a valence of $+1$ (gives up one electron), never with $+2$ (gives up two electrons). Figure 13-19 is a "picture" of how the lithium electron clouds should "look."

What is the l quantum number of the outer electron in lithium? According to all that we have learned so far, the ($N = 2$, $l = 0$) and ($N = 2$, $l = 1$) states should have the same energy. However, it can be seen by referring to Fig. 13-7 that the $l = 0$ state should be more strongly bound than the $l = 1$ state. This is because the lower angular momentum state ($l = 0$) has more of its electron wave near the nucleus than higher angular momentum states. In fact, all electron waves having l greater than zero have $\psi = 0$ when $r = 0$. This is also shown in Fig. 13-20 which compares the $l = 0, 1$, and 2 waves for the same N. That part of an electron wave near the nucleus sees a Z_{eff} almost as large as Z, whereas that part of the wave far from the nucleus sees a Z_{eff} of about 1. Hence a $l = 0$ wave sees a higher Z_{eff} than an $l = 1$ wave. This is why Z_{eff} is a function of l as well as N. This effect can cause an appreciable energy difference between the $l = 0$ and $l = 1$ or 2 "subshells." In fact, for $Z = 19$

Fig. 13-20. $N = 3$ hydrogen wave functions for $l = 0$, 1, and 2. Note that only the $l = 0$ wave is nonzero at the origin.

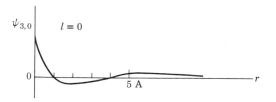

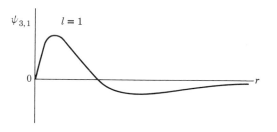

(potassium) the effect is so strong that the ($N = 4$, $l = 0$) energy level is lower than the ($N = 3$, $l = 2$) level. Table 13-1 shows the order in which energy levels occur. Another way to think of this effect is that a higher angular momentum orbit is more circular and consequently farther away from the nucleus than a lower angular momentum orbit. Hence the lower "l" states are more strongly bound.

$Z = 4$ (beryllium)

According to the exclusion principle there is room for two electrons in the ($N = 2$, $l = 0$) state. Because Z_{eff} for that part of the electron wave near the nucleus is now greater than for that of lithium, the ionization potential should be larger. The experimental value is 9.32 volts as compared to the 5.39 volts for lithium. However, for beryllium the second ionization potential is not much larger since this electron also comes from an $N = 2$ state. Thus beryllium has a valence of $+2$ in compounds.

$Z = 5$ (boron), $Z = 6$ (carbon), $Z = 7$ (nitrogen), $Z = 8$ (oxygen), $Z = 9$ (fluorine), and $Z = 10$ (neon)

These atoms are formed by filling up the ($N = 2$, $l = 1$) states. Since $l = 1$ occurs in three different states ($m_l = -1$, 0, $+1$), this ($N = 2$, $l = 1$) subshell can take up to 6 electrons. Boron would have 3 electrons in the $N = 2$ states, and consequently a valence of $+3$.* For oxygen and fluorine there is a new phenomenon called electron affinity. A single fluorine atom can take up an extra electron and become stable Fl⁻. This extra electron which sees a large Z_{eff} over part of its wave is bound with an energy of 4.2 ev. Thus the valence of fluorine is -1. The electron affinity for forming O⁻ is 2.2 volts. Oxygen and nitrogen commonly appear in chemical compounds with valences of -2 and -3, respectively. In neon the $N = 2$ states are all filled up and we have a closed shell. Because part of the $N = 2$ electron waves get fairly close to the nucleus (where now Z_{eff} goes up to 10), the ioniza-

---

* See page 362 in Chapter 14 for a more detailed discussion of why boron has valence 3, carbon has valence 4, etc.

tion potential is high, 21.6 volts. Thus we would expect neon, like helium, to be quite inert chemically.

$Z = 11$ (sodium) to $Z = 18$ (argon)

In sodium the exclusion principle forces the 11th electron to go into the $N = 3$ wave with a $Z_{\text{eff}} \simeq 1$ which is a much larger diameter wave than the $N = 2$ of the preceding element neon. Hence, the theory predicts that every time the outer electron is in a higher N orbital, the atomic size will be significantly larger. These abrupt increases in size are observed to occur for $Z = 3, 11, 19, \ldots$ as shown in Fig. 13-21. In the series of eight elements—sodium to argon—the ($N = 3$, $l = 0$) and ($N = 3$, $l = 1$) states are filled up in exactly the same way as the preceding 8 elements. The chemical properties are thus quite similar to the corresponding 8 preceding elements. This is the explanation of the "periodic system" of chemistry. So far we have explained the periodicity 2, 8, 8. Now we shall see why the next period must be 18 rather than 8.

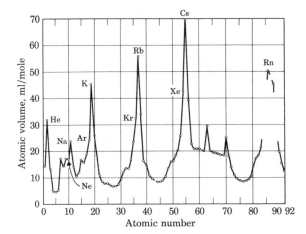

Fig. 13-21. Atomic volume as a function of Z.

$Z = 19$ (potassium) and up

We might expect that the next element would have its outer electron in the ($N = 3$, $l = 2$) state. However, as mentioned in the discussion of $Z = 3$ (lithium), the ($N = 4$, $l = 0$) wave sees a considerably larger Z_{eff} than the ($N = 3$, $l = 2$) wave because the $l = 0$ wave is most concentrated at $r = 0$ where the effective charge is the greatest. For this ($N = 4$, $l = 0$) wave, $Z_{\text{eff}} = 2.26$ and the binding energy is $13.6 \ Z_{\text{eff}}^2/4^2 = 4.34$ ev; whereas for ($N = 3$, $l = 2$), Z_{eff} is somewhat less than 1.7 corresponding to a binding energy less than 4.34 ev. If the 19th electron were put into the ($N = 3, l = 2$) state, it would quickly drop down to the lower energy state ($N = 4$, $l = 0$) emitting a photon equal to the energy difference.

When we get to $Z = 21$ (scandium), the ($N = 4, l = 0$) wave is all filled up and now the ($N = 4, l = 1$) competes with the ($N = 3, l = 2$) for the 21st electron. Now as we would expect the ($N = 3$) state is the lower one, so at scandium the $l = 2$ states of the $N = 3$ shell start filling up. Table 13-1

TABLE 13-1 ELECTRONIC STRUCTURE OF ATOMS*

Principal Quantum Number N		1	2		3			4	
Azimuthal Quantum Number l		0	0	1	0	1	2	0	1
Letter Designation of State		1s	2s	2p	3s	3p	3d	4s	4p
Z	Element	V_i volts							
1 H	Hydrogen	13.60	1						
2 He	Helium	24.58	2						
3 Li	Lithium	5.39		1					
4 Be	Beryllium	9.32		2					
5 B	Boron	8.30	Helium core	2	1				
6 C	Carbon	11.26		2	2				
7 N	Nitrogen	14.54		2	3				
8 O	Oxygen	13.61		2	4				
9 F	Fluorine	17.42		2	5				
10 Ne	Neon	21.56		2	6				
11 Na	Sodium	5.14				1			
12 Mg	Magnesium	7.64				2			
13 Al	Aluminum	5.98				2	1		
14 Si	Silicon	8.15				2	2		
15 P	Phosphorus	10.55	Neon core			2	3		
16 S	Sulfur	10.36				2	4		
17 Cl	Chlorine	13.01				2	5		
18 A	Argon	15.76				2	6		
19 K	Potassium	4.34							1
20 Ca	Calcium	6.11							2
21 Sc	Scandium	6.56						1	2
22 Ti	Titanium	6.83						2	2
23 V	Vanadium	6.74						3	2
24 Cr	Chromium	6.76						5	1
25 Mn	Manganese	7.43						5	2
26 Fe	Iron	7.90						6	2
27 Co	Cobalt	7.86	Argon core					7	2
28 Ni	Nickel	7.63						8	2
29 Cu	Copper	7.72						10	1
30 Zn	Zinc	9.39						10	2
31 Ga	Gallium	6.00						10	2	1
32 Ge	Germanium	7.88						10	2	2
33 As	Arsenic	9.81						10	2	3
34 Se	Selenium	9.75						10	2	4
35 Br	Bromine	11.84						10	2	5
36 Kr	Krypton	14.00						10	2	6

*From Charlotte E. Moore, *Atomic Energy Levels*, Vol. II, National Bureau of Standards Circular 467, Washington, 1952.

shows the electron structure of the elements up to $Z = 36$. The electron structures have been determined using quantum mechanics, chemistry, and spectroscopy for all the elements up to $Z = 104$. In fact, the theory is so good that the chemical properties of $Z = 105$, 106, etc., can be predicted in advance, before they are produced artificially.

To calculate precisely the ionization potentials and electron affinities requires an enormous amount of work; but we have the theory and it can be done in principle, thus explaining all of chemistry with one simple theory, the quantum mechanics of the spin-$\frac{1}{2}$ electron.

13-7 X-rays

The filling up of an empty hole

Each atom has a characteristic spectrum that is emitted when a small sample of the element is ionized by placing it in an electric arc or discharge. Unknown samples, whether on the earth or in stars, can be analyzed using the technology of spectroscopy. As in the case of hydrogen, the spectral lines correspond to quantum jumps between the various energy levels available for the outer one or two electrons. Since these energy levels are on the order of a few electron volts, the characteristic spectrum will consist of lines in the visible, ultraviolet, and infrared regions.

However, it is also possible for a heavy atom to emit much more energetic photons of hundreds or even thousands of electron volts energy. Such high-energy photons are called x-rays. An x-ray is emitted when an inner electron is missing from an atom. Then an outer electron will quickly jump down to replace the missing inner electron. Missing electrons can be produced by bombarding a sample with a beam of electrons that have been accelerated through several thousands of volts. Some of these beam electrons will collide with inner shell electrons of the sample and knock them out of their respective atoms. An electron of the $N = 1$ shell (the K shell in x-ray terminology) has an energy of $13.6\,Z_{\text{eff}}^2$ electron volts. In this instance, Z_{eff} is very close to $(Z - \frac{1}{2})$.

Example 7
What are the highest energy x-rays which can be emitted by copper and by uranium? Give the photon energies and wavelengths.

The highest energy x-rays occur when a free electron (zero energy) jumps all the way down to a vacancy in the K shell ($N = 1$).

For copper, $Z = 29$ and $hf = 13.6 \times (28.5)^2 = 11,000$ ev.

$$\lambda = \frac{12,390}{11,000} = 1.12 \text{ A}$$

For uranium, $Z = 92$ and $hf = 13.6 \times (91.5)^2 = 112,000$ ev.

$$\lambda = \frac{12,390}{112,000} = 0.11 \text{ A}$$

The above example illustrates a very reliable method that has been used to determine the Z of newly discovered elements. One merely determines the wavelengths of high-energy x-rays emitted from the unknown sample when it is bombarded with electrons.

Since x-rays have wavelengths comparable to the interatomic spacings in solids, they are also a very useful tool for the determination of the structure of solids. As mentioned in previous chapters, the periodically repeating planes of atoms in a crystal behave as the rulings of a diffraction grating. Thus if the x-ray wavelength is known, the interatomic spacings can be determined by measuring the x-ray diffraction angles (see Eq. 10-10 and Fig. 12-13).

Appendix 13-1

We can obtain Schroedinger's equation as follows. Referring to Eq. 3-10, we see that a pure sine wave as a function of time has the property that

$$\frac{\frac{d^2y}{dt^2}}{y} = -\frac{4\pi^2}{T^2}$$

where d^2y/dt^2 is the second derivative with respect to time (the acceleration) and T is the time of one complete sine wave (the period). Replacing time t with distance x, we see that a pure sine wave as a function of distance obeys the relation

$$\frac{\frac{d^2y}{dx^2}}{y} = -\frac{4\pi^2}{\lambda^2}$$

or

$$\frac{d^2y}{dx^2} = -\frac{4\pi^2}{\lambda^2}y$$

Then a particle wave of wavelength λ must obey the equation

$$\frac{d^2\psi}{dx^2} = -\frac{4\pi^2}{\lambda^2}\psi$$

Now we substitute the right-hand side of Eq. 13-3 for λ and obtain

$$\frac{d^2\psi}{dx^2} = -\frac{4\pi^2 m}{h^2}(W - U)\psi$$

which is the Schroedinger equation. In calculus the second derivative $d^2\psi/dx^2$ is a measure of the curvature. If it is negative, it means the slope is decreasing, or that ψ is curving toward the x-axis. If $d^2\psi/dx^2$ is positive, then ψ is curving away from the x-axis. Mathematically there is always a solution to the Schroedinger equation, whether or not the right-hand side is positive or negative. This means ψ can exist even when $(W - U)$ or the kinetic energy is negative.

Appendix 13-2

We shall make an approximate calculation of the radius and energy of the hydrogen atom using the modern wave theory of the electron rather than the older, semiclassical theory of Bohr. We start by noting that the electron wave function consists of a standing wave which is maximum in the center and drops off at the "edges" of the atom. Although this standing wave does not have any exact wavelength, we will assign it an average wavelength, λ_{av}, and note that the lowest order standing wave would be $\frac{1}{2}\lambda_{av}$ across the "box" or atom. Hence

$$\tfrac{1}{2}\lambda_{av} = 2R_{av} \qquad \text{or} \quad \lambda_{av} = 4R_{av} \tag{13-13}$$

where $2R_{av}$ is the average diameter of the atom. We bring in quantum theory by using the de Broglie relation as expressed in Eq. 13-3:

$$\lambda_{av} = \frac{h}{\sqrt{2m\text{KE}_{av}}}$$

Both sides of this equation can be expressed in terms of R_{av}. For the left-hand side use Eq. 13-13, and for the right-hand side, substitute $(Ze^2/2R_{av})$ for KE_{av} (as given by Eq. 13-8). Then one obtains

$$(4R_{av}) = \frac{h}{\sqrt{2m(Ze^2/2R_{av})}}$$

Squaring both sides gives

$$16R_{av}^2 = \frac{h^2}{mZe^2/R_{av}}$$

Solving for R_{av} gives

$$R_{av} = \frac{h^2}{16mZe^2} \qquad (13\text{-}14)$$

Note that this approximate result is close to the Bohr theory result $h^2/4\pi^2mZe^2$ given in Eq. 13-12.

To get the binding energy we use

$$W = KE + U = KE + (-Ze^2/R)$$

and we again make the approximation that $KE_{av} = Ze^2/2R_{av}$ to obtain

$$W = -\frac{Ze^2}{2R_{av}}$$

Now substitute the result for R_{av} in Eq. 13-14 to obtain

$$W = \frac{8mZ^2e^4}{h^2}$$

This agrees with the exact answer ($W = -2\pi^2mZ^2e^4/h^2$) within a factor of 2.5. This is a reasonable agreement for such an approximate calculation.

Problems

1. What is the minimum energy that can be absorbed by an H atom that starts out in its ground state?

2. What is the ground state Bohr radius in He^+?

3. A one-dimensional "box" is 10^{-8} cm long. What are the wavelengths of the electron standing waves corresponding to the three lowest possible momenta?

4. Make qualitative sketches of ψ_4 and ψ_5 for the potential energy diagram shown in Fig. 13-2.

5. Using the ideas of both *relativity* and *quantum mechanics,* calculate the de Broglie wavelength of an electron traveling at $v = \frac{4}{5}c$.

6. In the ground state of the Bohr model of the hydrogen atom
  (a) What is the kinetic energy in ev?
  (b) What is the potential energy in ev?
  (c) What is the binding energy in ev?
  (d) If the electron in the hydrogen atom was at rest, but still at the distance $R = \frac{h^2}{4\pi^2me^2}$, what would be the energy required to remove the electron?

7. What is the number of electrons in
   (a) a $N = 2$ closed shell.
   (b) a $N = 3$ closed shell.
   (c) a $N = 4$ closed shell.

8. What are the maximum number of electrons in a $N = 6, l = 2$ subshell?

9. In the Bohr theory of the hydrogen atom, which is greater in magnitude: $(Ke + U)$, KE, or U?

10. At $Z = 56$ (barium) the $N = 6, l = 0$ subshell becomes filled up. In the next element ($Z = 57$, lanthanum), the $N = 4, l = 3$ subshell starts getting filled up. How many electrons will it take to completely fill up the $N = 4, l = 3$ subshell? (These elements are called the rare earths. They all have similar chemical properties because they all have the same two outer electrons ($N = 6, l = 0$).)

11. A sample of hydrogen gas is excited to $N = 5$. What is the total number of lines that can appear in the emission spectrum of this gas?

12. Assume the ground state for the outer electron of a certain atom is ($N = 3, l = 0$) with $Z_{eff} = 3.2$. Assume the next highest energy level for this atom is ($N = 3, l = 1$) with $Z_{eff} = 1.6$. What energy photon is emitted in a transition from the ($N = 3, l = 1$) level to the ground state?

13. The following three lines are observed in the absorption spectrum of Element X: $f_1 = 2.2 \times 10^{15}$ cps, $f_2 = 3.0 \times 10^{15}$ cps, and $f_3 = 3.5 \times 10^{15}$ cps. (An absorption spectrum appears as dark lines in a continuous spectrum, and is obtained by passing a continuous spectrum through gas in the ground state. Only those frequencies are absorbed which correspond to transitions from the ground state to higher energy levels; that is, each of the above three lines is a transition from the ground state to a higher energy level.)
   (a) Will the above three lines also appear in the emission spectrum of Element X?
   (b) List the frequencies of three other lines appearing in the emission spectrum of Element X.

14. The absorption spectrum of hydrogen is obtained by passing a continuous spectrum of radiation through hydrogen gas in the ground state. What are the photon energies in the absorption spectrum of hydrogen? What are the wavelengths of the dark lines?

15. The highest energy x-rays emitted by an unknown sample have a wavelength of 2.16 A. Which element is this sample?

16. The energy required to completely ionize helium is 79 ev.
   (a) How much energy is required to remove the second electron after the first one has already been removed?
   (b) How much energy is required to remove only one of the two electrons?

17. A certain atom has all of its electrons in closed shells except for one outer electron. Suppose the only possible energy levels for this outer electron are -1, -4, -7, and -10 ev.

    (a) What is the ionization potential?

    (b) How many different lines could appear in the emission spectrum? (Do not count the same frequency more than once.)

    (c) What would be the wavelengths of the lines in the emission spectrum?

18. The ionization potential of a certain atom is 9 volts. The first two excited states of this atom are $E_2 = -8$ ev and $E_3 = -6$ ev.

    (a) What is the ground state energy E_1 in ev?

    (b) Give the photon energies in ev of 3 lines in the spectrum of this element. Only use lines with λ greater than 3000A.

19. Arrange the following orbitals in order of increasing energy (write the lowest or most strongly bound energy level first). Assume the inner shells are all filled. $(N = 3, l = 1)$; $(N = 3, l = 2)$; $(N = 4, l = 0)$; $(N = 4, l = 1)$.

20. A particle of mass m and positive charge q is in a circular orbit of radius R about a heavy particle of mass M and charge $-Q$. The only appreciable force between them is an attractive electrostatic force.

    (a) Write down an equation for the force between them.

    (b) What is the potential energy of q in terms of its mass m and its velocity v?

    (c) What is the total energy (KE + U) in terms of m and v?

    (d) Using the Bohr postulate to quantize the angular momentum, give an expression for v in terms of R, m, h, and the quantum number N.

    (e) What is the quantized radius in terms of m, h, Q, q, and N?

    (f) What is the total energy in terms of m, h, Q, q, and N?

21. Find a line in the He$^+$ spectrum having the same wavelength as a line in the hydrogen spectrum. What is this wavelength?

22. A mu-mesic atom consists of a nucleus of charge Z with a captured μ^- meson (a particle 207 times as heavy as an electron in the ground state.

    (a) What is the binding energy of a μ^- captured by a proton?

    (b) What is the radius of the $N = 1$ Bohr orbit in Part (a)?

    (c) What energy photon is emitted when the above μ^- jumps from $N = 2$ to the ground state?

23. Consider a lead $(Z = 88)$ mu-mesic atom. Is the $N = 1$ Bohr orbit for the mu-meson inside or outside the nucleus? The diameter of the lead nucleus is 1.6×10^{-12} cm.

24. Consider element Q, a hypothetical atom of valence $+1$. The binding energy of the outer electron is 3.2 ev. It is also known that the energy levels for 3 excited states of the outer electron are -1.0 ev, -1.4 ev, and -2.0 ev.

    (a) What is the energy level of the ground state in ev?

    (b) List all the lines that should appear in the emission spectrum of element Q. Specify the photon energies in ev.

25. The most intense x-ray line is due to the transition of an electron in the $N = 2$ shell to fill the hole in the $N = 1$ shell. Assume Z_{eff} for the $(N = 2)$ state is $(Z - 3)$. What is the wavelength of this line for copper?

26. The attractive force between a neutron (mass M) and an electron (mass m) is given by $F = GMm/R^2$. Let us consider the lowest possible Bohr orbit for this electron circling around the neutron.

    (a) Write down the centripetal force in terms of m, R, and v where v is the electron circular velocity.

    (b) What is the kinetic energy in terms of G, M, m, and R?

    (c) What is the potential energy in terms of G, M, m, and R?

    (d) What is the total energy in terms of G, M, m, and R?

    (e) Write down an equation expressing the Bohr postulate for quantization of the orbits.

    (f) What is the $N = 1$ radius in terms of h, G, M, and m? What is the numerical value?

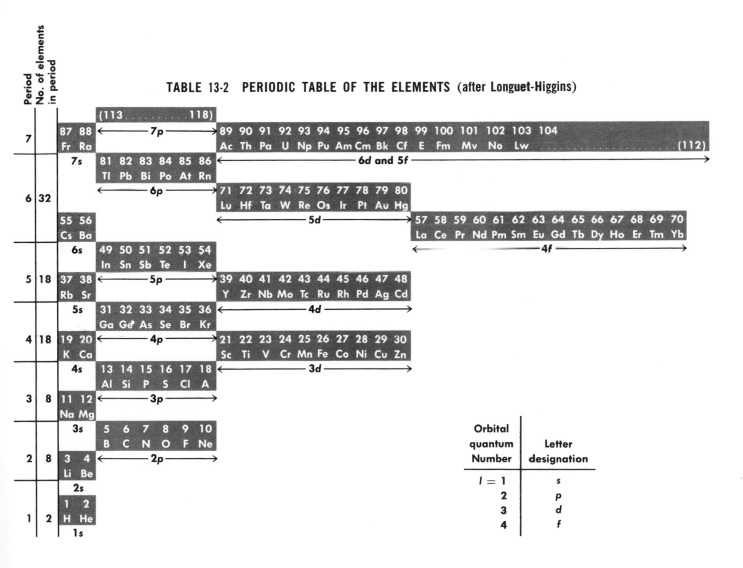

TABLE 13-2 PERIODIC TABLE OF THE ELEMENTS (after Longuet-Higgins)

Orbital quantum Number	Letter designation
$l = 1$	s
2	p
3	d
4	f

Linus Pauling

The structure of matter

Chapter **14**

The structure of matter

The explanation of the structure and properties of matter involves an understanding of the interactions among atoms, electrons, and molecules at small distances. As we have seen, such interactions are explained by the quantum theory. Thus we might expect some of the "peculiar" aspects of the quantum theory to show up in our study of the structure of matter. Indeed, in this chapter we will find examples of "peculiar" quantum phenomena showing up on a large-scale as well as on an atomic scale.

14-1 Molecular Theory

Electron stealing vs. electron sharing

We shall discuss two different mechanisms that cause atoms to bind together into molecules: ionic binding and covalent binding.

Ionic binding

Ionic binding is due to Coulomb's law. For example, there will be an attractive force between positive and negative singly charged ions. If the electrostatic energy of attraction e^2/R exceeds the amount of energy needed to form the two ions, a stable molecule is possible. As an example, consider a molecule of potassium chloride, KCl. The energy needed to form the two ions K^+ and Cl^- is 0.52 ev. This is because the ionization potential of potassium is 4.34 volts and the electron affinity of chlorine is 3.82 ev. By electron affinity we mean that 3.82 ev of energy is given off when an electron is added to a chlorine atom. The net energy required to form the two ions is then 4.34 minus 3.82, or 0.52 ev. If the two ions can get close enough so that e^2/R is greater than 0.52 ev, a stable molecule will be formed. This electrostatic potential energy can be estimated by putting R equal to the sum of the radii of the two ions. K^+ and Cl^- both have a radius of about 1.5 A. Then the e^2/R contribution would be 4.8 ev. Thus we would expect the energy of formation of a KCl molecule to be approximately 4.8 minus 0.52 ev. The measured value of 4.4 ev checks quite well with our crude estimate. The binding of most inorganic compounds is due to ionic binding which can be thought of as electron stealing.

14

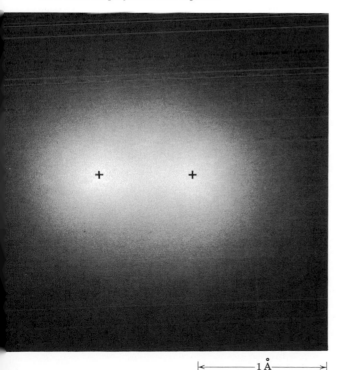

Fig. 14-1. The electron charge density in H_2. The amount of whiteness is proportional to the square of the electron wave function projected onto a plane.

|←————1 Å————→|

Covalent binding

Another very prevalent mechanism for molecular binding that occurs in most organic molecules is called covalent binding. A covalent bond is sharing of electrons by two atoms. The simplest example of covalent binding is the hydrogen molecule. First we shall consider the ionized hydrogen molecule H_2^+. This consists of two protons surrounded by an electron cloud. The binding energy of the electron in the presence of two protons is of course larger than in the presence of one proton only. On the other hand, the electrostatic repulsion of the two protons tends to oppose the binding. However, since the electron wave gets in close and tends to concentrate between the two protons, the effect of the electrostatic attraction of the electron to the two protons dominates. The binding energy of the hydrogen atom to the hydrogen ion in H_2^+ is 2.65 ev; that is, $(2.65 + 13.6)$ ev is required to dissociate completely an H_2^+ ion into two protons and one electron.

According to the Pauli exclusion principle there is room for a second electron to fit in the same electron wave as the first electron. This system of two electrons and protons is the neutral hydrogen molecule. Here the electron wave function is somewhat more spread out than that of the single electron in H_2^+ because of the electrostatic repulsion between the two electrons. The binding energy of the two hydrogen atoms in the neutral H_2 molecule is 4.48 ev. The electron cloud (square of the wave function) for the two electrons of the H_2 molecule would look like Fig. 14-1 if "it could be seen by eye."

Carbon atoms usually form covalent bonds. The carbon atom has a tendency to share with four other electrons in an attempt to fill up its $N = 2$, $l = 1$ shell. The simplest such case is CH_4 (methane) shown in Fig. 14-2. As with the hydrogen molecule, the electron waves tend to concentrate between the positive charges where they will make the greatest contribution to the binding energy. Since these four electron clouds are mutually repulsive, the lowest energy configuration is achieved when they are the farthest from each other as in Fig. 14-2. The shapes of electron waves in molecules are determined by the Schroedinger equation along with the condition that the energy levels be as low as possible.

Fig. 14-2. A qualitative drawing showing the electron wave structure of methane. The hydrogen nuclei are at the corners of a regular tetrahedron, or at 4 of the 8 corners of a cube. The electrons are concentrated along the carbon-hydrogen lines.

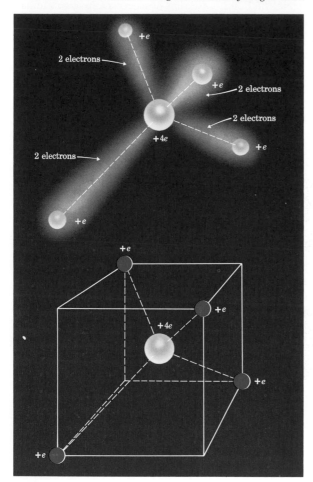

Hybridization

We see that when forming molecules, the four $N = 2$ electrons of carbon can extend out and play equal roles giving carbon a valence of 4. But how could this be, when the first two $N = 2$ carbon electrons are more strongly bound and in an $l = 0$ closed subshell? Similarly, why should boron ($Z = 5$) with its one $l = 1$ electron not have a valence of 1 rather than 3? The answer is that atoms in molecules are not the same as when they are alone. In the pure atom, there is a distinction between the $l = 0$ and $l = 1$ electrons. As emphasized in Chapter 13, the $l = 0$ electrons are indeed more tightly bound than the $l = 1$ electrons. But, molecules are not pure atoms. The CH_4 molecule, for example, has in addition to the carbon nucleus, four other centers of positive charge. With a system of five atomic nuclei, the solution to Schroedinger's equation for standing waves is now more complicated, and gives the result for the ground state shown in Fig. 14-2. This phenomenon of putting the $l = 0$ and $l = 1$ electrons on an equal footing in molecule formation is called hybridization.

It is not difficult to see how our familiar solutions to the Schroedinger equation can generate an electron standing wave that extends to one side as do those of Fig. 14-2. This can be done for the four $N = 2$ electrons of carbon by describing each outer electron wave function as a mixture of the $l = 0$ and $l = 1$ waves of carbon. An $l = 0$ wave function by itself is shown schematically in Fig. 14-3a. Here the value of ψ must be of the same sign over the entire central region (we choose the positive sign). The ($N = 2$, $l = 1$) wave function will be of opposite signs in the two ends of the "dumbbell" as indicated in Fig. 14-3b. In looking at the sum (Fig. 14-3c), we see that the upper parts add constructively and the lower parts of opposite sign tend to cancel each other out. The result is a lopsided lobe which extends out in one direction from the carbon nucleus. Chemists call this an *sp orbital* (s stands for $l = 0$ and p stands for $l = 1$).

The study of organic chemistry is concerned with molecules containing carbon atoms and covalent bonds. Carbon atoms have the special characteristic that they do not mind sharing electrons with each other. Thus many of the mole-

Fig. 14-3. Diagram showing how addition of $l = 0$ and $l = 1$ wave functions give a electron cloud projecting out mainly to one side. (a) represents the ($N = 2$, $l = 0$) spherical wave function. (b) represents the ($N = 2$, $l = 1$) dumbell-shaped wave function. (c) is the sum of the two above wave functions. Here the lower lobe of ψ_{21} is mainly cancelled out by ψ_{20} because these parts are of opposite sign.

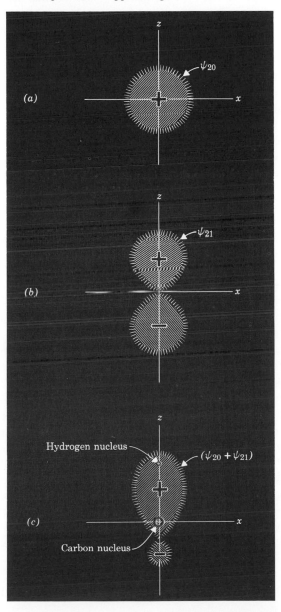

cules in organic chemistry contain long chains of carbon atoms. These long chains often contain thousands of atoms. The opening figure for this chapter shows a small section of such a molecule. Many of them occur in nature as products of living processes. With the help of quantum mechanics and advanced techniques of physics and chemistry, it is now possible to determine the exact structure of giant protein molecules that are basic to life. In fact, it has recently become easier to understand how some of these complicated molecules are able to reproduce themselves. A more complete understanding of the "secret" of life is, of course, one of the problems for the future. Perhaps some day scientists will be able to synthesize new forms of life out of completely inanimate matter.

14-2 Crystalline Solids

Super molecules

Most compounds and elements when in the solid state occur with the atoms in a periodic lattice that appears as a pure crystal by eye. The same mechanisms that bind atoms together into molecules can also bind them into an unlimited periodic structure or "super molecule." Many of these substances do not appear to the eye as obvious crystals because they are made up of many tiny crystals (polycrystalline structure). Some solids and liquids, especially at low temperatures, exhibit "peculiar" large-scale properties characteristic of the quantum theory. The study of the properties and phenomena of solids and liquids is called solid state physics and is at present one of the major fields of physics research.

Just as there are ionic and covalent molecules, there are also ionic and covalent crystals. Figure 14-4 shows the structure of the NaCl ionic crystal. Notice that each Na^+ ion has 6 Cl^- ions for its nearest neighbors. This type of spatial configuration of Na^+ and Cl^- ions has the lowest energy (it gives off the most heat during formation) of all possible other configurations. This explains the tendency for NaCl and many other substances to form pure crystals as they are cooled below their "freezing points." As the temperature is

Fig. 14-4. The crystal structure of NaCl. The small circles indicate the positions of the centers of the Na and Cl atoms. (*a*) Centers of atom. (*b*) Entire atoms.

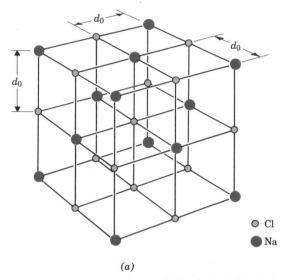

d_0

○ Cl
● Na

(*a*)

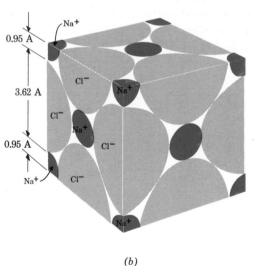

0.95 A

3.62 A

0.95 A

Na⁺
Cl⁻
Na⁺
Cl⁻
Na⁺
Cl⁻
Na⁺
Na⁺

(*b*)

raised, the thermal kinetic energy eventually becomes large enough to overcome the binding into a regular crystal, and the crystal melts.

Metallic binding

There is another type of binding called metallic binding which occurs only for elements having few enough outer electrons to satisfy the following description. This type of binding occurs when the atoms are pushed together closer than the size of the electron cloud of the outer electrons. Because of the exclusion principle, such a configuration will tend to raise the outer electrons to higher energies. However, in the case of the substances called metals this configuration still has a lower energy than if the atoms were held farther apart. Let us consider a substance such as lithium, potassium, or sodium which has only one outer electron. Figure 14-5 shows the outer electron cloud of the *free* lithium atom. The positions of the neighboring nuclei of the lithium lattice are shown as small *x*'s. If the atoms are crowded together so that their inner closed shells are touching, the neighboring nuclei will be inside what was the outer electron cloud for the free atom. In such a case the outer electron is attracted by the neighboring nuclei, which both increases its binding energy and spreads out its size even more. This permits it to be near even more remote neighbors, which in turn "pull out" the electron cloud even more. As one might expect, the end result is that each outer electron wave function gets uniformly spread out over the entire crystal! Thus a single electron cloud can be as large as the Empire State building or as long as the longest wire.

To a first approximation, the attractive forces of the nuclei on an outer electron can be averaged out to correspond to a uniform attractive potential energy, the magnitude of which we shall call U_0. A plot of this averaged potential energy, called a potential well, is shown in Fig. 14-6. Each outer electron is described by a standing wave confined to this potential well. We now see that the hypothetical example in the previous chapter of an electron trapped in a box is not so hypothetical after all.

We can begin to see that quantum theory provides a

Fig. 14-5. Outer electron cloud of the free lithium atom. The positions of the neighboring lithium nuclei in metallic lithium are indicated as x's. Note that the neighboring lithium nuclei are inside the outer electron cloud.

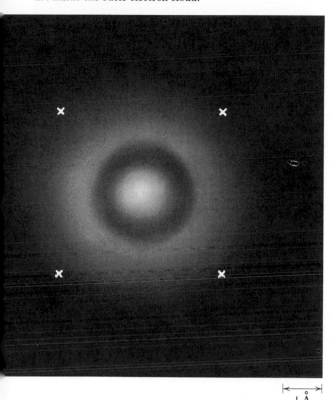

$$\overset{\longleftrightarrow}{1\text{ Å}}$$

Energy of Fermi level

Q.1: Does the wavelength of the lowest energy conduction electron depend on the size of the piece of metal?

reasonable explanation of why metals conduct electricity and why other substances do not (or almost do not). That metals contain at least one "free" electron per atom is due in part to the wave nature of the electrons. These "free" or conduction electrons are not bound to any particular atom and are free to flow anywhere in the metal as discussed on page 367. In ionic and covalent crystals the outer electrons are bound to their particular atoms; hence these crystals generally do not conduct electricity. They are called insulators. The fact that pure metallic crystals can have free electrons should be considered as a large-scale quantum mechanical phenomenon. Classically each electron would belong to its own atom.

14-3 Fermi Gas of Electrons

A peculiar gas imbedded in solid matter

We have learned that a metal of volume V can be considered as a box of volume V holding N electrons. Because of the exclusion principle only two of these electrons are permitted in each of the states specified by Eq. 13-1. All N of these electrons try to crowd into the lowest energy states, forming what is known as a Fermi gas. Such a gas has certain interesting, nonclassical properties first pointed out by Enrico Fermi. The N electrons will fill up all the energy states from the lowest state to a state of kinetic energy KE_0 called the Fermi level. If N is known, one should be able to calculate KE_0. The calculation of KE_0 in terms of $\mathfrak{N}$, the number of electrons per unit volume, is presented in the Appendix to this chapter. The result is

$$KE_0 = \frac{h^2}{8m}\left(\frac{3}{\pi}\,\mathfrak{N}\right)^{2/3} \tag{14-1}$$

where $\mathfrak{N}$ is the number of free electrons per cubic centimeter. As one might expect, the result is independent of the particular shape or volume of the piece of metal. It depends only on how tightly the free electrons are crowded together.

Example 1

The density of lithium is 0.534 gm/cm³. What is the Fermi level of the conduction electrons in lithium in ev?

Fig. 14-6. Solid curve is the approximate potential energy of an outer electron as it crosses the surface of a metal.

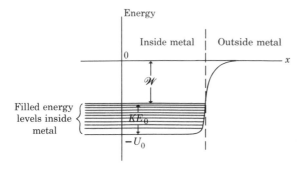

Lithium has only one outer electron, hence $\mathfrak{N}$ is the number of atoms per cubic centimeter. Since the atomic weight of lithium is 6.94, there will be $N_0 = 6.02 \times 10^{23}$ atoms in 6.94 gm of lithium. Then $N = 0.534/6.94 \times 6.02 \times 10^{23} = 4.63 \times 10^{22}$ free electrons per cubic centimeter and

$$KE_0 = \frac{h^2}{8m}\left(\frac{3 \times 4.63 \times 10^{22}}{\pi}\right)^{2/3} = 7.55 \times 10^{-12} \text{ erg}$$

$$KE_0 = 4.7 \text{ ev}$$

The approximate potential seen by an electron at the edge of a metal is shown in Fig. 14-6. We define zero energy as the energy of a free electron at rest outside the metal. The energy levels of the electron Fermi gas are indicated as fine horizontal lines starting at $-U_0$ and going up an energy interval KE_0 from the bottom of the potential well. The minimum energy required to remove an electron from the metal is then $U_0 - KE_0$. By definition this is the work function $\mathfrak{W}$ defined in the section on photoelectric effect. Although an electron at the Fermi level has kinetic energy $= KE_0$, its total energy is

$$W = KE + U$$
$$= KE_0 + (-U_0)$$
$$= -\mathfrak{W}$$

Hence the Fermi level is at $-\mathfrak{W}$ on the energy scale. Actually, one has a sharply defined work function only at absolute zero. At a temperature T degrees absolute, the electrons are in thermal equilibrium, so that a significant number of them should have some thermal energy in addition to their Fermi energy. As shown in Chapter 6 the average thermal energy per particle in a classical gas is $(\frac{3}{2})kT$. In a Fermi gas, only the particles with kinetic energies close to KE_0 can have thermal energy. Thus there will be some electrons with kinetic energies a little higher than KE_0; at room temperature kT is 0.025 ev, whereas KE_0 and $\mathfrak{W}$ are on the order of several electron volts.

Contact potential

Whenever two dissimilar metals are joined, a potential difference appears between them. Now that we know about

Ans. 1: Yes. In the one-dimensional case the relation is $N\left(\dfrac{\lambda_N}{2}\right) = L$ where L is the length of the "box".

Fig. 14-7. When two dissimilar metals are joined, electrons will flow until the Fermi levels line up as in (c).

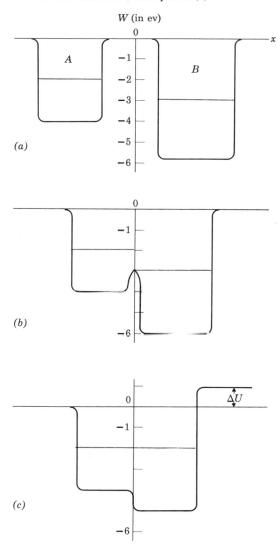

W (in ev)

(a)

(b)

(c)

Q.2: In Fig. 14-7(c) which side is more positive, A or B?

Fermi levels, we can explain this phenomenon with the use of potential well diagrams. Consider two different metals A and B as in Fig. 14-7a. The Fermi levels of A and B are at -2 and -3 ev respectively, and the electron potential energies inside the two metals are -4 and -6 ev respectively—all with respect to the energy of an electron at rest just outside the metal. Figure 14-7b shows the situation just as the two metals are first brought into contact. Now electrons in A are free to move to B, where lower energy states are available. But as electrons move into B, it rapidly acquires a negative charge with respect to A. Now more work must be done to bring a negative electron to the negatively charged metal B; in other words, the entire potential energy diagram of B is raised with respect to A. This process continues until the Fermi levels meet as shown in Fig. 14-7c. This equilibrium situation is achieved after a very small fraction of the conduction electrons have moved from A to B. As seen in Fig. 14-7c, the potential energy difference is ΔU which must be equal to the initial energy difference of the Fermi levels; that is, if the Fermi levels of two metals differ by 1 volt, then, when brought together, there will be a potential difference of 1 volt between the two metals.

14-4 Electrical Conductivity

Quantum mechanics on a large scale

We have seen that in a perfect metallic crystal lattice the outer electrons behave as free electrons in a box. Since these electrons can carry electrical current, we would expect the electrical resistance of a perfect metal to be zero. However, real metals have impurities and lattice imperfections. Since the condition for obtaining free electrons was that the lattice be perfect with no impurities, a "free" electron can interact and lose energy to imperfections and impurities. The electrical resistance in ohms depends on the mean free path for electron collisions with the imperfections and impurities. From this theory of electrical resistance, we can easily derive Ohm's law as was done in Chapter 9 on page 219. Ohm's law states that the resistance is completely independent of the value of the current and only dependent on the temperature.

The increase of electrical resistance with temperature is easily seen according to this theory of electrical conduction. An intrinsic source of lattice imperfection is the vibrational motion of the atoms due to the fact that they are not at absolute zero. Hence we have predicted that the resistance of a pure metal should increase with increased thermal motion of the atoms. The theory predicts the resistance should approach zero as the temperature approaches absolute zero. This prediction agrees with experiment.

Superconductivity

The fact that a pure metal can have zero resistance or infinite conductivity at absolute zero should not be confused with a different quantum phenomenon called superconductivity. Superconductivity is infinite conductivity at temperatures several degrees above absolute zero. Actually quite a few metals have this strange property of superconductivity. Once a circular current has been started up in a superconductor, it should keep going by itself until the cooling system breaks down. Such currents have kept going by themselves for years in the laboratory. The quantum mechanical explanation of superconductivity is one of the current problems in theoretical solid state physics. Recently, considerable progress has been made in the understanding of this most remarkable phenomenon. A very brief description of the theory goes as follows. Below a certain temperature, the disturbance of the lattice by a conduction electron is greater than the thermal motion of the lattice disturbing the electron. The disturbance of the lattice by electron A will show an effect on the motion of electron B. The net effect is an effective attractive force between electrons A and B, which in some materials is greater than the electrostatic repulsive force. Hence, if both electrons are set in motion in the same direction (a net current), this will be the lowest energy state available for the electrons and they must stay in that state because there is no lower state available for them; consequently, there will be a permanent net electron current in their direction of motion.

Ans. 2: Electrons left side A making it more positive in net charge.

14-5 Semiconductors

Freedom increases with temperature

There are some nonmetallic covalent crystals, such as silicon and germanium, where the wave functions of the outer electrons are not quite pulled out over the entire crystal, but are only pulled out as far as the atom's nearest neighbor. However, for silicon and germanium, the first excited state of an outer electron of the atom consists of an electron cloud large enough to meet the condition that it be pulled out over the entire crystal. As we have seen, whenever the size of the electron cloud becomes so great that it envelops several of the nuclei of the crystal, the wave is pulled out over the entire crystal. Silicon and germanium with all their electrons in the lowest energy states are covalent crystals or insulators. However, if some of the outer electrons can be excited into the next highest energy state (called the conduction band), these excited electrons suddenly become free electrons and the crystal can now conduct electricity. In germanium 0.72 ev is required for an outer electron to be excited to the conduction band. At room temperature ($kT = 0.025$ ev) only an extremely small fraction of electrons will have this much thermal energy. However, even though the fraction is very small, at least there will be some conduction electrons, with the result that the conductivity of germanium will be many times that of an insulator or of germanium at absolute zero. This fact explains why germanium is called a semiconductor.

The presence of a small amount of impurity can increase the conductivity of a semiconductor enormously. For example, a few parts per million of arsenic can increase the conductivity of germanium by a factor of 1000 at room temperature. A comparison of the electron structures (Table 13-1) of arsenic and germanium makes this seem reasonable. Arsenic has one more electron than germanium and, because of the Pauli principle, this electron must already be in the next highest energy state. So the outer electron of an arsenic atom imbedded in a germanium crystal should essentially be in the conduction band. Germanium is deliberately manufactured with a controlled amount of arsenic impurity. This

Q.3: Can a superconductor with a small amount of impurity still have *exactly* zero resistance?

is called *n*-type germanium (*"n"* for negative). Also germanium crystals can be grown with gallium impurity. The gallium atom captures an electron away from a neighboring germanium atom leaving an electron hole which can move around from one germanium atom to another. This hole behaves as a positive carrier of electric current. Germanium seeded with gallium is called *p*-type (*"p"* for positive).

A *p-n* junction consists of a semiconductor having regions of *p*- and *n*-type impurities adjacent to each other. Then the semiconductor has the unusual property that its electrical resistance is hundreds of times larger in one direction than the other. Such a device is called a crystal diode and has electrical characteristics similar to those of a diode vacuum tube; that is, if a positive voltage is applied to the *p*-type side of a *p-n* junction, a large current will flow, whereas if a negative voltage is applied, a small current will flow. As with vacuum tube diodes, a *p-n* junction has a low resistance if the voltage is applied in one direction, and a high resistance if applied in the other direction. This effect can be understood by referring to Fig. 14-8. In Fig. 14-8*a* are two separate samples of germanium, one seeded with arsenic donors (*n*-type) and the other with gallium acceptors (*p*-type). Then there will be a difference in Fermi levels depending on the amount of seeding or doping. Let us consider an example where the difference in Fermi levels is 0.2 ev. In Fig. 14-8*b*, where the two samples are joined, the Fermi levels will line up and a contact potential of 0.2 volt will be developed as explained in Section 14-3. If the two sides are connected by a wire, there will be equal and opposite electron drift currents, I_p and I_n. I_p is very small because there are very few electrons which have the 0.72 ev of thermal energy to get up into the conduction band. On the *n*-type side there are many electrons in the conduction band due to the arsenic donors, however, very few of them have the 0.2 ev of kinetic energy required to climb up the potential hill which is of that height. Now if the *p*-type side of the diagram is lowered (by applying a positive voltage) a larger number of the conduction electrons will have enough energy to climb the reduced potential hill. I_n will increase greatly and the net current will be large as shown in Fig. 14-8*c*. However, if the *p*-type side is raised

Ans. 3: Yes. Otherwise such currents would not last undiminished for years as has been observed. Any laboratory sample will have a small amount of impurity. Also a small amount of impurity does not negate the quantum mechanical explanation.

Fig. 14-8. Electron currents in n-p junction. (a) and (b) No applied voltage, (c) Forward voltage, and (d) Back voltage. Note that applied voltage V causes a difference in Fermi levels eV. Net current I is conventional positive current.

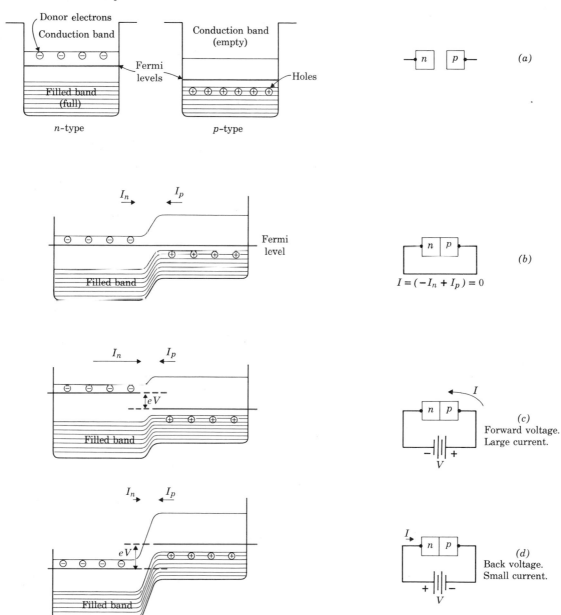

as in Fig. 14-8d, the hill is raised higher than 0.2 ev and there are fewer electrons able to make it, with the result that the net current is small.

The theory of semiconductors was so well understood in 1949 that J. Bardeen, W. H. Brattain, and W. Shockley invented a crystal that should behave as a triode electron tube. They then proceeded to manufacture these semiconductor crystals, now called transistors. This development of vast technological importance never would have been possible without an understanding of the theory of conductors and semiconductors. A transistor consists of a *p-n-p* or a *n-p-n* junction with three electrical connections. We will not take the time here to go into the detailed explanation of the theory of the transistor; however, the reader can get some idea of the principles involved by referring again to Fig. 14-8 and imagining a third piece of germanium at the right with a fixed voltage applied across the three pieces. In Fig. 14-8c when an additional positive voltage is applied to the middle piece, a large current is available to flow to the right, whereas when a negative voltage is applied to the center, only a small current could make its way through to the right. So here, as in the triode, we have the situation where a small change in signal voltage can cause large changes of current in the external circuit. In this case the "signal" voltage was applied to the center element of a *n-p-n* transistor, whereas in the vacuum tube triode it is applied to the control grid. By now transistors have replaced triodes in many electronic circuits. They have the advantage that no heated cathode is required and they can be run on small voltages and small batteries.

Another important property of semiconductors is photoconductivity. This mechanism can be thought of as the photoelectric effect completely imbedded in a solid. Photons in the visible spectrum (or even in the infrared) can be absorbed by the ground state outer electrons. The electron now has the additional energy of the photon, which is sufficient to raise it to the conduction band. When light is shined on a photoconductive cell, its electrical resistance abruptly decreases. A most practical photo detector is a *p-n* junction of high resistance (back-voltage applied). Now if light is shined on it, the change in current is especially striking; the electrons excited by the light have enough energy to pass the small barrier.

Another of the many applications of semiconductors is the solar battery. If light shines on a crystal diode, both holes and free electrons are created. The currents of these will not be symmetric however because of the enormous differences in resistances for flows in the two directions across a *p-n* junction. Thus more negative charge will flow one way and more positive charge the other, and a potential difference will be produced.

14-6 Superfluidity

The easy way to fill a glass

Another strange quantum mechanical phenomenon that occurs near absolute zero is the superfluidity of liquid helium. As helium gas is cooled down it liquefies at 4.2° absolute. As the liquid is cooled down further, it suddenly changes its properties at 2.2° absolute. Then large-scale phenomena occur that are completely contrary to common experience. For example, a partially filled vessel that is open at the top will quickly empty itself of this strange form of liquid helium (called helium II). The explanation is that the liquid crawls up the inside surface of the vessel (no matter how tall it is) over the rim and down the outside. For the same reason the reverse phenomenon also occurs (see Fig. 14-9). If an empty glass is partially immersed, the helium will quickly creep up the glass as shown until the beaker is filled to the same level. As the temperature is decreased below 2.2° K, the amount of helium II increases. At absolute zero it would be all helium II. A strange property of pure liquid helium II is that it cannot exert forces on anything. A high pressure firehose shooting a stream of this liquid could not even knock over a coin balanced on edge. The liquid would freely flow around the coin without exerting any net force on the coin. Could a fish swim in liquid helium II? No, because it would freeze. But even a hypothetical nonfreezing fish could not swim because it would have nothing to push against. All it could do would be to obey Newton's first law of motion. Physicists express these strange properties mathe-

Fig. 14-9. Arrows represent surface film creep of liquid helium II into empty vessel. The surface acts as a syphon.

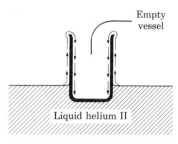

Empty vessel

Liquid helium II

Q.4: The usual symbol for a *p-n* junction diode is ⊣◀⊢. The convention is that the "easy" direction for current flow is along the arrow. Then which is the "*p*" and which is the "*n*" side?

matically by saying the viscosity of helium II is zero. The intriguing problem is why is the viscosity zero? Like superconductivity, the peculiar properties of liquid helium are now actively under study. Considerable progress has been made on the theoretical understanding of the superfluidity of liquid helium II.

14-7 Lasers

Stimulated atoms work together

Until the discovery of lasers in 1960, man was unable to shine a spot of light on the moon and detect it, or, at the other extreme, to obtain a tiny spot of concentrated light no more than a wavelength in diameter. Laser beams of about 5×10^8 watts power over very short periods of time have been achieved. And since this energy can be concentrated in a spot the size of a wavelength of light, enormous temperatures can be achieved which will instantly vaporize any substance. Ordinary sources of light emit independent photons from individual atoms in the source. Using lenses it is impossible to focus all the energy from such a source into a spot smaller than the size of the source itself. However, if one could have a pure, continuous sine wave source of light, a simple lens will focus to a spot size of about one wavelength. It is possible to obtain such a pure source of light by making use of what is called stimulated emission. There are two mechanisms by which an excited atom can emit its photon. One is the usual random process called spontaneous emission, and the other is stimulated emission. In this process an excited atom in the presence of external radiation from other similar atoms can be stimulated to emit its photon *in phase* with the external radiation. In other words, if an atom is hit by a photon like the one it would emit, the atom will emit it and will do so in phase with the "parent" photon.

Hence if a source of excited atoms can be continuously supplied between two parallel mirrors, there will be a buildup of photons all in phase with each other bouncing back and forth; that is, a continuous sine wave of electromagnetic radiation at the characteristic frequency of the atomic radiation. In practice, a partially reflecting mirror is used so that

Ans. 4: The "*p*" is on the right side and "*n*" on the left.

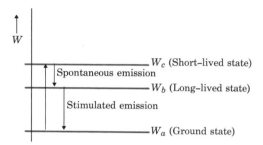

Fig. 14-10. Energy level diagram for typical laser.

a continuous stream of pure sine wave light leaks out. The only problem left is how to supply the continuous source of excited atoms. This can be done by what is called optical pumping. Light of a higher frequency from an outside source is continuously supplied to a sample of atoms which have energy levels similar to those in Fig. 14-10. This higher frequency corresponds to transitions from the ground state W_a to W_c. Now the excited atoms will either re-emit the same frequency photon by dropping back down to the ground state, or else emit a lower frequency photon by dropping down to W_b. In certain materials, there is a higher probability to drop down quickly to W_b rather than all the way to the ground state, and, when at W_b, the probability is then relatively low to give a spontaneous emission down to W_a. Such a material is ruby crystal (chromium ions trapped in aluminum oxide). It is possible by fast optical pumping to get more of the chromium atoms in state W_b than in the ground state. Then the first few spontaneously emitted photons corresponding to $W_b - W_a$ will quickly stimulate other excited atoms to emit their photons in phase.

For chromium ions the frequency $f = \dfrac{W_b - W_a}{h}$ corresponds to red light. Since 1960 many other substances, both solids and gases, have been used to emit laser beams over a wide range of frequencies in the infrared and visible spectrum. The excited states can be produced by electrical and chemical means as well as by optical pumping. Even certain types of *p-n* junctions will emit laser beams if opposite crystal surfaces are made reflecting by polishing or cleaving. Here the basic mechanism is photon emission by a conduction electron dropping down into a hole in the valence band. This transition can proceed via stimulated emission as well as spontaneous emission.

14-8 Barrier Penetration

Leaking through a wall

Thermionic emission

If an electric field is applied which tends to pull electrons away from the metal, it is found that a steady current of

Fig. 14-11. Same as Fig. 14-6 except for external electric field.

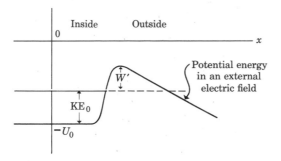

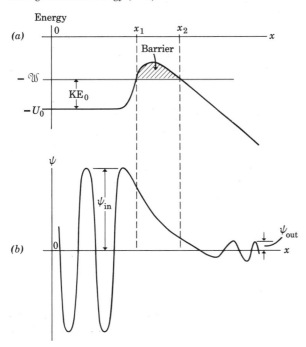

Fig. 14-12. (*a*) The same potential energy as in Fig. 14-11. (*b*) The corresponding wave function of an inside electron having the Fermi energy ($-\mathcal{W}$).

electrons will actually leave the metal. The potential energy due to a uniform field E is $U = -eEx$. The combined potential energy curve seen by a conduction electron is shown in Fig. 14-11. We can see that those few electrons that have thermal energy greater than W' can escape from their metallic prison. This phenomenon is called thermionic emission. We would expect that a small increase in temperature would cause a large increase in electron emission. This is why the cathodes in electronic tubes are heated. But even if the cathodes are cooled to absolute zero, still some electrons are emitted! This is a quite different phenomenon and is discussed in the next paragraph.

Field emission

It is observed that a smaller electron current is still emitted from a cathode even at very low temperatures where no electrons can have thermal energies as high as W'. This phenomenon is called field emission and is an example of an important quantum mechanical phenomenon that blatantly violates classical physics. The striking phenomenon referred to is the quantum mechanical penetration of a potential barrier. In this example we have a potential barrier of height W' (see Fig. 14-12). Classically an electron of kinetic energy KE_0 inside the metal would have zero kinetic energy at the position x_1. The attractive potential of the metal would then pull it back in. Classically no electron could ever penetrate the slightest amount into the barrier. In the region between x_1 and x_2 a classical electron would have negative kinetic energy which is classically impossible. However, we know from the Schroedinger equation that the electron wave could still exist in this region. It must curve away from the x-axis as shown in Fig. 14-12. Note that in this figure there is some probability of finding the electron outside the metal. According to quantum mechanics the chance that a given electron gets through the barrier must be ψ^2_{out}/ψ^2_{in} for each time the electron collides with the barrier. A classical example of a potential energy barrier would be a marble rolling inside a bowl. If the marble is released inside the bowl with its center at the height of the rim, it will roll back and forth and never get out. However, according to quantum mechanics there is an extremely small probability [about one chance in $10^{(10^{29})}$] that the marble will escape.

Actually there is an example of barrier penetration in classical physics. Since the Schroedinger equation is of the same form as the wave equation for light or water waves, we might expect to find an example of barrier penetration in optics. The surface of a piece of glass presents a barrier to light inside the glass trying to get out. If the light strikes the surface at an angle of incidence greater than the critical angle (angles forbidden by Snell's law), then the light beam cannot get past the barrier and is consequently totally reflected back into the glass as shown in Fig. 14-13a. However, if another piece of glass is brought near (within a wavelength or two) as in Fig. 14-13b, some light will penetrate the barrier and continue on in the second piece of glass. The ripple tank analogy of this is shown in Fig. 14-14.

Alpha decay

It is this same barrier penetration mechanism that explains the natural radioactive decay of some of the heavy elements into an α-particle (a helium nucleus) and a residual nucleus. The α-particle is strongly bound by the nuclear force inside the nucleus. The potential which binds the α-particle is the deep "well" shown in Fig. 14-15. Outside the nuclear radius R, the potential energy is the electrostatic potential energy $Q_1 Q_2 / r$. Q_1 is the charge of the α-particle and Q_2 is the charge of the residual nucleus. However, in spite of the large barrier, the α-particles do eventually manage to penetrate to the outside. For example, there is a 50% chance of barrier penetration or radioactive decay in 4.5 billion years for a U^{238} nucleus. Thus the radioactive halflife of U^{238} is 4.5×10^9 years. Since the earth is about this same age, there is still plenty of U^{238} around. All the isotopes having Z greater than 92 have high α-particle energies and consequently smaller barriers, and thus much shorter halflives. That is why none of them now occur naturally on the earth.

We now have a simple quantum mechanical understanding of the phenomenon of radioactivity. It is because of the probabilistic nature of quantum mechanics. We see why it is impossible to predict just when a given U^{238} nucleus will decay

Fig. 14-13. Barrier penetration of light. When second plate of glass is brought near the first, some of the light can then escape from the first into the second.

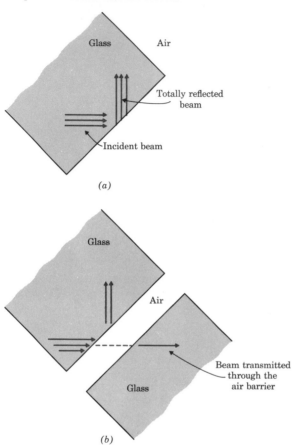

(a)

(b)

Fig. 14-14. Barrier penetration in a ripple tank. In (*a*) waves are totally reflected from a gap of deeper water. As this gap is narrowed in (*b*) and (*c*), a transmitted wave appears. The transmitted wave increases in intensity as the gap decreases. (Courtesy Educational Services Inc.)

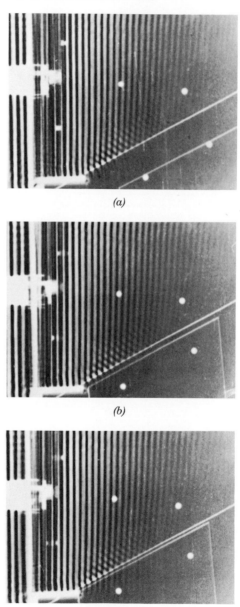

(a)

(b)

(c)

Fig. 14-15. The potential energy of an alpha particle "trapped" in a nucleus of radius R. After escaping from the nucleus its kinetic energy would be $W\alpha$.

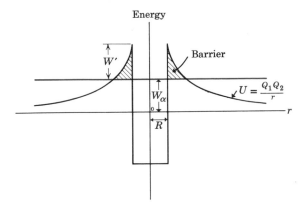

Energy

no matter how accurately it is measured. For the same reason, it is impossible to predict just where the next electron will strike the screen in electron diffraction. The subject of radioactivity and other aspects of nuclear physics are discussed in more detail in the next chapter.

Example 2

The radius of the U^{238} nucleus is 8.6×10^{-13} cm. When it decays it emits an α-particle of energy 4.2 Mev. What potential barrier in Mev does this α-particle see when inside the nucleus?

From Fig. 14-15 we see that the height of the barrier is Q_1Q_2/R minus the energy of the α-particle. Thus

$$W' = \frac{Q_1Q_2}{R} - 4.2 \text{ Mev}$$

$$Q_1 = 2e$$

$$Q_2 = 90e$$

$$R = 8.6 \times 10^{-13} \text{ cm}$$

$$\frac{Q_1Q_2}{R} = 4.82 \times 10^{-5} \text{ ergs} - 30.1 \text{ Mev}$$

$$W' = 25.9 \text{ Mev}$$

Appendix 14-1

Calculation of the Fermi energy

Consider an empty cubical box of volume $L \times L \times L$ cubic centimeters at a temperature of absolute zero. If a single electron is put into this box it will quickly radiate down to the lowest energy level. A second electron would also drop down to this lowest energy level. Because of the Pauli exclusion principle, a third electron must occupy the next highest or second energy level. A fifth electron will occupy the third energy level and so on. If the box contains a total of N electrons, the Fermi level has the kinetic energy of the $(N/2)$-th energy level. Let us now consider the electron waves corresponding to each energy level as we did for the one-dimensional box on page 322. In this three-dimensional case a given electron wave is specified by three integers or quantum numbers: N_x, N_y, and N_z. As explained on page 323, the quantum number N_x is the number of standing waves in the x-direction and has the value

$$N_x = \frac{2L}{h} P_x$$

where P_x is the x-component of the electron momentum. If $\text{KE}_0 = P_0^2/2m$ stands for the kinetic energy at the Fermi level, then

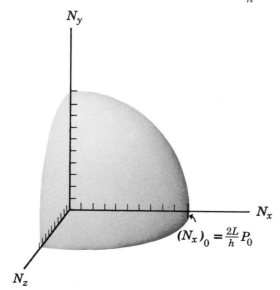

Fig. 14-16. Upper quadrant of sphere of radius $R = \dfrac{2L}{h} P_0$.

N_y

N_x

N_z

$(N_x)_0 = \dfrac{2L}{h} P_0$

P_0 is the maximum possible value of P_x for any of the N electrons in the box. The highest value of N_x still corresponding to a filled state is then $(N_x)_0 = 2LP_0/h$. The same is true for N_y and N_z. To obtain the total number of states that are filled we have to count up all possible combinations of three integers N_x, N_y, and N_z up to the above limit of $2LP_0/h$. This counting would be extremely tedious without the aid of Fig. 14-16. In this figure, for convenience, we number at 1-cm intervals N_x along the x-axis, N_y along the y-axis, and N_z along the z-axis. Then each possible state (or combination of the three integers) is represented as a unique point in space. These points make up a lattice of 1-cm cubes. Note that there are the same number of 1-cm cubes as of these points. Now use the fact that the filled states will be inside a radius $R = 2LP_0/h$. The total number of filled states will then numerically be the number of 1-cm cubes enclosed by the spherical surface shown in Fig. 14-16. Since this volume is one-eighth of an entire sphere,

$$\text{total number of states} = \frac{1}{8} \times \frac{4}{3}\pi R^3 = \frac{1}{6}\pi\left(\frac{2LP_0}{h}\right)^3 = \frac{4\pi L^3 P_0^3}{3h^3}$$

Since there are two electrons per state, the total number of electrons is

$$N = \frac{8\pi V P_0^3}{3h^3}$$

where $V = L^3$ is the volume of the box.

Solving for P_0 gives

$$P_0 = \left(\frac{3h^3}{8\pi}\frac{N}{V}\right)^{1/3}$$

Now substitute this into $\text{KE}_0 = P_0^2/2m$ and we obtain

$$\text{KE}_0 = \frac{h^2}{8m}\left(\frac{3}{\pi}\mathfrak{N}\right)^{2/3}$$

where

$$\mathfrak{N} = \frac{N}{V}$$

Problems

1. Which should change faster with temperature, the resistance of a pure metal or of a semiconductor?

2. Will the resistance of a semiconductor increase or decrease with temperature?

3. The depth of the potential well in tungsten metal is 9 ev and the maximum kinetic energy of the conduction electrons is 5 ev. What is the work function in ev? What is the Fermi level in ev?

4. A 1-cm diameter vessel contains a 5-cm column of liquid helium II. How long after the top is removed will it take before all the liquid creeps out of the vessel? The creep velocity is 50 cm/sec and the film thickness is 10^{-5} cm.

5. In a metal the depth of the potential well is 11 ev. The work function is 4 ev.

(a) What is the total energy (KE + U) of a conduction electron at the Fermi level? (Give correct sign.)

(b) When an electron enters the surface of this metal, how much kinetic energy does it gain?

6. How many conduction electrons are there per gram of sodium? How many conduction electrons are there per gram of germanium seeded with four parts per million of arsenic?

7. Solar radiation at the earth's surface is 2 calories/min/cm². What must be the area of a 20% efficient solar battery in order to generate 100 watts?

8. An electron with KE = 3 ev strikes a metal. As it enters the surface its KE increases to 8 ev. What is the depth of the potential well? (Assume the electron penetrates into the metal before undergoing a collision.)

9. Metal A has $U_0 = 4$ and $KE_0 = 3$ ev. Metal B has $U_0 = 3.5$ and $KE_0 = 2$ ev. When connected together electrons will flow from one to the other until the Fermi levels are equalized. Which of the two obtains a positive contact potential with respect to the other?

10. What is the heat of formation in calories of one mole of KCl gas starting with K atoms and Cl atoms?

11. The work function of lithium metal is 2.36 ev. What is U_0 for lithium?

12. The distance between the two protons in H_2^+ is 1.06 A. What would be the distance between two protons bound together by a μ^- meson (207 times the electron mass)?

13. In a certain heavy nucleus an α-particle collides with the potential barrier 10^{22} times per second and $\psi_{out}/\psi_{in} = 10^{-15}$. What is the probability that this nucleus would decay in a period of one year?

14. Two different metals which happen to have identical potential wells have Fermi levels differing by 3 ev.

(a) What will be the contact potential between these two metals?

(b) Will the metal with the lowest Fermi level be positive or negative relative to the other?

15. Consider a hypothetical substance, Metal A. When photons of 7 ev energy are absorbed by this metal, photoelectrons having energies up to 3 ev emitted. The density of conduction electrons in Metal A is such that the conduction electrons have kinetic energies up to

5 ev inside the metal (neglecting any thermal energy they may also have).

(a) What is the Fermi level energy for Metal A?

(b) What is the work function for Metal A?

(c) What is the depth of the potential well for Metal A?

(d) How much kinetic energy does an electron lose as it leaves the surface of Metal A?

(e) What is the photoelectric threshold in ev; that is, photons of energy less than this will never eject photoelectrons from Metal A.

16. A certain metal has an attractive potential U_0 for electrons. A certain electron outside the metal has a wavelength $\lambda = 10$ A. When it reaches the metal, the wavelength of this electron decreases to 4 A. What is the value of U_0 in electron volts?

Prob. 17

17. If two metals I and II are joined together, the potential energy diagram is as shown in the figure.

(a) What is the maximum kinetic energy of the conduction electrons while in metal II?

(b) What is the work function in II?

(c) What is the potential energy in II?

(d) What is the ratio of electron wavelengths for an electron at the Fermi level?

$$\frac{\lambda_I}{\lambda_{II}} = ?$$

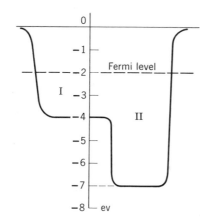

18. The quantum numbers for a certain electron in a cubical box of side L are N_x, N_y, and N_z.

(a) Write down equations for the x-, y-, and z-components of the momentum of this electron in terms of L, m, h and the quantum numbers.

(b) What is the energy of this electron in terms of h, m, L, N_x, N_y, and N_z? (*Hint:* $P^2 = P_x{}^2 + P_y{}^2 + P_z{}^2$.)

19. The solution to Part (b) of Problem 12 is $W = h^2/8mL^2 \times (N_x{}^2 + N_y{}^2 + N_z{}^2)$. Calculate the five lowest energy levels in electron volts for a box with $L = 10^{-7}$ cm.

20. Repeat Problem 18 for a rectangular box of sides L_x, L_y, and L_z.

21. Assume the potential energy curve between an electron and a negatively charged speck of dust of 10^{-4} cm radius would be as in Fig. *a*. Consider an electron inside the sphere which has a wave function as in Fig. *b*.

(a) What is the electron wavelength inside the sphere?

(b) What is the electron velocity inside the sphere?

(c) What is the probability per collision for the electron to escape from the sphere?

(d) Assuming the electron loses no energy, what is the mean-life for this electron to escape from the sphere?

Prob. 21

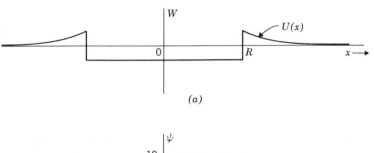

(a)

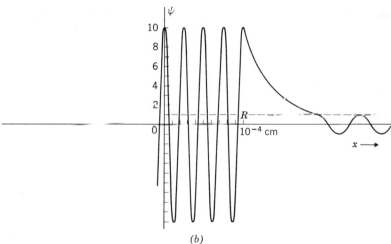

(b)

Los Alamos Scientific Laboratory

Nuclear physics

Chapter 15

Nuclear physics

15-1 Properties of Nuclei

Miniaturized atoms

This chapter is mainly devoted to the understanding of the atomic nucleus and its structure. As we shall see, protons and neutrons are very similar. For this reason the term "nucleon" is used for either a proton or a neutron. We would hope to explain nuclear structure in a manner similar to atomic structure; namely, as a structure of nucleons bound into definite orbits analogous to the structure of electron orbits in the atom. Ideally we would like to be able, in principle, to calculate the mass of a given nucleus, its spin and magnetic moment, the binding energies of its nucleons, and the energies of its excited states. In order to meet this goal, the force between two nucleons must be known exactly. Unfortunately, the nucleon-nucleon force is not as simple as the electric force, and so far it is not known with very great accuracy. Nevertheless, considerable progress has been made in understanding the structure of the nucleus and many nuclear properties can be predicted at least approximately.

Before discussing our present knowledge of the nucleon-nucleon force and the structure of the nucleus, we shall review some of the experimental knowledge of nuclear properties. First, just how small is the nucleus? Nuclear size can be determined by bombarding atoms with high-energy electrons and counting how many of them score direct hits.* The result is that a nucleus of mass number A (number of protons plus neutrons) has nearly all of its nucleons tightly packed within a radius

Nuclear radius

$$R = 1.2 \times 10^{-13} A^{1/3} \text{ cm} \tag{15-1}$$

The experiments show that this equation holds quite well for all but the very smallest nuclei. Note that since the volume is proportional to R^3, it is proportional to the first power of A, the number of nucleons. Then according to Eq. 15-1 all nuclei should have the same density independent of size.

---

*For details on how nuclear sizes are determined see Section 15-5 and Appendixes 15-1 and 15-2.

Example 1

What is the density of nuclear matter in nucleons per cubic centimeter and grams per cubic centimeter? The number of particles per cubic centimeter is

$$N = \frac{A}{V}$$

and according to Eq. 15-1 the nuclear volume is

$$V = \frac{4\pi}{3} R^3 = \frac{4\pi}{3} (1.2 \times 10^{-13})^3 A$$

Thus

$$N = \frac{A}{\dfrac{4\pi}{3} \times 1.2^3 \times 10^{-39} A}$$

$$= 1.38 \times 10^{38} \text{ nucleons per cm}^3 \qquad (15\text{-}2)$$

The density is this number times 1.67×10^{-24} gm, the mass of one nucleon. Thus

$$D = 2.3 \times 10^{14} \text{ gm/cm}^3$$

According to this, 1 cm³ of nuclear matter would contain 250 million tons.

As with electrons in atomic structure, the nucleons are bound together by attractive forces which must be strong enough to overcome the repulsive Coulomb force between protons. This is called a nuclear force and is discussed in Section 15-2. Experimentally, in light nuclei the binding energy per nucleon increases with increasing A (see Fig. 15-1). We might expect such an effect as the result of addition of forces; a single nucleon is more strongly bound if attracted by several others rather than by just one or two. However, above $A = 80$ the binding energy per nucleon gradually decreases with increasing A. This suggests that the attractive nuclear force is of short range (about the diameter of a single nucleon). Past this range the electrostatic repulsive force dominates; that is, when two protons are more than 2.5×10^{-13} cm apart, the force is repulsive rather than attractive. Nuclei in the region $A = 80$ are the most tightly bound.

Two consequences of this behavior of the binding energy as a function of A are nuclear fusion and nuclear fission. First consider what happens when an electron and proton are

Q.1: Give a formula for the cross-sectional area of a nucleus of mass number A.

Fig. 15-1. Experimental values of binding energy per nucleon as a function of the mass number A.

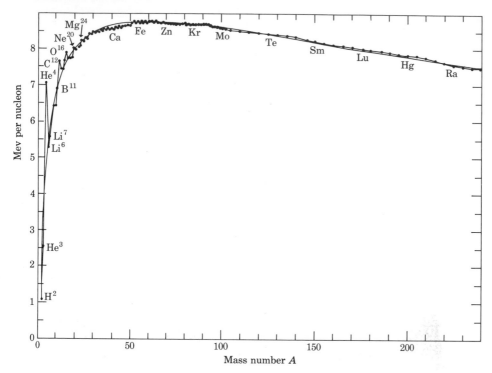

brought together. Then 13.6 ev of energy is released with the result that the mass of the hydrogen atom is 13.6 ev* less than the mass of a free electron plus proton. Similarly, two light nuclei will have more mass or rest energy than will their sum. If they can be brought together they will "fuse" into their sum with a release of energy corresponding to the mass difference. This process is called nuclear fusion. In Section 15-5 we will see that this mass difference can be larger than one-half of 1%. On the other hand, if a *heavy* nucleus is split into two smaller nuclei, the two pieces will have less mass than does their parent by as much as one-tenth of 1%. Thus

Ans. 1: Area $= \pi R^2$
$$= \pi (1.2 \times 10^{-13} A^{1/3})^2 \text{ cm}^2$$
$$= 4.52 \times 10^{-26} A^{2/3} \text{ cm}^2.$$

*It is common practice in physics to use energy as a unit of mass. The corresponding value of mass can always be obtained from the relation $M = W/c^2$.

there is a tendency for a heavy nucleus to fission into two smaller nuclei with a release of energy. The energy of an A-bomb and of a nuclear reactor is the energy released in nuclear fission. The energy of an H-bomb is energy released in fusion.

Alpha decay (see Section 14-8) can be thought of as a lopsided fission where the parent nucleus M splits into a small α-particle and a large residual nucleus M'. Alpha decay is possible only if in the reaction

$$M \to M' + \alpha$$

the mass M is greater than M' plus the α-particle mass. Then the nucleus will be radioactive and can alpha decay. It turns out that $M > (M' + M_\alpha)$ for all nuclei with $Z > 82$ (lead). Above $Z = 92$ (uranium) the alpha decay halflives become significantly shorter than the age of the earth. This is why no elements of atomic number greater than 92 occur naturally on the earth. Such elements, however, can be produced artificially by man. For example, plutonium ($Z = 94$) can be produced from uranium in nuclear reactors; this process has now become sufficiently common so that the price is about \$15 per gram. So far elements up to $Z = 104$ have been produced, but at much higher prices and usually in very minute quantities. It is expected that radiochemists will eventually succeed in producing extremely minute quantities of new elements even beyond $Z = 104$.

Just as electrons in an atom can be excited to higher energy "orbits" or energy levels, so nucleons in a nucleus can be excited to higher energy "orbits." Thus each nucleus has a set of energy levels above its ground state. See Fig. 15-2 for the energy levels and gamma ray transitions of phosphorus-31. Note that this nuclear energy level diagram has the same features as the atomic energy level diagram of Fig. 13-13. Such an excited nucleus can drop to a lower energy level by emitting a photon (called gamma ray by nuclear physicists). For example, if a neutron hits a U^{238} nucleus (the superscript means that $A = 238$), a $(U^{239})^*$ nucleus is formed, where the star means some excited state of the U^{239} isotope. Then

$$(U^{239})^* \to U^{239} + \gamma$$

Fig. 15-2. Energy levels and gamma transitions in P^{31}. Each horizontal line represents an excited state of the P^{31} nucleus. (Reprinted from *Reviews of Modern Physics,* October 1957.)

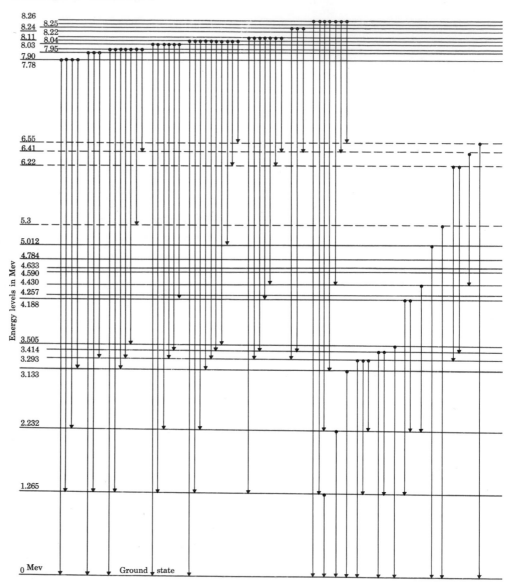

where γ stands for the emitted gamma ray. Of course, each excited state has its own definite energy level, spin, and magnetic moment. The measurements of these plus other quantities for the many excited states of over 1000 different isotopes keep the experimental nuclear physicist and radiochemist busy. The results of such measurements have both practical applications and improve our understanding of nuclear structures and nuclear forces.

15-2   Radioactive Decay

When it goes, nobody knows

A large class of physical phenomena exhibit what is called exponential decay. So far we have studied two such phenomena: (1) alpha decay of a heavy nucleus; and (2) emission of a gamma ray or photon from an excited atom (Chapter 13) or nucleus (Section 15-1). In each of these cases there is a fixed probability of decay per unit time.

We recall from Section 14-8 that the U^{238} nucleus decays according to the laws of probability as a consequence of the wave nature of matter. The probability that the α-particle be found outside the nucleus is proportional to $(\psi_{out})^2$ (see Fig. 14-12). The nature of probability is such that if a given nucleus has by rare coincidence managed to survive many halflives without decaying, this previous history will in no way influence its future chance of decay. The same is true for coin tossing. If you happen to toss five heads in a row, the probability is still one-half that the sixth toss will also be a head. We can never predict when a given nucleus will decay. All nuclei of the same kind always have exactly the same probability for decay no matter how long they have lived. For example, half the nuclei of a radioactive isotope of one-year halflife will decay in the first year, but an individual nucleus that survived the first year will still have a 50-50 chance of decaying in the second year. If it survives the first two years, its probability of decay during the third year is still one-half.

Let us now study in more detail the decay of U^{238}. The probability that any given U^{238} nucleus decay in one year is *prob.* $= 1/(6.5 \times 10^9)$. In a period of Δt years the probability

Q.2: What is the probability that a radioactive nucleus should survive 10 halflives?

decay is

$$prob. = \frac{\Delta t}{6.5 \times 10^9} \tag{15-3}$$

Using this probability we can calculate the average length of time a U^{238} nucleus will live. The result is the value in the denominator of Eq. 15-3, or 6.5×10^9 years. This quantity is defined as the meanlife τ. In terms of the meanlife, the probability for decay in a time Δt is

$$prob. = \frac{\Delta t}{\tau} \tag{15-4}$$

Example 2

Derive an approximate formula for the meanlife due to alpha decay in terms of the ratio $\left(\frac{\psi_{out}}{\psi_{in}}\right)$.

In Section 14-8 it is shown that the probability of escape each time the alpha particle collides with the barrier is

$$\frac{prob.}{coll.} = \left(\frac{\psi_{out}}{\psi_{in}}\right)^2$$

The number of collisions in a time Δt is the distance traveled ($v\,\Delta t$) divided by the diameter of the nucleus ($2R$),

$$coll. \simeq \frac{v\,\Delta t}{2R}$$

If we multiply the two above equations by each other, we obtain

$$prob. \simeq \frac{v\,\Delta t}{2R}\left(\frac{\psi_{out}}{\psi_{in}}\right)^2$$

or

$$\frac{\Delta t}{prob.} \simeq \frac{2R}{v}\left(\frac{\psi_{in}}{\psi_{out}}\right)^2$$

From Eq. 15-4, $\Delta t/prob. = \tau$, hence the above quantity is the meanlife; namely,

$$\tau \simeq \frac{2R}{v}\left(\frac{\psi_{in}}{\psi_{out}}\right)^2$$

This equation demonstrates how quantum mechanics can be used to calculate lifetimes as well as the other properties of atoms and nuclei. Present quantum mechanical theory gives as full an explanation of radioactive decay as it does of anything else. As we shall now see, Eq. 15-4 holds only

Ans. 2: It is $(\frac{1}{2})^{10}$ or 1 chance in 1024.

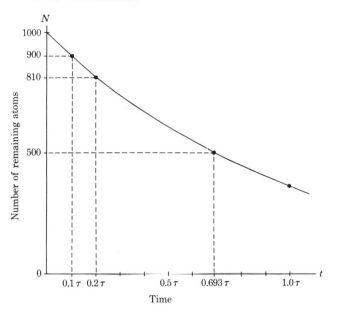

when Δt is much less than τ. According to this equation, if $\Delta t = \frac{1}{10}\tau$, then each nucleus has one chance in ten of decaying. If we start with 1000 U²³⁸ atoms, about 100 of them will decay in $\tau/10 = 6.5 \times 10^8$ years. In the next period of 6.5×10^8 years about 90 of the remaining 900 will decay. In the third interval of 6.5×10^8 years, about 81 of the remaining 810 will decay, etc. The number of atoms remaining after a time t is plotted in Fig. 15-3.

In order to calculate this curve precisely, we must take Δt to be very small. Then the number of atoms remaining after a time Δt is

$$N \approx N_0\left(1 - \frac{\Delta t}{\tau}\right)$$

where N_0 is the initial number of unstable particles of meanlife τ. After n intervals of time Δt, where $t = n\,\Delta t$

$$N \approx N_0\left(1 - \frac{\Delta t}{\tau}\right)^n$$
$$\approx N_0\left(1 - \frac{\Delta t}{\tau}\right)^{t/\Delta t}$$

Now let

$$x = -\frac{\Delta t}{\tau}$$

then

$$N \approx N_0(1 + x)^{-t/x\tau}$$
$$\approx N_0[(1 + x)^{1/x}]^{-t/\tau}$$

This relation becomes exact in the limit as Δt (or as x) approaches zero. The limit $(1 + x)^{1/x}$ as x approaches zero is a famous number in mathematics. It is

$$\lim_{x \to 0}[(1 + x)^{1/x}] = e$$

where $e = 2.718 \ldots$. Hence

$$N = N_0 e^{-t/\tau} \tag{15-5}$$

What must be the value of t/τ for half of the particles to have decayed? For this particular value of t that we have previously called the halflife we shall use the symbol T. According to Eq. 15-5 we have

$$e^{-T/\tau} = \tfrac{1}{2}$$

Q.3: Is the average lifetime of a radioactive nucleus equal to the halflife of that isotope?

A table of logarithms shows that

$$e^{-0.693} = \tfrac{1}{2}$$

Thus $\quad \dfrac{T}{\tau} = 0.693$

or $\quad\quad T = 0.693\tau$ is the halflife $\quad\quad\quad\quad$ (15-6)

Note that for U^{238} the halflife is $T = 0.693 \times 6.5 \times 10^9$ years $= 4.5$ billion years. If we substitute the quantity $T/0.693$ for τ in Eq. 15-5, we obtain

$$\frac{N}{N_0} = e^{-0.693\,t/T} = (e^{-0.693})^{t/T}$$

or

$$\frac{N}{N_0} = \left(\frac{1}{2}\right)^{t/T} \quad\quad\quad\quad (15\text{-}7)$$

At the end of two halflives ($t = 2T$), one-fourth of the particles remain. At the end of three halflives, one-eighth remain, etc. Equation 15-7 is plotted in Fig. 15-4. These decay curves (Figs. 15-3 and 15-4) are called exponential decay curves.

For gamma emission, the probability of a nucleus in an excited state emitting a photon can be calculated by using quantum mechanics. Thus each excited state of the phosphorus nucleus in Fig. 15-2 has a definite halflife. A similar type calculation gives the halflife or probability that an $N = 2$, $l = 1$ hydrogen atom will emit a photon and "decay" down to a ground state hydrogen atom. The result is that the probability of photon emission is $prob. = \Delta t / 1.6 \times 10^{-9}$ for Δt in seconds. Thus the halflife of a sample of excited hydrogen gas with all the atoms in the $N = 2$, $l = 1$ state is 1.1×10^{-9} sec.

15-3 Radioisotopes

"Man-made" nuclei

New isotopes that do not occur in nature can be produced easily by bombarding naturally occurring isotopes with neutrons. Unlike charged particles, neutrons lose little energy while traversing matter and can easily penetrate to the atomic nucleus. Quite often when a neutron hits a nucleus it

Fig. 15-4. Radioactive decay curve. The number of atoms remaining as a function of time over a time interval of four halflives.

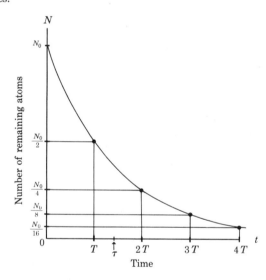

Number of remaining atoms

Time

Ans. 3: No. The average lifetime is the meanlife which is longer than the halflife.

will be absorbed. Then a new isotope containing one additional neutron is formed in an excited energy level. The newly formed excited nucleus will then usually gamma decay to the ground state. Quite often the ground state is also radioactive. Those isotopes whose ground states are radioactive are called radioisotopes. As an example, let us consider the bombardment of U^{238} with neutrons. When U^{238} absorbs a neutron it forms the radioisotope U^{239} in an excited state which we shall denote as $(U^{239})^*$.

$$N + U^{238} \rightarrow (U^{239})^*$$

Then the excited nucleus quickly gamma decays to its ground state:

$$(U^{239})^* \rightarrow U^{239} + \gamma$$

The new radioisotope U^{239} does not occur in nature, and beta decays with a halflife of 24 min to form a residual nucleus having $Z = 93$ (neptunium):

$$U^{239} \rightarrow Np^{239} + e^- + \bar{\nu}$$

The symbols $e^- + \bar{\nu}$ and the mechanism of beta decay are discussed further in Chapter 16. It turns out that the residual nucleus Np^{239} is also unstable to beta decay. It decays with a halflife of 2.3 days to plutonium ($Z = 94$):

$$Np^{239} \rightarrow Pu^{239} + e^- + \nu$$

In January 1934 the idea occurred to Enrico Fermi that he might be able to produce artificial radioactivity by bombarding different substances with neutrons. Fermi and his Rome group systematically bombarded every element they could obtain in order of increasing atomic number. By May 1934 they had worked their way up to uranium. However, not only did they observe the U^{239} beta decay halflife but also a complicated mixture of other halflives. They felt that they had probably produced one or more trans-uranic elements by the beta decay of U^{239} to a nucleus of $Z = 93$ and that perhaps this also beta decayed into $Z = 94$, etc. At the time Fermi said that he did not completely understand his results. The reason his results were so complicated is that, unknown to Fermi, some of the excited U^{239} nuclei fissioned,

Q.4: A sample of U^{238} is bombarded by neutrons for 30 min. During bombardment the amount of U^{239} is increasing. After bombardment does the amount of U^{239} follow a pure exponential decay curve? Does the amount of Pu^{239} in the sample follow a pure exponential decay curve?

and the many possible fission products made up a mixture of many different halflives. If Fermi had completely understood his results, the whole world, including Hitler, would have known back in 1934 about the possibility of an atomic bomb.

In December 1938, Fermi was awarded the Nobel prize for his work in discovering these new isotopes and how to produce them. Since it is the custom for one's family to accompany the prize winner to Sweden, Fermi had the golden opportunity to get his family out of Fascist Italy. After receiving his award, Fermi headed for the United States. Shortly after arriving in the United States, Fermi learned about fission and did the pioneer work that led to the atomic bomb. The famous letter of Einstein to President Roosevelt pointing out the significance of Fermi's work is shown in Fig. 15-5.

The use of radioisotopes in industry, medicine, agriculture, and research is widespread and of great economic importance. Small samples of biological and industrial material can be "tagged" with radioisotopes. Biological and industrial processes can then be studied by tracing the tagged materials with radiation detectors. For example, genetic material in a chromosome can be tagged. The resulting information on gene location, chromosome duplication, and combination is invaluable to the science of genetics.

15-4   Biological Effects of Radiation

What you don't see can hurt you

When charged particles pass through matter, they collide with atomic electrons leaving a trail of ions behind. Consequently, molecules of living matter are disrupted or damaged by the passage of charged particles. The nature and effects of this damage are of more interest to biologists than physicists. However, because of the widespread interest and importance of the subject, we shall summarize some of the main points. The effects of radiation are important to those interested in civilian defense, military planning, foreign policy, nuclear power, nuclear weapons testing, radioisotope applications, nuclear research, etc.

Ans. 4: The U^{239} follows a pure exponential decay curve. However, as the U^{239} is decaying, the amount of Pu^{239} is at first increasing rather than decreasing.

Fig. 15-5. Letter to President Roosevelt from Albert Einstein.

August 2nd, 1939

F.D. Roosevelt,
President of the United States,
White House
Washington, D.C.

Sir:

Some recent work by E.Fermi and L. Szilard, which has been communicated to me in manuscript, leads me to expect that the element uranium may be turned into a new and important source of energy in the immediate future. Certain aspects of the situation which has arisen seem to call for watchfulness and, if necessary, quick action on the part of the Administration. I believe therefore that it is my duty to bring to your attention the following facts and recommendations:

In the course of the last four months it has been made probable - through the work of Joliot in France as well as Fermi and Szilard in America - that it may become possible to set up a nuclear chain reaction in a large mass of uranium, by which vast amounts of power and large quantities of new radium-like elements would be generated. Now it appears almost certain that this could be achieved in the immediate future.

This new phenomenon would also lead to the construction of bombs, and it is conceivable - though much less certain - that extremely powerful bombs of a new type may thus be constructed. A single bomb of this type, carried by boat and exploded in a port, might very well destroy the whole port together with some of the surrounding territory. However,

I understand that Germany has actually stopped the sale of uranium from the Czechoslovakian mines which she has taken over. That she should have taken such early action might perhaps be understood on the ground that the son of the German Under-Secretary of State, von Weizsäcker, is attached to the Kaiser-Wilhelm-Institut in Berlin where some of the American work on uranium is now being repeated.

Yours very truly,

A. Einstein

(Albert Einstein)

From Franklin D. Roosevelt Library

A practical unit of radiation dose for monitoring purposes is the rad. A rad is so defined that if 1 gm of biological matter receives a radiation dose of 1 rad, then 100 ergs of energy will be dissipated (energy lost by the ionizing particles).

1 rad liberates 100 ergs/gm

The rad is a modification of an older unit called the roentgen.

1r (roentgen) liberates 83 ergs/gm

Let us now estimate the radiation dose contributed by the cosmic rays in one year. Experiments show that a fast-moving charged particle loses about 2 Mev in traveling through 1 cm of water. The cosmic ray flux at sea level liberates about 4 ergs/gm of water per year. Thus the natural radiation level due to cosmic rays is about 0.04 rad per year, or 3 rads per human lifetime (70 years). A comparable amount of radiation comes from natural radioactive isotopes contained in soil and rock. Thus the total background dose due to natural sources is about 0.1 rad per year, or 7 rads per lifetime.

Lethal dose

If a human receives 400 rads over his entire body in a short period of time, the biological damage is so great that there is about a 50% chance the individual will die. This amount of radiation is called the lethal dose.

Lethal dose = 400 rads

A dose of about 200 rads will cause radiation sickness. As some of the body tissues have recuperation ability, an individual receiving 400 rads over several years will probably survive, although his health may be adversely affected.

"Permissible" dose

Various industrial and research jobs involve work with radioactive materials. In 1957 the International Commission on Radiological Protection established 5 rads per year (100 millirads per week) as the maximum dosage for radiation workers.

Fig. 15-6. (*a*) Photograph of diffraction pattern obtained by shining monochromatic light on a black disk. (*b*) Plot of intensity obtained from the photograph in (*a*) as a function of diffraction angle.

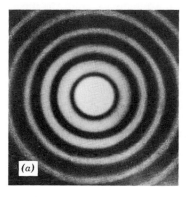

(*a*)

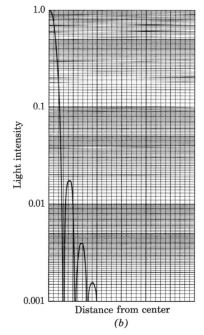

Light intensity

Distance from center

(*b*)

Permissible dose for radiation workers $= 5$ rads per year

Note that for a 50-year working life, this corresponds to about forty times the dose received from natural radiation.

The maximum permissible dose for a large population is usually taken to be 10% of the dose for the occupationally exposed people.

Permissible dose for average citizens
$$= 0.5 \text{ rad per year or } 30 \text{ rads per lifetime}$$

This is about four times the dose received from natural radiation. In the United States a major source of radiation comparable to the natural background radiation is the medical use of x-rays. The radiation level due to nuclear weapons tests will remain below that of natural background as long as the 5 nuclear powers refrain from testing in the atmosphere.

A common misconception is that radiation below the permissible dose does no harm. The fact is that any radiation does biological damage. A chromosome mutated by radiation will reproduce itself in the same mutated form. There are many uncertainties in the science of genetics, but it is certain that natural radiation contributes to the natural mutation rate, and that any increase in radiation, no matter how small, will increase this rate. The natural mutation rate in humans is so high that it results in serious abnormalities in about 3% of all births. Thus nuclear warfare is to be feared, not only for the unprecedented damage it can do to the belligerents but for the accompanying increase in world-wide radiation level which would affect all people in all nations.

15-5   Diffraction Scattering

The most powerful microscope

If the wave nature of matter is correct, we must expect to observe classical diffraction patterns whenever a beam of particles of short enough wavelength strikes an atomic nucleus. As shown in Fig. 15-6, when a beam of light of wavelength λ strikes a black disk of radius R, there is a resulting circular diffraction pattern where the position of the first

$$\sin \theta_{\min} = 0.61 \frac{\lambda}{R} \qquad (15\text{-}8)$$

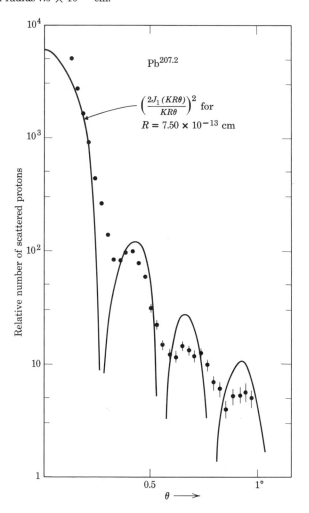

Fig. 15-7. Results of scattering 19 Bev protons on lead. Number of scattered protons is plotted vs. scattering angle. The smooth curve is the classical diffraction pattern corresponding to light of the same wavelength shining on a disk of radius 7.5×10^{-13} cm.

This phenomenon is basically the same as single-slit diffraction described in Problem 26, Ch. 10. Those photons which strike the black disk are removed from the incoming beam. Similarly, when a proton beam is aimed at a large nucleus like lead, those protons which score direct hits are lost from the beam. Hence the lead nucleus is mathematically the same as a black disk of radius R illuminated by an incoming beam of wavelength $\lambda = h/P$ where P is the proton momentum. Whether the particles are photons or protons, the solution to the wave equation is the same as that of the classical problem of light waves diffracted by a black disk; that is, Eq. 15-8 must apply to high-energy protons as well. All we need do is to replace λ in Eq. 15-8 by h/P to obtain

$$\sin \theta_{\min} = 0.61 \frac{h}{PR} \qquad (15\text{-}9)$$

The proton diffraction pattern is observed by measuring the number of elastically scattered protons as a function of θ. The result for the scattering of 19 Bev protons on lead is shown in Fig. 15-7. The points show the results of measurements made at 30 different angles ranging from 0.11° to 1.03°. The smooth curve is the standard intensity pattern of light diffracted by a black disk assuming light of the same wavelength as the protons. The departures of the proton results from the smooth curve are not because of any violation of the wave nature of matter, but because the lead nucleus does not behave as a simple black disk with sharp edges. A black disk with partially transparent edges can give a smooth curve which would agree with all the experimental points.

Equation 15-9 can be used to determine the size of any nucleus by solving for R:

$$R = 0.61 \frac{h}{P \sin \theta_{\min}} \qquad (15\text{-}10)$$

For example, in Fig. 15-7, $\theta_{\min}$ occurs at 0.24° which when substituted into Eq. 15-10 gives $R = 7.5 \times 10^{-13}$ cm for the radius of the lead nucleus. Even scattering of protons against protons gives a diffraction pattern, but in this case the pattern departs even more from that of the black disk with sharp edges because the proton is increasingly transparent from its center to its edge. No second maximum is seen in the proton-proton scattering pattern; however, the effective radius of the proton can be obtained by measuring the width of the central maximum. The result is 1.1×10^{-13} cm for the optical radius of the proton.

But what about possible structure of the proton? Does it have a hard, "black" core as does the atom? A small black disk or core superimposed on a large semitransparent disk will give a complex diffraction pattern: the central maximum will be a superposition of two central maxima of the two corresponding widths. So far no such effect has been clearly seen. An upper limit on the size of a possible core can be obtained from Eq. 15-10 by inserting the highest value of $P \sin \theta$ observed so far. The highest energy proton-proton scattering was done at the Brookhaven AGS utilizing the full energy of the accelerator which is the highest in the world. A target containing hydrogen was bombarded by a beam of 32 Bev protons. The scattered and recoil protons each went off at 15° from the beam direction, each with half the beam energy. The product $Pc \sin \theta$ is then 4 Bev or $P \sin \theta = 2.13 \times 10^{-13}$ gm cm/sec which when inserted into Eq. 15-10 gives a value of 0.19×10^{-13} cm. Actually a core of about half this size should give enough of its central maximum to be detectable. Hence we conclude that the proton has no core as large as 0.1×10^{-13} cm. We see that the AGS and other high energy accelerators can correctly be thought of as super microscopes which can measure distances down to a fraction of a fermi. (The fermi is a unit of length equal to 10^{-13} cm.) This is a billion times more resolution than the best optical microscope. In order to see structure in the proton with a resolution better than 0.1×10^{-13} cm, higher energy accelerators must be built because the above value of $P \sin \theta$ is the highest that can be obtained by existing accelerators.

Q.5: Would the central diffraction peak of a proton "core" be wider or narrower than that of the semitransparent disk?

15-6 The Nucleon-Nucleon Force

The elementary force of the nuclear "atom"

The main goal of physics is to explain all physical phenomena by means of a small number of simple, fundamental principles. Since material objects are made up of electrons and nuclei, our approach so far has been to study the fundamental interactions of electrons, nuclei, and photons. We saw in the last chapter that this approach has met with great success. It has given a complete (although difficult to calculate) explanation of the structure and interactions of matter. In fact, the modern theory of quantum electrodynamics is so good that when applied to atomic physics it predicts results more accurately than can be measured. So far no discrepancy between experiment and theory has been found despite the fact that some of the experimental results are more accurate than one part in ten million.

On the other hand, quantum electrodynamics cannot explain the structure of the atomic nucleus which we now know to be made up of protons and neutrons. A new, fundamental force law is needed to explain what holds the protons so tightly together within the nucleus. This force must be even stronger than the electrostatic force in order to overcome the electrostatic repulsion of the protons. This new force, called the nuclear force or strong interaction, is roughly a hundred times stronger than the electrostatic force.

Except for a weak electrostatic repulsion, the strong proton-proton, proton-neutron, and neutron-neutron nuclear forces are all the same and are called the nucleon-nucleon force. Although the detailed form of this force is still not known, a crude plot of the potential energy between two nucleons is shown in Fig. 15-8. The e^2/r electric potential energy between two protons is shown for comparison (dashed curve). This nuclear force also has the peculiar feature that it looks like Fig. 15-8 if the nuclear spins are parallel; on the other hand, when the spins are antiparallel, the nuclear force is considerably weaker. As we shall see in the next section, the depth of the potential well shown in Fig. 15-8 can be determined from the binding energy of the deuteron. More detailed information on the shape of this potential

Fig. 15-8. Potential energy diagram of the elementary nucleon-nucleon force.

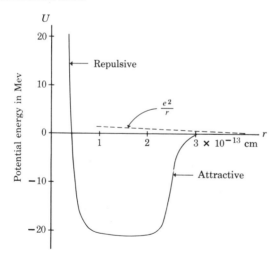

Ans. 5: In Eq. 15-9, as R decreases, θ_{min} increases; hence the diffraction pattern is wider.

comes from the proton-proton and neutron-proton scattering experiments. A proton-proton scattering is shown in Fig. 3-2.

15-7 Structure of the Nucleus

Orbits of nucleons

In atomic physics, the only atom we were able to analyze easily was the one containing just one proton and one electron, the hydrogen atom. We encounter the same situation in nuclear structure: the easiest nucleus for us to analyze is made up of just one proton and one neutron, the deuteron. In the deuteron, the neutron and the proton are bound together with an energy of 2.22 Mev.

Example 3

The masses of the deuteron, proton, and neutron are 1875.49 Mev, 938.21 Mev, and 939.50 Mev, respectively. What is the binding energy of the deuterium nucleus?

The total binding energy is the sum of the masses of the individual nucleons minus the mass of the nucleus. Thus binding energy = $(M_P + M_N) - M_D = 2.22$ Mev.

As with the hydrogen atom, we should be able to calculate this binding energy once the force between the two particles is known. The problem is merely one of finding the lowest order wave function corresponding to the potential energy of Fig. 15-8. To first approximation the potential energy can be drawn as a "square well" of radius $r_0 = 2.3 \times 10^{-13}$ cm. This is shown in Fig. 15-9a. The red horizontal line is the energy W corresponding to the lowest order standing wave shown in Fig. 15-9b.

We shall now make a calculation. We have the choice of calculating the lowest energy level W knowing the depth of the well, or else of calculating the depth of the well knowing W (the measured binding energy). The latter is what was actually done historically, and is thus what we chose to do here. We will start with the knowledge of the binding energy and the range r_0 of the nuclear force. Our goal is to obtain from this information the depth of the potential well which is a measure of the strength of the nuclear force. We start by noting that the lowest order standing wave is about a half-

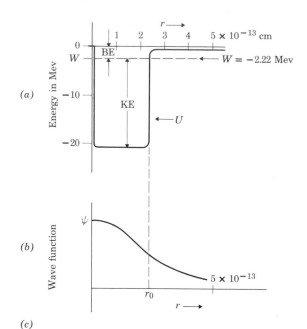

(a)

(b)

(c)

wavelength across the deuteron; that is,

$$\frac{\lambda}{2} = 2r_0$$

$$\lambda = 4r_0$$

The momentum of the proton or neutron will be $p = h/\lambda$. While inside the well, the momentum of the proton or neutron will be

$$p = \frac{h}{4r_0}$$

and the kinetic energy will be

$$\text{KE} = \frac{p^2}{2M} = \frac{h^2}{32Mr_0^2}$$

The total kinetic energy is twice this, or

$$\text{KE}_{\text{total}} = \frac{h^2}{16Mr_0^2}$$

If we use $r_0 = 2.3 \times 10^{-13}$ cm, we obtain

$$\text{KE}_{\text{total}} = 3.00 \times 10^{-5} \text{ ergs} = 18.7 \text{ Mev}$$

If U_0 is the depth of the well, we see from Fig. 15-7a that

$$U_0 = \text{KE}_{\text{total}} + \text{BE}$$
$$= 18.7 + 2.2 = 19.9 \text{ Mev}$$

The results of many experiments such as neutron-proton scattering and the binding energy of the deuteron are used to determine the nucleon-nucleon force (Fig. 15-8). Note that the deuteron wave function extends out beyond $r = 5 \times 10^{-13}$ cm. Thus the deuteron is less dense than heavier nuclei.

◀ **Fig. 15-9.** (a) The approximate potential well for the neutron-proton force. (b) The lowest energy wavefunction. The corresponding energy of this state is $W = -2.22$ Mev. (c) Distribution of nuclear matter in the deuterium nucleus. The amount of whiteness is proportional to the square of the nucleon wave function projected onto a plane.

Heavy nuclei

The reason for the high density of nucleons in heavy nuclei can be understood by the following. Start with a large number of free nucleons with a distance s between adjacent particles. Now slowly push them together (decrease s). Suddenly when s becomes less than 2.5×10^{-13} cm the nucleons feel the strong attractive force of their neighbors, and their binding energy increases correspondingly. On the other hand, we saw in Chapter 14 that when free electrons are crowded closely together, their average kinetic energy must increase because of the Pauli exclusion principle (see Eq. 14-1). Since protons and neutrons are also spin-$\frac{1}{2}$ particles, they must also obey the Pauli principle. Thus the effect of the exclusion principle is to decrease the binding energy as s is decreased. Fortunately, the nucleon-nucleon attractive force is just barely strong enough to allow a "happy medium" between these two effects; that is, there exists a value of s, where the binding energy is maximum (if the nucleon-nucleon force had been 30% weaker, the effect of the exclusion principle would dominate and no nuclei could ever exist). The value of s that gives the maximum binding energy determines the size of the nucleus. The experimental result is $s = 1.9 \times 10^{-13}$ cm as determined from Eq. 15-2.

Let us now consider a single neutron inside of a heavy nucleus. The neutron sees an attractive force averaged over all the other nucleons in the nucleus. In Chapter 14 an analogous situation was the potential seen by a free electron inside of a metal. The average potential energy curve seen by our neutron is shown in Fig. 15-10. It is about 42 Mev deep for all medium and large nuclei. Adding more nucleons will not increase the strength of the net force on a given nucleon because, as pointed out before, a nucleon is only attracted by its nearest neighbors. Actually about $A/2$ neutrons are crowded into this potential well. Because of the Pauli principle they will occupy different states or energy levels up to the Fermi level.

Example 4

What is the Fermi level for $A/2$ neutrons which are trapped in the potential shown in Fig. 15-10?

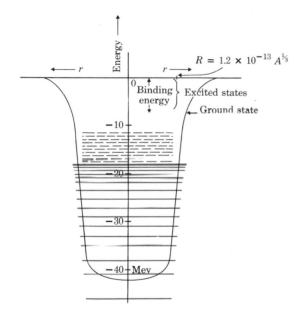

Fig. 15-10. Average nuclear potential seen by a neutron in a nucleus of radius R. The heavy red lines are the occupied states. The dashed red lines are unoccupied or excited states.

Q.6: If in a large nucleus the nucleon spacing s is halved, how does this effect the maximum kinetic energy of the nucleons?

According to Eq. 14-1 the highest energy neutron has kinetic energy

$$\mathrm{KE}_0 = \frac{h^2}{8M} \left(\frac{3}{\pi} \mathfrak{N} \right)^{2/3}$$

where $\mathfrak{N}$ is the number of neutrons per cubic centimeter. The number of neutrons per cubic centimeter would be approximately one-half the result obtained in Eq. 15-2. Thus

$$\mathrm{KE}_0 = \frac{h^2}{8M} \left(\frac{3}{\pi} \times 0.69 \times 10^{38} \right)^{2/3}$$

$$= \frac{(6.62 \times 10^{-27})^2}{8 \times 1.67 \times 10^{-24}} \times (66 \times 10^{36})^{2/3} = 54 \times 10^{-6} \mathrm{\ erg}$$

$$\mathrm{KE}_0 = 33.7 \mathrm{\ Mev}$$

We see from the example that the highest occupied neutron energy level is about 34 Mev above the bottom of the potential well, or about 8 Mev below zero energy. Thus a minimum of 8 Mev is needed to remove a single neutron from a typical nucleus. This is in agreement with the data shown in Fig. 15-1. We have only discussed neutrons; however, the protons see just about the same average potential and have just about the same energy levels.

From this simplified "square-well" model of nuclear structure we can predict many things, such as the energies of the excited states (the dashed lines in Fig. 15-10). We would get the result that for a nucleus like phosphorus ($A = 31$), the average spacing between the lowest excited states is about 1 Mev. This checks with the experimental result (see Fig. 15-2). As the diameter of the nucleus or of the potential well is increased, higher order standing waves of the same wavelength (or KE) will fit in; that is, a larger nucleus contains a larger number of permissible energy levels. And since the depth of the well stays the same, the average spacing between energy levels must become less for larger nuclei. For a nucleus-like lead we would predict a spacing between excited states significantly less than 1 Mev. This also agrees with experiment. Heavy nuclei have many closely spaced excited states, while light nuclei have few widely spaced excited states. Not only can the energy values of excited states be crudely predicted, but so can the angular momenta. This is discussed in the next paragraph where an important refinement will be made to our potential well model.

Ans. 6: If s is halved, the particle density $\mathfrak{N}$ increases by a factor of 8, and the Fermi level increases by a factor of $8^{2/3}$ or 4.

Fig. 15-11. Relative spacing of nuclear energy levels taking into account spin-orbit interaction (force is larger if spin and orbital angular momentum are parallel). As with the hydrogen atom, each energy level or "shell" contains subshells corresponding to the quantum number m. The total number of nucleons required to fill these subshells is given in the right-hand column.

N	l	j	No. nucleons per energy shell
4	0	1/2	2
5	2	3/2	4
6	4	7/2	8
5	2	5/2	6
7	6	11/2	12
6	4	9/2	10
5	3	5/2	6
4	1	1/2	2
7	6	13/2	14
4	1	3/2	4
5	3	7/2	8
6	5	9/2	10
3	0	1/2	2
6	5	11/2	12
4	2	3/2	4
4	2	5/2	6
5	4	7/2	8
5	4	9/2	10
3	1	1/2	2
3	1	3/2	4
4	3	5/2	6
4	3	7/2	8
2	0	1/2	2
3	2	3/2	4
3	2	5/2	6
2	1	1/2	2
2	1	3/2	4
1	0	1/2	2

(Energy levels — vertical axis label; the lowest group is braced and labeled 50.)

The shell model

Each of the neutron energy levels shown in Fig. 15-10 corresponds to a definite wave function or "orbit" of definite energy and angular momentum. This is true for both the occupied levels and the higher levels or excited states. These energies and angular momenta can be calculated theoretically and agree quite well with experiment. Figure 15-11 shows the calculated energy and angular momentum of each of the states. These energy levels are those of the potential well shown in Fig. 15-10 modified to take into account the observation that the force on a nucleon is greater if its spin and orbital angular momentum are in the same direction. The sum of the spin and orbital angular momentum is the total angular momentum j. We see that a nucleus that has the $N = 5$, $l = 4$, $j = \frac{9}{2}$ shell completely filled has a total of 50 neutrons (or 50 protons). We also note from Fig. 15-11 that there is a large energy gap between this energy shell and the next highest one. Thus we would expect that nuclei having 50 neutrons ($A-Z = 50$) or 50 protons ($Z = 50$) would be strongly bound and particularly stable. This checks experimentally. For example tin ($Z = 50$) has 10 stable isotopes, which is more than any other element. Also the natural abundance of nuclei with 50 neutrons or 50 protons is significantly higher than those with 51 neutrons or protons. For these reasons 50, along with 2, 8, 20, 82, and 126, is called a magic number. In its present form the shell model does not explain all the observed properties of all the nuclei. However, it has been quite successful in predicting, among other things, the angular momenta of most nuclei and it shows great promise of further improvement.

15-8 Nuclear Fission

"Fission is really of no significance to anybody except people." D. H. Wilkinson

This quotation is an observation that there is nothing of fundamental importance to be learned from fission. As we observed in Section 15-1, the increasing contribution of the electrostatic repulsion in the very large nuclei makes them less strongly bound than nuclei of half the size. Thus if one could slice a big uranium nucleus in two, the resulting two groups of nucleons left over would reorder themselves into more tightly bound nuclei—releasing energy while in the process of reordering. We see that spontaneous fission is permitted by the conservation of energy. However, in the case of the naturally occurring nuclei, the potential barrier (see Section 14-8) is so great that the probability for spontaneous fission is even smaller than that for α decay. For example, the halflife of U^{238} due to spontaneous fission alone is 8×10^{15} years. This is more than one million times the age of the earth. On the other hand, if such a nucleus is hit by a neutron, it can be excited to a higher energy level closer to the top of the electrostatic potential barrier and thus would have higher probability for fission. Also, such an excited state can have high angular momentum and be egg-shaped. The far ends find it much easier to penetrate the barrier because they have already partially penetrated it, hence the barrier is reduced even more when the nucleus is egg-shaped. When U^{238} captures a neutron it becomes U^{239} in an excited state, and the halflives of some of these states for spontaneous fission are just fractions of a second. When U^{235} or Pu^{239} captures a slow neutron, states are formed which have extremely short lifetimes for fission. The mass difference between the uranium nucleus and typical fission products is such that 200 Mev of energy is released in the average uranium fission.

Example 5

How many ergs of energy are released by the fission of 1 gm of uranium fuel? The rest mass of a uranium nucleus is 2.2×10^5 Mev.

The fraction of mass converted to energy is then 200 Mev divided by 2.2×10^5 Mev or 9×10^{-4}. Thus, almost one-tenth of 1% of uranium mass gets converted to usable energy. Since 1 gm of any substance has $Mc^2 = 9 \times 10^{20}$ ergs, 1 gm of uranium releases

$$W = 9 \times 10^{-4} \times 9 \times 10^{20} \text{ ergs} = 8.1 \times 10^{17} \text{ ergs} \qquad (15\text{-}8)$$

Comparing the above result to the 2.9×10^{11} ergs released by burning 1 gm of coal, we see that uranium fission fuel is almost three million times as "efficient." On the other hand, 1 gm of uranium is more expensive than 1 gm of coal. However, the cost per erg is 400 times more for coal than for uranium fuel. This explains the rapidly increasing use of nuclear reactors for power production.

Nuclear fission can be made self-sustaining by a chain reaction process. Each fission releases 2 or 3 neutrons. Then, if one of these neutrons manages to induce fission in another uranium nucleus, the process is self-sustaining. An assembly of fissionable material that meets this criterion is called a critical assembly. The first of these, called a nuclear pile, was constructed by Enrico Fermi in a squash court at the University of Chicago. A bronze plaque is mounted at the site which reads "On December 2, 1942 man achieved here the first self-sustaining chain reaction and thereby initiated the controlled release of nuclear energy."

A mass of U^{235} or Pu^{239} can also be made supercritical. Here the neutrons from one fission induce more than one secondary fission. Since neutrons travel with velocities greater than 10^8 cm/sec, a supercritical assembly can be all used up (or blown apart) in much less than a thousandth of a second. This device is called an A-bomb. The standard method for making a sphere of plutonium supercritical is the implosion technique. A subcritical sphere of plutonium is surrounded by chemical explosives. When these go off, the plutonium sphere is momentarily compressed. Because its density is then significantly increased, it will absorb neutrons at a rate faster than the rate it loses neutrons to the outside. This is the condition for supercriticality. Apparently the explosion of an A-bomb can be made reasonably efficient (most of the plutonium is consumed rather than blown apart). Chemical energies are such that one ton of TNT releases 10^9 calories, or 4×10^{16} ergs. We see from Eq. 15-8 that an A-bomb consuming 1 kg of plutonium or U^{235} has an energy release of 8×10^{20} ergs or 20,000 times as much. This is called a 20-kiloton A-bomb. Thus present-day megaton bombs are roughly a million times more powerful than conventional TNT "blockbusters." Not only is the energy release a million times worse but also each gram of plutonium or U^{235} consumed must result in a gram of fission products, which are all initially radioactive. This is an extremely large amount of radioactivity.

Q.7: What would be the weight of plutonium used up in a 1 megaton fission explosion?

15-9 Nuclear Fusion

$$2 + 2 = 3.975$$

On a scale where the deuterium mass is 2, the observed value of the helium mass would be 3.975. Thus when two deuterons are joined together into helium, six-tenths of 1% of the original rest mass is converted into energy. We see that if this fusion process could be used for energy production, it would be about six times more efficient than fission of uranium. Furthermore, there is an unlimited and inexpensive supply of deuterium in the water of the lakes and oceans, which is not so for other fuels. The world's supply of gas and oil will be depleted in a few decades. Even the supply of coal and uranium will last only a few centuries at most. The big stumbling block to obtaining unlimited energy from "sea water" is Coulomb's law. The electrostatic repulsion between two deuterons at room temperature does not permit them to get within the range of each other's short-range, attractive nuclear force.

Example 6

Assume two deuterons must get as close as 10^{-12} cm in order for the nuclear force to overcome the repulsive electrostatic force. What is the height in million electron volts of the electrostatic potential barrier?

$$U = \frac{e^2}{r} = \frac{(4.8 \times 10^{-10})^2}{10^{-12}} = 2.3 \times 10^{-7} \text{ erg} = 0.14 \text{ Mev}$$

Example 7

To what temperature must deuterium be heated so that the average kinetic energy per deuteron is 0.14 Mev?

$$\overline{\text{KE}} = \tfrac{3}{2}kT = \tfrac{3}{2} \times 1.38 \times 10^{-16} \, T = 2.3 \times 10^{-7} \text{ erg}$$
$$T = 1.1 \times 10^{9\circ} \text{ C}$$

From this example we see that if deuterium could be heated to a billion degrees, fusion would take place. Because of barrier penetration, the temperature need not be this high. Nuclear reactions that require temperatures on the order of millions of degrees are called thermonuclear reactions. The temperatures momentarily obtained in an A-bomb explosion are high enough to ignite a mixture of deuterium, tritium,

Ans. 7: It would be 50 times as much as in a 20 kiloton explosion; namely, 50 kg.

and Li[6]. Once the thermonuclear reactions are started, the extra energy they release can sustain the high temperature until much of the material is quickly "burned." Then we have what is called an H-bomb. The thermonuclear fuel for an H-bomb is very inexpensive, and there is no limit to how much can be used in an individual bomb. H-bombs of about 20-megatons yield (equivalent to 2×10^7 tons of TNT) probably cost about one million dollars each.

Controlled fusion

In order to obtain usable power from fusion, we must have some control over the thermonuclear reactions. We must have some way of generating and sustaining temperatures of many millions of degrees. One of the technical problems is that of confining the high temperature gas or plasma in such a way that the container walls will not melt. A large amount of effort has gone into solving this technological problem, with little success so far. Strong magnetic fields are used in an attempt to keep the plasma ions from the walls. At the present stage of development it is difficult to predict whether this approach will ever be economically successful. Some experts in the field predict success by the end of the century. Another possible approach to thermonuclear power is to find a practical way of tapping the energy release of H-bombs.

Stellar energy

Our own sun is smaller than the average star. Yet it continuously radiates 4×10^{23} kilowatts of energy out into space and has been doing so for billions of years. Such an enormous amount of energy could be supplied only by converting mass into energy, as in the fusion processes. It is believed that one major source of the sun's energy is the conversion of hydrogen into helium by the following sequence of fusion processes:

$$H^1 + H^1 \rightarrow H^2 + e^+ + \nu$$

The symbol H^1 stands for hydrogen and ν for neutrino, a neutral particle of zero rest mass. The above reaction involving a neutrino is called a weak interaction and is discussed in Chapter 16. The above reaction is followed by

Q.8: In a gas of deuterium at a temperature of 1.1×10^8 degrees is it possible for any of the deuterons to have kinetic energy above 0.14 Mev?

$$H^2 + H^1 \rightarrow He^3 + \gamma$$
$$He^3 + He^3 \rightarrow He^4 + H^1 + H^1$$

In the above sequence three hydrogen atoms are used to make He^3. Then two such He^3 eventually combine to form He^4 along with two hydrogens. The net result is that four hydrogen atoms combined to form one He^4 with an over-all energy release of 28.5 Mev.

15-10   Cosmic Rays

The poor man's high-energy accelerator

About 30 ergs per square meter of starlight strike the earth every second. However, an additional, but invisible, 40 ergs per square meter coming from outer space also strike the earth every second. This somewhat larger flux of energy is in the form of high-energy particles, mainly protons and alpha-particles. These primary particles and the secondaries they produce in the earth's atmosphere are called cosmic rays. The number of primary particles striking one square meter of the earth's upper atmosphere and having energy greater than 5 Bev is about 1500 per second. Actually, most of these cosmic ray primaries are below 10 Bev, although some have been observed as high as 10^9 Bev. Cosmic rays have been and still are a useful tool for the study of high-energy interactions. However, the modern high-energy accelerators permit much more controlled and accurate determinations than do most experiments that use the low intensity, uncontrolled cosmic ray particles, although several of the more recent elementary particles (see Chapter 16) were first discovered by using cosmic rays. Now they all are produced by the high-energy accelerators and can be studied in much greater detail.

There are two problems to be explained: (1) where do the cosmic rays come from, and (2) how do these particles get accelerated to such high energies? In 1949 Fermi proposed what is considered to be a major mechanism for accelerating the cosmic ray particles. Observations show that there are moving interstellar gas clouds that have associated magnetic fields (moving charges produce an associated mag-

Ans. 8: Yes. The energy distribution will be exponential with a mean value $\overline{KE} = 0.014$ Mev. A small fraction of the particles can have 10 times the mean energy. (It is like the probability for a nucleus to survive 10 meanlives.)

netic field). Fermi proposed that the "collisions" of the faster moving cosmic ray particles with these regions of magnetic field will tend, on the average, to accelerate the cosmic ray particles. Charged particles moving randomly about encounter the magnetic fields carried by drifting gas clouds, and are "reflected" by them with increased energy if the cloud is initially drifting toward the particle and with decreased energy if the cloud is drifting away. Since more collisions per unit time will occur with the cloud moving toward the particle than when it is moving away, there should be a net gain in energy by the particle, on the average. This is the same mechanism as that which gives rise to the equipartition of energy discussed on page 138.

More recently clues have been obtained as to the source of the cosmic ray particles, which are then further accelerated by the Fermi mechanism as they randomly drift through space. With the help of modern radio telescopes, it has been determined that the Crab Nebula (see Fig. 15-12) contains a high intensity of cosmic ray particles that are leaking out into space. It has been estimated that the number of such particles is so great that they could account for as much as a third of the cosmic ray intensity. Actually, these particular cosmic ray particles have not reached us yet. Because of their repeated Fermi-type collisions with regions of magnetic field, these particles have not had sufficient time to diffuse the 3500 light years distance to the earth. We know that most of these particles were emitted within a period of a few years because the Crab Nebula is the visible remains of a giant supernova, which occurred on July 4, 1054 A.D. The explosion was so spectacular that it was first noticed during the daylight hours. This enormous nuclear explosion was well recorded by the Chinese and Japanese. In fact, there are even crude records of it by the Navaho Indians in an Arizona canyon. But not a single mention of it has been found in any European chronicle. Perhaps Europe had a stretch of bad weather during the peak days of the explosion.

As its hydrogen is consumed, the density of a star changes and the thermonuclear processes can become unstable, as in an H-bomb. A supernova is a gigantic, uncontrolled thermonuclear explosion of an entire star. Supernovae in our galaxy

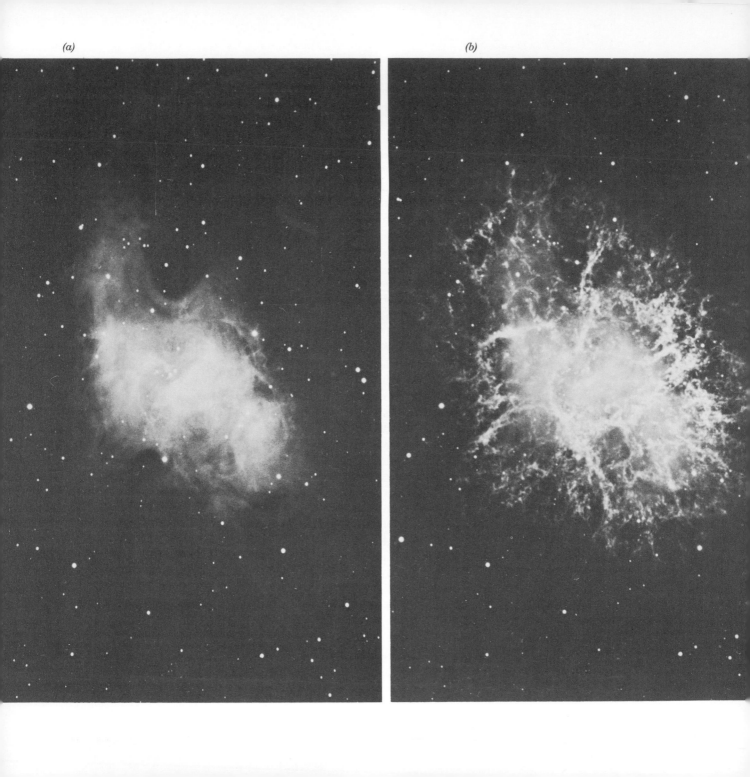

(b)

seem to occur every few hundred years. It seems reasonable that much of our present cosmic rays were emitted by ancient supernovae. The more rare very high-energy cosmic rays may have their source in the newly discovered quasars.

15-11 Social Responsibility of Scientists

The second duty

Whether we like it or not, basic physics research affects more than just our understanding of the physical world. For example, fission and fusion have had such a great impact on our culture that many of us feel the very survival of civilization as we know it is dangerously hanging by a thread. As expressed by E. B. White: "The bomb has given us a few years of grace without war and now it promises us a millenium of oblivion."

Since the development of the atomic bomb in 1945, a small, but quite significant fraction of the world's scientists, have felt some social responsibility for at least pointing out to the public and the politicians the extreme risks involved in policies that might lead to nuclear war. One of the activities of a group of these scientists is the publication of a monthly magazine of science and public affairs, *Bulletin of the Atomic Scientists* (see Fig. 15-13). This sense of responsibility on the scientists' part helps explain the unusual amount of political activity we find among part of the scientific community. We might have expected that the campaign to stop the testing of nuclear weapons would have been initiated by the social scientists, or by the religious leaders, rather than by the scientists.

Some people resent this "meddling" of scientists into politics. There is the valid objection that a scientist who is high in the public esteem for his scientific accomplishments might make rash, oversimplified statements about matters outside his field of competence, thus confusing and misleading the public. Because of his prestige and reputation for careful thinking, a scientist who speaks publicly on related political matters must exercise great care. Just as his scientific opinions are based on careful thought and study of the relevant subjects, so should his political opinions be based on

Fig. 15-12. Two views of the Crab Nebula taken in different kinds of light. View (a) is a continuous spectrum of light emitted by the radiation of high energy particles accelerating in magnetic fields. View (b) is taken in light corresponding to the emission spectrum of hydrogen. This view shows the locations of the gas clouds rather than the cosmic ray particles.

careful thought and study. But once this condition is met, the scientist who delves into the relationship between science and politics is performing a valuable public service.

Such a scientist also feels he is performing one of his duties as a scientist. This "second duty" of the scientist is expressed by John M. Fowler in his book *Fallout*.

Public opinion on the great nuclear questions remains largely unformed and uninformed. To present to the public the raw materials from which this opinion can be forged is both the privilege and the duty of the scientist. For in our world of complex knowledge and burgeoning technology, scientists have not one but two essential duties: first, the traditional duty of seeking the truth; secondly, the duty to communicate to all who need it the knowledge gained in their search.

Fig. 15-13. Front cover of two successive issues of the Bulletin of the Atomic Scientists. Serving as a warning that there is little time left, the hands of the clock were moved forward as a consequence of the first Soviet H-bomb explosion of Aug. 8, 1953.

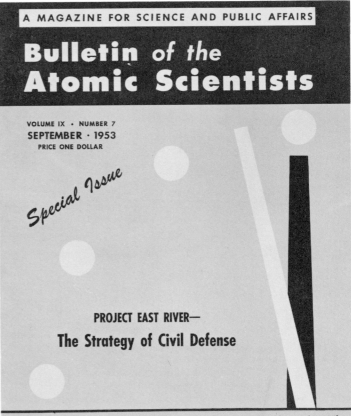

A MAGAZINE FOR SCIENCE AND PUBLIC AFFAIRS

Bulletin *of the* **Atomic Scientists**

VOLUME IX · NUMBER 6
JULY · 1953
PRICE FIFTY CENTS

Eleanor Bontecou: President Eisenhower's "Security" Program
Reuel Denney: Reactors of the Imagination
Wayne Morse: Wanted—Fair Procedures

A MAGAZINE FOR SCIENCE AND PUBLIC AFFAIRS

Bulletin *of the* **Atomic Scientists**

VOLUME IX · NUMBER 7
SEPTEMBER · 1953
PRICE ONE DOLLAR

Special Issue

PROJECT EAST RIVER—
The Strategy of Civil Defense

Special Editor: Ralph Lapp · *Contributors:* Val Peterson · Alexander Wiley
Ramsay Potts, Jr. · Joseph McLean · Robert Stokley · David F. Cavers · Henry
Parkman · Roland Sawyer · Horatio Bond · Donald Monson, and Others

Some scientists feel this "second duty" more strongly than others. However, as long as political judgments must be based on technical knowledge and scientific judgment, there will be scientists who feel it their duty to help contribute toward political judgments as well. Certainly it would be dangerous to society for a politician who has no scientific background to make such vital decisions alone.

Appendix 15-1 *Determination of Nuclear Sizes Using Neutron Beams*

A neutron will travel in a straight line until it hits an atomic nucleus. Then it will interact and be lost from the original neutron beam. The effective area of the atomic nuclei in a sheet of material such as a copper plate can be determined by measuring how many neutrons are lost (the quantity $(N - N')$ in Fig. 15-14). The probability of losing a neutron is the ratio of the total nuclear area to the area of the plate.

Fig. 15-14. Beam of neutrons striking plate of area $\mathcal{C}$.

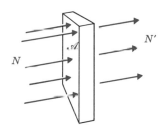

$$prob. = \frac{N_a \sigma}{\mathcal{C}}$$

where N_a is the total number of atoms in the plate, σ is the effective area of each atomic nucleus, and $\mathcal{C}$ is the area of the plate. The number of neutrons lost will be this probability times the number of neutrons hitting the plate:

$$(N - N') = \left(\frac{N_a \sigma}{\mathcal{C}}\right)N$$

Solving for σ gives

$$\sigma = \frac{N - N'}{N \times N_a}\mathcal{C} \quad \text{is the area of the nucleus}$$

N and N' can be measured by a special neutron counter similar to a geiger counter. The total number of atoms N_a is obtained by weighing the plate

$$N_a = \frac{M}{A}N_0$$

where M is the mass of the plate, A is the atomic number, and N_0 is Avogadro's number. The quantity σ is called the nuclear cross section and for high energy neutrons is πR^2, where R is the nuclear radius.

Appendix 15-2

Determination of Nuclear Size Using High Energy Electrons

As shown in Fig. 15-15 if a high energy electron passes by a nucleus of charge Ze at a distance b, it will be deflected toward the nucleus by an angle θ. Because of Coulomb's law, the closer the approach to the nucleus, the stronger will be the force, and the greater will be the deflection. An equation can be derived giving θ in terms of b. However, if the electron gets so close that it penetrates inside the nucleus, it no longer sees a strong Coulomb force (we recall that the electric field inside a uniformly charged spherical shell is zero). Hence the maximum angle of deflection occurs when $b = R$, the nuclear radius.* We see that the nuclear radius can be determined by measuring $\theta_{\max}$. All we need is the formula relating b to θ. An approximate derivation goes as follows.

Let Δp be the change in momentum of the electron due to the Coulomb force. From Newton's definition of force we have

$$\Delta p = F\,\Delta t$$
$$= \frac{Ze^2}{b^2}\,\Delta t$$

We make the approximation that the force acts full strength when the electron is within a distance b of its point of nearest approach; that is, $\Delta t = 2b/v$. Then

$$\Delta p \approx \frac{Ze^2}{b^2}\left(\frac{2b}{v}\right)$$

From the figure we see that $\Delta p = p \tan \theta$. Substituting this into the left-hand side of the above equation gives

$$p \tan \theta \approx \frac{2Ze^2}{vb}$$

or

$$b \approx \frac{Ze^2}{pv\,\dfrac{\tan \theta}{2}}$$

Our approximations turn out quite good this time since the exact relation is

$$b = \frac{Ze^2}{pv \tan \dfrac{\theta}{2}}$$

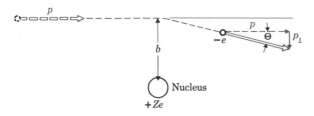

Fig. 15-15. Electron of momentum p scattered by atomic nucleus of charge Ze.

---

*Actually, electrons penetrating inside the nucleus will still experience some force due to charge inside the nucleus, and hence be deflected somewhat more than $\theta_{\max}$. The limit $\theta_{\max}$ is thus a fuzzy limit. But if we measure the shape and extent of this "fuzziness" in the scattering distribution, we can calculate back and determine the charge distribution inside the nucleus. This has even been done for high-energy electron-proton scattering with the result that no charged core to the proton has been found.

The nuclear radius in terms of θ_{max} is then

$$R = \frac{Ze^2}{W \tan \frac{\theta_{max}}{2}}$$

where W is the electron energy.

Problems

1. Is the mass of an atomic nucleus less or greater than that of the particles of which it is composed?

2. A radioactive isotope with a one year halflife is produced and an individual nucleus doesn't decay in the first year. What then is the probability that this nucleus will decay sometime during the second year? If this particular nucleus happens to survive the first two years without decaying, what will be the probability that it will decay sometime during the third year?

3. If the halflife of radium is 1600 years, what fraction of a sample of radium will have decayed after 3200 years?

4. What is the halflife of the isotope whose decay curve is as shown?

5. Which is longer, three halflives or two meanlives?

6. What per cent of a radioactive sample decays during one meanlife? During two meanlives?

7. In a certain heavy nucleus an alpha particle collides with the potential barrier 10^{99} times per second and

$$\left(\frac{\psi_{out}}{\psi_{in}}\right) = 10^{-14}.$$

    (a) What is the probability that this nucleus would decay in one second?
    (b) Approximately how long would the average nucleus of this type live? (Give answers to the nearest power of 10.)

8. Consider a sample of 1000 radioactive nuclei with a halflife T. Approximately how many will be left after a time $T/2$? (*Hint:* the same fraction decays during the first half-halflife as the second.)

9. The diagram represents the potential well for a proton in an iron ($Z = 26$) nucleus. In the ground state all the proton energy levels are filled up to W_0.
    (a) Which letter represents the kinetic energy of the Fermi level?
    (b) If a proton of energy W_0 inside the nucleus is given additional energy of the amount $(a + d)$, what will be its kinetic energy when it gets far away from the nucleus?
    (c) Write a formula for the quantity "a" in terms of R and any other quantities you wish to use.

Prob. 4

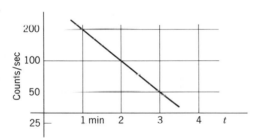

Prob. 9

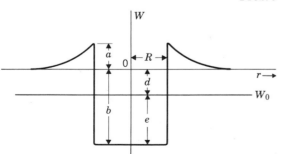

(d) If we go to a smaller nucleus such as carbon, the quantity "d" will be larger, smaller, the same in magnitude. If we go to a larger nucleus such as lead, the quantity "d" will be larger, smaller, the same in magnitude.

(e) If we go to a nucleus with twice as many nucleons as iron, by what factor will R increase?

(f) Assume that in this double-sized nucleus the density of nuclear matter stays the same. Does quantity "e" increase, decrease, or remain the same?

10. The sun has a diameter of 864,000 miles and an average density of 1.41. What would be the diameter of the sun if it had the same mass but the density of nuclear matter?

11. In a thermonuclear bomb, 18 kg of explosive can give an energy release equivalent to one million tons of TNT. One ton of TNT releases 10^9 calories. How many grams of this thermonuclear explosive get converted to energy?

12. A big nucleus X contains 204 nucleons and has a binding energy per nucleon (BE/A) equal to 8 Mev. Assume that the rest energy of one free proton or one free neutron is 940 Mev.

(a) Find the rest energy (or rest mass) of nucleus X.

(b) Nucleus X emits an α-particle (BE/A) = 7 Mev and changes into nucleus Y which has (BE/A) of 8.1 Mev. Find the amount of energy released (in the form of KE of α and Y) in this process.

13. According to Fig. 15-11 a nucleus having $Z = 50$ (tin) would be particularly stable. From this figure determine another element that should be particularly stable.

14. A sample of radioactive material contains 10^{12} radioactive atoms. If the halflife is 1 hr, how many of these atoms will decay in 1 sec? (*Hint:* What is the relation between the probability that an atom decays in 1 sec and its meanlife? What is the relation between the meanlife and the halflife?)

15. One rad will liberate 1.2 statcoulombs of positive and 1.2 statcoulombs of negative ions per cubic centimeter of air. If the density of air is 1.3×10^{-3} gm/cm^3, how many electron volts are required for each ion pair?

16. We shall consider the following photonuclear reaction:

$$\gamma + Cu^{63} \rightarrow (Cu^{63})^* \qquad \text{followed by} \qquad (Cu^{63})^* \rightarrow Ni^{62} + P.$$

The highest energy proton in a copper nucleus sees a potential well as shown above. For this proton KE + U = -8 Mev.

(a) What is the kinetic energy of this proton?

(b) If this particular proton absorbs a 15 Mev gamma ray, what will be the barrier height for it? (How much more energy is needed to get over the top of the barrier?)

(c) What is the potential energy of this proton while inside the nucleus?

Prob. 16

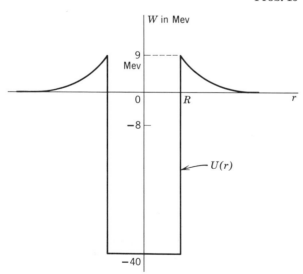

(d) What is the binding energy of this proton?

(e) One could now calculate $\dfrac{\psi_{\text{out}}}{\psi_{\text{in}}}$ for such an excited proton. Assume the result is 10^{-8} and that the proton makes 10^{22} collisions per second with the barrier. What then is the meanlife for the decay $(Cu^{63})^* \rightarrow Ni^{62} + P$?

(f) What would be the KE of the decay proton after it is far from the nucleus?

(g) Suppose the proton has instead absorbed a lower energy gamma ray. Would the meanlife be greater, less or the same as your previous result?

(h) What is the energy threshold for this photonuclear reaction; that is, what is the lowest possible energy gamma ray that could still emit a proton?

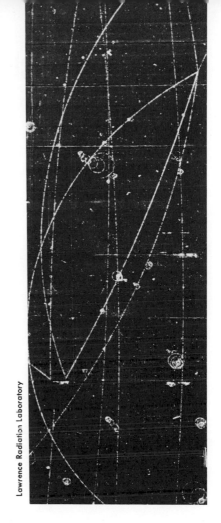

Lawrence Radiation Laboratory

Particle physics

Chapter 16

Particle physics

16-1 Introduction

Peeling the onion

The main goal of physics is to explain all physical phenomena by means of a small number of simple, fundamental principles. Since all matter is made up of just a few different kinds of elementary particles, the countless number of physical phenomena and properties of matter can be explained by the few simple properties of a small number of elementary particles.

In man's search for the elementary particles, he first found that compounds were made up of "elementary" molecules. Then it was learned that molecules are made up of "elementary" atoms. Many years later these "elementary" atoms were discovered to be made up of "elementary" nuclei and orbital electrons. These successive probings to learn what is truly elementary are like peeling away successive layers of an onion. The latest stage in this peeling down of ordinary matter was the discovery that all nuclei are made up of protons and neutrons.

Have we finally reached the core of the onion? Are the proton and neutron made up of even smaller elementary particles? According to our present knowledge the proton, neutron, electron, and photon are all thought of as elementary particles. Now that we have gotten down to the basic building blocks of ordinary matter and explained its structure and properties in terms of fundamental principles, this would seem like an appropriate place to end the book. However, since the purpose of this book is to cover fundamental topics of physical reality, before closing we must ask the question: are there any other physically real elementary particles that happen not to appear in ordinary matter? The answer is an emphatic yes! Since 1933 physicists have discovered over 190 other elementary particles. As far as we know, none of these new elementary particles can be broken down into smaller particles.* They are all considered elementary in the sense that they are structureless (cannot be explained as a system of other elementary particles). For

---

*However, some physicists have recently proposed that the proton, neutron, and all other strongly interacting particles are made up of an even more basic particle called the quark.

example, there are several reasons why the neutron cannot be explained as made up of an electron plus a proton bound together. One reason is that the mass of a neutron exceeds that of a proton plus an electron by about 1.5 electron masses. A system of electron and proton bound together must have total mass less than the sum of the two (as does the hydrogen atom). Another reason is that the spins of the neutron, proton, and electron are all $\frac{1}{2}$. According to quantum mechanics there is no way to obtain a resultant angular momentum of $\frac{1}{2}$ unit from two spin-$\frac{1}{2}$ particles. Another reason why any "elementary" particle cannot contain an electron in its structure is given by the uncertainty principle. According to this principle, the electron momentum would be at least as large as Planck's constant divided by the diameter of the "elementary" particle. Since elementary particles have diameters of about 10^{-13} cm, an electron "bound" in an elementary particle would have

$$ P = \frac{h}{10^{-13}} = 6 \times 10^{-14} \text{ gm cm/sec} $$

An electron of such momentum would have a kinetic energy of about 100 Mev and would immediately escape.

Most of the new elementary particles that we have not yet covered are unstable—they decay or transform into other elementary particles of smaller mass. Before discussing these new particles, it is necessary to learn more about beta decay and what is called the weak interaction.

16-2   Beta Decay and the Weak Interaction

Slow decay

We recall from Chapter 4 that there are only four basic interactions: the strong or nuclear, the electromagnetic, the gravitational, and the weak interaction. The strong interaction was featured in Chapter 15. Chapters 7 and 8 were devoted to the electromagnetic interaction, and Chapter 4 to the gravitational interaction. In this section we finally take up the weak interaction.

If it were not for the law of conservation of heavy particles, all the matter in the universe would decay into electrons and

Q.1: What would be the minimum momentum of a nucleon "bound" in an elementary particle?

Fig. 16-1. R. P. Feynman at "work." Professor Feynman is one of the main contributors to the modern theory of quantum electrodynamics and the universal Fermi interaction. (Photo by H. Y. Chiu.)

neutrinos in less than a thousandth of a second. This is a consequence of what is called the weak interaction. The weak interaction can be thought of as a universal disease attacking all particles with the same strength. It tries to transform elementary particles into electrons and neutrinos as a final product. Historically these decay electrons are called beta rays.

Since beta decay always involves neutrinos, a description of the neutrino is essential. A neutrino is an elementary particle that has no charge and no rest mass. Furthermore, the interaction between a neutrino and anything else is so weak as to be almost unobservable. If a beam of 10^{12} neutrinos were shot at the earth, all but one would pass through the earth completely unaffected. So far this particle sounds like a figment of the imagination. However, the neutrino is not a hoax dreamed up by theoretical physicists. In the last few years the evidence, both experimental and theoretical, has become so convincing that no competent scientist questions the existence of the neutrino.

As an elementary particle, the neutrino is in some ways analogous to a photon. Since a photon has zero rest mass, its energy is $W = Pc$ where P is the photon momentum. Because of relativity, this same relation must also hold for neutrinos (see page 308). Furthermore, as with electrons, protons, and neutrons, the neutrino has a spin of $\frac{1}{2}$ (an intrinsic angular momentum of $\frac{1}{2}h/2\pi$).

In 1958 great progress was made in understanding the weak interaction. A specific interaction that can transform particles into electrons and neutrinos was proposed. This specific interaction is called the universal Fermi interaction. For example, the new theory gives an accurate prediction of the muon (a new particle discussed in Section 16-4) lifetime.

One example of a weak interaction is the beta decay of a free neutron by the process

$$N \rightarrow P + e^- + \bar{\nu}_e$$

with a halflife of 12 min and energy release of 0.8 Mev. The neutron-proton mass difference is 1.3 Mev. Since the rest mass of the electron is 0.5 Mev there is 0.8 Mev left over which goes into the kinetic energy of the electron plus the

Ans. 1: $P \approx \Delta P \approx \dfrac{h}{R}$ is independent of the mass of the particle. The nucleon momentum would be the same as that of an electron confined to $R = 10^{-13}$ cm. However the nucleon energy would be 1837 times smaller because $\mathrm{KE} = \dfrac{P^2}{2M}$.

kinetic energy of the antineutrino. (The symbol $\bar{\nu}_e$ is used for the antineutrino that accompanies an electron.) The difference between the neutrino and the antineutrino is explained in the next section. Note that the decay electron and antineutrino can share the 0.8 Mev of kinetic energy in any way they choose. Thus, in a large group of neutron decays, electrons may be found having values of kinetic energy anywhere from zero to 0.8 Mev. If the neutron decayed into a proton and electron only, the electron would always have the full kinetic energy of 0.8 Mev. Experimentally the electron energy can be determined by measuring its radius of curvature in a magnetic field (see Fig. 8-12). The experimental result is that beta rays rarely have their maximum permissible kinetic energy. Historically this "puzzle" of what happened to the "missing" kinetic energy was the reason the neutrino was "invented" in the first place.

In order to preserve conservation of energy, W. Pauli proposed in 1930 that the missing energy may have been carried off by an undetectable, light, neutral particle. Shortly thereafter E. Fermi named this particle the neutrino (little neutral one) and worked out the theory of beta decay that is similar to the modern, more universal theory of weak interactions, called the universal Fermi interaction. Another check on the existence of the neutrino would be to measure its momentum. In a cloud chamber picture of an individual beta decay, the electron energy can be measured. This measurement determines not only what the neutrino energy should be but also its momentum according to the relation $P = W/c$. The neutrino momentum can then be determined independently by measuring the electron momentum and the recoil momentum of the residual nucleus. These determinations of neutrino energy and momentum always checked with the theoretical prediction $P = W/c$. Not only does the "invention" of the neutrino save the laws of conservation of energy and momentum but it also saves the law of conservation of angular momentum. In Section 16-1 we observed that quantum mechanics does not permit two spin-$\frac{1}{2}$ particles (such as $P + e^-$) to have a total angular momentum the same as that of the neutron which is also spin-$\frac{1}{2}$. However, the three decay products of the neutron,

Q.2: In the beta decay of a neutron, suppose the decay electron and proton have the same momentum. What is the ratio of their energies according to classical mechanics?

all of which are spin-$\frac{1}{2}$, can add up to give a total angular momentum of $\frac{1}{2}(h/2\pi)$.

The evidence for neutrinos as decay products was quite satisfying. However, the physicists wanted to be more satisfied by observing a direct interaction starting with a beam of neutrinos. According to the Fermi theory the following reaction should take place for antineutrinos which have sufficient energy to make up the mass difference:

$$\bar{\nu}_e + P \rightarrow N + e^+$$

where e^+ is a positron (positive electron). However, the predicted occurrence of this direct neutrino interaction was close to zero. The only possibility of observing this reaction would be to obtain an extremely intense beam of antineutrinos. The recent development of the nuclear reactor made this possible. Each fission in a nuclear reactor leads to several beta decays, and hence several antineutrinos. Such an experiment would require a high-power nuclear reactor and an elaborate large-sized detector. A group from Los Alamos undertook this experiment using the detector shown in Fig. 16-2, with the result that the elusive neutrino was first trapped by man in 1956.

Because of the law of conservation of heavy particles, the free proton is immune to the weak interactions. Since there is no heavy particle of mass less than the proton, there is nothing available for it to decay into.

Beta decay

The question now arises: how can neutrons inside of atomic nuclei be stable if the halflife of the free neutron is 12 min? One way of answering this "paradox" is that the binding energy of the neutron to the nucleus effectively lowers its mass below that of a proton in the corresponding situation. If an atomic nucleus by virtue of its binding energy has less mass than any possible combination of decay products, then because of the law of conservation of energy such a nucleus will be stable. This is why only certain isotopes are stable and all others are radioactive. For example, hydrogen and deuterium are stable, but tritium (one proton and two neutrons) is not.

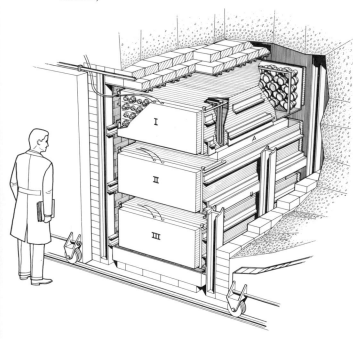

Fig. 16-2. Cutaway drawing of neutrino detector in wall of nuclear reactor. Tanks I, II, and III each contain 370 gallons of liquid scintillator and are each monitored by 110 photomultiplier tubes. The positron in the reaction $\bar{\nu} + P \rightarrow N + e^+$ will produce distinctive light pulses in the liquid scintillator. (Courtesy Los Alamos Scientific Laboratory and Dr. F. Reines.)

Ans. 2: $\dfrac{\text{KE}_e}{\text{KE}_P} = \dfrac{P_e^2}{2M_e} \div \dfrac{P_P^2}{2M_P} = \dfrac{M_P}{M_e} = 1837.$

Example 1

The nuclear masses of tritium and helium 3 are

$$M_{\mathrm{H}^3} = 2805.205 \ \mathrm{Mev}$$

and

$$M_{\mathrm{He}^3} = 2804.676 \ \mathrm{Mev}$$

respectively. The electron mass is

$$M_{e^-} = 0.511 \ \mathrm{Mev}$$

Will tritium beta decay, and if so, what will be the maximum possible beta ray energy?

If the mass of tritium happens to be greater than that of He³ plus an electron, then the reaction

$$\mathrm{H}^3 \rightarrow \mathrm{He}^3 + e^- + \bar{\nu}$$

is not forbidden by the conservation of energy. The kinetic energy shared by the e^- and the $\bar{\nu}$ would be the mass difference

$$M_{\mathrm{H}^3} - (M_{\mathrm{He}^3} + M_{e^-}) = 2805.205 - (2804.676 + 0.511)$$
$$= 0.018 \ \mathrm{Mev}$$

Hence the decay electrons from a sample of tritium will have energies ranging all the way from zero to this value of 0.018 Mev.

Tritium can be produced by bombarding heavy water with neutrons. The decay electrons or beta rays from tritium are observed to have kinetic energies up to 0.018 Mev with a halflife of 12 years.

Another example of beta decay is the isotope C¹⁴ which occurs naturally in small amounts. It is formed by atmospheric absorption of neutrons produced by cosmic rays and by H-bomb explosions. The C¹⁴ nucleus is of slightly greater mass than N¹⁴ plus an electron. Then, because of the weak interaction, the following decay process must occur:

$$\mathrm{C}^{14} \rightarrow \mathrm{N}^{14} + e^- + \bar{\nu}_e$$

In this case the halflife is $T = 5000$ years. On the other hand, the stable isotope C¹² is of smaller mass than any possible combination of byproducts containing 12 nucleons. According to the law of conservation of energy there is nothing for C¹² to decay to and it thereby is saved from the weak interaction. Most of the radioisotopes that can be produced beta decay, emitting either a positron or electron along with a neutrino or antineutrino respectively.

Q.3: Would the C¹⁴ nucleus be stable if it were 0.4 Mev heavier than N¹⁴?

16-3 Antimatter

The antiworld

The relativistic quantum theory of spin-$\frac{1}{2}$ particles not only gives us the exclusion principle but it also predicts the existence of what is called an antiparticle. The antiparticle of a given particle should have exactly the same mass, but opposite charge. Also an antiparticle can be annihilated by its corresponding particle. Then the two rest masses are directly converted into energy, in the form of other particles such as photons. The first antiparticle known to man was the positron which was discovered in 1933 in a cloud chamber exposed to the cosmic rays. The first positron was found by accident even though its existence had already been predicted by the relativistic quantum theory. The positron or positive electron has the same mass, but opposite charge, as that of an electron. When a positron comes to rest in matter, it is quickly annihilated by an electron usually into two photons:

$$e^+ + e^- \rightarrow 2\gamma$$

Then each photon must have an energy of 0.51 Mev, which is the rest mass of an electron. Positrons are easily produced by a process called pair production (a high-energy photon strikes a nucleus and is completely converted into an electron-positron pair).

$$\gamma \rightarrow e^+ + e^-$$

The above is one of many examples of direct conversion of energy into rest mass. As mentioned in the previous section, both electrons and positrons can be products of beta decay. In beta decay a positron is always produced along with a neutrino.

The antiparticle of a proton is called the antiproton or negative proton $\bar{P}$. The standard notation for an antiparticle is a bar over the symbol. Hence $\bar{P}$ stands for antiproton and $\bar{N}$ for antineutron. The antielectron (the positron) would be $\bar{e}^-$, but for this the usual convention is e^+. After the discovery of the positron in 1933, many physicists felt that there must also be an antiproton. According to theory, antiprotons could

Ans. 3: Yes, because then it could not beta decay into ($N^{14} + e^-$) which would be heavier by 0.1 Mev.

be produced by bombarding nuclei with protons of six billion electron volts kinetic energy. One of the production reactions should be then

$$P + P \rightarrow P + P + \bar{P} + P$$

The 6 Bev is directly converted into the rest mass energy of a proton-antiproton pair in addition to the kinetic energy of the final particles. The possibility of discovering the antiproton was one of the main considerations that led the United States AEC to build the Bevatron, a high-energy proton accelerator located in Berkeley, California. The Bevatron, shown in Fig. 16-3, accelerates protons to a kinetic

Fig. 16-3. The Bevatron, located at the University of California, Berkeley, California. (Courtesy Lawrence Radiation Laboratory.)

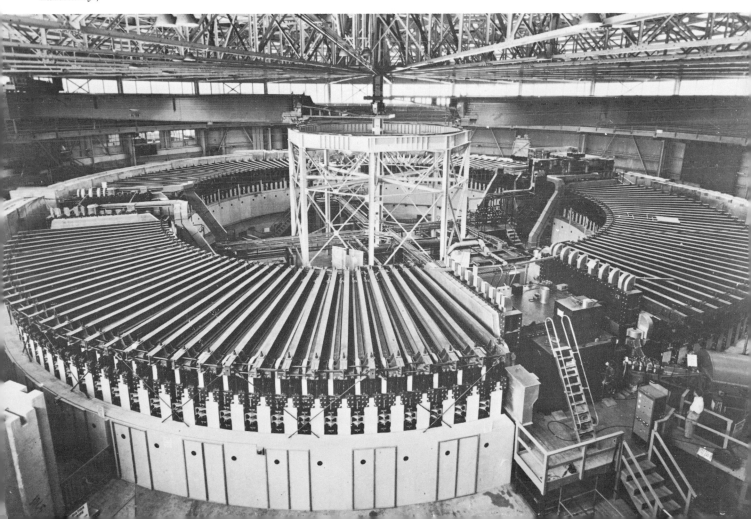

energy of 6.2 Bev, which is just barely enough to produce nucleon-antinucleon pairs. In 1955, the second year of Bevatron operation, the antiproton was discovered. The antineutron $\overline{N}$ was discovered a year later. Since the neutron is neutral in charge, so must the antineutron be neutral. However, the antineutron is quickly annihilated by either a neutron or a proton. With antinucleon annihilation, the annihilation products are usually pions (see Section 16-6). In Fig. 16-4 an antiproton enters a liquid hydrogen bubble chamber, slows down to rest, and is annihilated by one of the hydrogen nuclei. In this particular photograph, the rest mass of the antiproton and proton is converted into five pions:

$$\overline{P} + P \rightarrow \pi^+ + \pi^+ + \pi^- + \pi^- + \pi^0$$

The question arises why all hydrogen atoms are made up of positive protons and negative electrons in preference to negative protons (antiprotons) and positive electrons (positrons). Such a "reversed" hydrogen atom is called an antihydrogen atom. Matter made of antinucleons and orbital positrons is called antimatter. By general symmetry considerations one would expect half the atoms in the universe to be antimatter. It is difficult to understand why there should be a preference to positive charge over negative charge. On the other hand, if there were any antimatter on earth, or even in our galaxy, it would not last very long. It would quickly annihilate away releasing energy with more than 1000 times the efficiency of an H-bomb. At present there is speculation that some galaxies may be made of antimatter, but so far there is no sufficient evidence.

As already mentioned, the antiparticle of the neutrino is the antineutrino $\overline{\nu}$. According to the theory, the photon must be its own antiparticle. Counting the antiparticles, our list

Fig. 16-4. Antiproton coming to rest in a liquid hydrogen bubble chamber. The antiproton is annihilated by a proton. The annihilation products are five pions: two positive, two negative, and one neutral. One of the positive pions also comes to rest and then decays into a μ^+. The μ^+ comes to rest and decays into a position. (Courtesy liquid hydrogen bubble chamber group of the Lawrence Radiation Laboratory.)

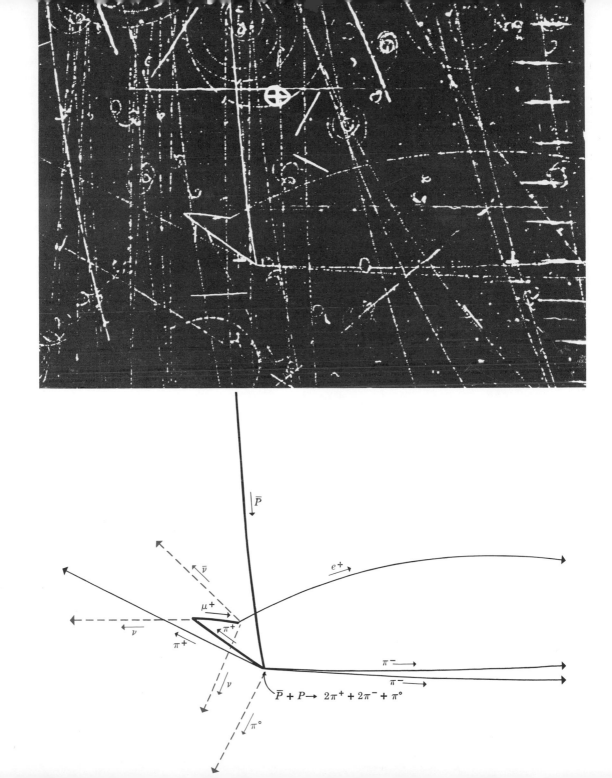

$\bar{P}$

$e+$

$\bar{\nu}$

$\mu+$

$\pi+$

ν

$\pi+$

ν

π^-

π^-

$\bar{P} + P \rightarrow 2\pi^+ + 2\pi^- + \pi^\circ$

π°

of elementary particles has now grown to nine (γ, ν, $\bar{\nu}$, e^-, e^+, P, $\bar{P}$, N, $\bar{N}$).

16-4   Antiparticle Symmetry

Is the antiworld any different?

Suppose during the course of any physical experiment that all the particles were suddenly changed to their corresponding antiparticles. Would the experiment proceed to give the same results? Up until 1957 physicists had assumed that antiparticles should obey exactly the same laws of physics as their counterparts. In principle there should be no way to tell whether a certain physical system is built out of antimatter or ordinary matter. We shall call this fundamental symmetry principle antiparticle symmetry. Theoretical physicists usually call it the law of charge conjugation invariance. Charge conjugation is a mathematical operation that changes every particle to its antiparticle while leaving everything else the same. The charge conjugate of a hydrogen atom is antihydrogen. The principle of antiparticle symmetry predicts, for example, that the spectrum emitted by antihydrogen gas should be exactly the same as that from ordinary hydrogen. Since antiparticles are hard to produce (antihydrogen has never been produced), some aspects of the principle of antiparticle symmetry are difficult to check by experiment.

Actually, in 1957 physicists were suddenly shocked to learn that the principle of antiparticle symmetry is violated by the weak interaction. The nature of this violation is discussed in Section 16-6.

16-5   Conservation of Leptons

Weak particles

Not only does the weak interaction cause a heavy particle to emit an electron and neutrino, but in some cases a muon and neutrino instead. The muon is an elementary particle which is the same as the electron in all respects except for its rest mass. It happens to have a rest mass 207 times that of the electron. The muon should be thought of as a heavy electron. However no other heavy electrons occur in nature. Just why there should be two kinds of "electrons" nobody knows. One of the current problems in theoretical physics is to explain why nature provides us with two and only two types of electrons. It is also a puzzle why the muon should be so heavy.

Muons are not uncommon. They were discovered in 1936 in the cosmic rays. Actually at sea level the cosmic rays consist mainly of muons and electrons in the ratio of 4 to 1 muons to electrons. Just as the e^+ is the antiparticle of the e^-, so is the μ^+ the antiparticle of the μ^-. The reason why ordinary matter is not made of muons as well as electrons is because muons can decay by the weak interaction into electrons with a halflife of 1.5×10^{-6} sec:

$$\mu^- \to \nu_\mu + e^- + \overline{\nu}_e$$

Note that we are giving subscripts e and μ to the decay neutrinos. This is because in 1963 a group at the Brookhaven AGS discovered that there are two distinct kinds of neutrinos: the ν_μ which is associated with the muon, and the ν_e which is associated with the electron. The ν_μ should be thought of as the neutral muon and the ν_e as the neutral electron. We could have used the notation μ° for ν_μ and e° for ν_e. As a class, muons, electrons, and their corresponding neutrinos are called leptons. Leptons are involved in weak interactions. It is observed that whenever leptons are created, they are created in pairs. This observation has led to a new conservation law called the conservation of leptons. Actually there are two independent conservation of lepton laws, one for electrons and one for muons. The electron e^- and its neutrino ν_e, are given electron lepton number $+1$ and their antiparticles, e^+ and $\overline{\nu}_e$, are given electron lepton number -1. In any closed system the total electron lepton number must be conserved. For example, in neutron decay

$$N \to P + e^- + \overline{\nu}_e$$

Lepton no.: $(0) \to (0) + (1) + (-1)$

the total lepton number before decay is zero. Hence the total lepton number on the right-hand side must also be zero. We see that it is $(1) + (-1) = 0$. Similarly in a beta decay,

Q.4: Counting all charges, what is the total number of leptons presently known?

where a positron is emitted, the neutrino must be of opposite lepton number. Since the positron lepton number is (-1), the emitted neutrino must have lepton number $(+1)$; hence it must be ν_e rather than $\bar{\nu}_e$.

Our final example of conservation of leptons is the decay of the muon:

$$\mu^- \rightarrow \nu_\mu + e^- + \bar{\nu}_e$$

Muon lepton no.: $(1) \rightarrow (1) + (0) + (0)$

Electron lepton no.: $(0) \rightarrow (0) + (1) + (-1)$

The initial muon lepton number is $+1$. This value of $+1$ is conserved if the muon decays into a neutral muon ν_μ. Electric charge is conserved by emitting an electron, but then to conserve the electron lepton number, an antineutrino, $\bar{\nu}_e$, must also be emitted.

In summary, there are 8 leptons: $(\mu^-, \nu_\mu, e^-, \nu_e)$ and their antiparticles $(\mu^+, \bar{\nu}_\mu, e^+, \bar{\nu}_e)$ and these particles occur in pairs in weak interactions.

16-6 The Hadrons

Strong particles

Just as the leptons are the particles of the weak interaction, the hadrons are the particles of the strong or nuclear interaction. The electromagnetic interaction cuts across these boundaries and interacts with any charged particle as well as the photon. It interacts with charged leptons as well as charged hadrons.

The most familiar hadrons are our friends the proton and neutron. But since 1947 many other unstable hadrons have been discovered. Twenty one of them (see Fig. 16-8) have long enough halflives so that their tracks can be seen in nuclear emulsions or bubble chambers. Most of these 21 have halflives of about 10^{-10} sec. A particle of velocity $v = c$ will travel 3 cm in 10^{-10} sec. In spite of the fact that certain unifying properties and groupings among these particles will become evident, the amount of data presented here may seem overwhelming, making us wonder whether there is something more elementary and simpler than the elementary particles. This indeed is the central question physics is now

Ans. 4: Eight.

facing. In this book we have now reached the outer frontier of human knowledge and understanding of the physical world. The remaining sections of the book attempt to give some feeling for our present-day "pioneer" exploration of the physical world.

There are two types of hadrons called mesons and baryons. The mesons have spin 0, 1, 2, or some whole integer. The baryons have half-integer spin; namely, $\frac{1}{2}$, $\frac{3}{2}$, $\frac{5}{2}$, etc. The baryons obey the law of conservation of baryons and always have a proton or antiproton as a final decay product. All baryons such as the proton and neutron have baryon number $+1$ and all antibaryons such as $\overline{P}$ and $\overline{N}$ have baryon number -1. For a closed system, the total baryon numbers must remain constant. Mesons have zero baryon numbers.

The mesons

The two longest-lived mesons are the pion and the K meson. The pion mass is about one-seventh that of a proton and the K meson is about one-half a proton in mass.

Pions have spin zero and occur with negative, positive, and neutral charge (π^-, π^+, π^0). The π^- is the antiparticle of the π^+. As with the photon, the π^0 is its own antiparticle. The "reason" for the existence of pions is easier to understand theoretically than that for muons. In fact, in 1936, eleven years before its discovery, the pion was predicted by H. Yukawa. Yukawa tried to explain the strong nuclear force in the same way that quantum electrodynamics explains the electric force. In quantum electrodynamics the electric force is explained in terms of an electric charge continually emitting and reabsorbing virtual quanta (photons). Yukawa invented a new type of virtual quantum to explain the strong, short-range nuclear force. The quantum theory specifies the mass of this new type of particle (quantum) in terms of the range of the nuclear force. This can be seen crudely using the uncertainty principle. If the range of these virtual quanta is R, the uncertainty principle says

$$\Delta p R \approx h$$

where Δp is the uncertainty in momentum which will be on the order of $m_\pi v$. Hence

Q.5: The muon is often called the mu meson. According to our definition of meson, is the muon a meson?

$$(m_\pi v)R \approx h$$

or

$$m_\pi \approx \frac{h}{Rv}$$

This uncertainty is smallest for $v = c$. Then the above equation says

$$m_\pi \approx \frac{h}{Rc}$$

The more accurate quantum-field theory calculation of Yukawa gave

$$m_\pi = \frac{h}{2\pi Rc}$$

and predicted a mass which agrees well with the measured mass of the pion. In addition to predicting the correct mass, Yukawa also predicted that pions would interact strongly with nucleons. For example, pions are easily produced by collisions of nucleons. In this case kinetic energy of the nucleons is directly converted into rest mass. Some pion production reactions are

$$P + P \rightarrow P + N + \pi^+$$
$$P + N \rightarrow P + P + \pi^-$$
$$\gamma + P \rightarrow N + \pi^+$$
$$\gamma + P \rightarrow P + \pi^0$$
$$\gamma + N \rightarrow P + \pi^-$$

Protons of several hundred million electron volts are needed to produce pions. Such proton beams are provided by synchrocyclotrons such as the one shown in Fig. 9-13.

In 1947 the pion was first discovered in the cosmic rays by examining the tracks they made in nuclear emulsions (see Fig. 16-5). A year later the first man-made pions were detected at the Berkeley synchrocyclotron. The charged pions decay by the weak interaction as follows

$$\pi^+ \quad \begin{matrix} \nearrow \mu^+ + \nu_\mu \\ \text{or} \\ \searrow e^+ + \nu_e \end{matrix}$$

$$\pi^- \quad \begin{matrix} \nearrow \mu^- + \bar{\nu}_\mu \\ \text{or} \\ \searrow e^- + \bar{\nu}_e \end{matrix}$$

Fig. 16-5. Drawing of track left by stopping pion in nuclear emulsion. The decay chain $\pi^+ \rightarrow \mu^+ + \nu$ followed by $\mu^+ \rightarrow e^+ + \nu + \nu$ is seen (except for the neutrinos). Grains of a photographic emulsion can be sensitized by collisions with a charged particle as well as by light.

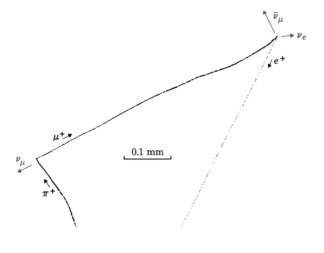

0.1 mm

Ans. 5: No. Mesons are hadrons and hadrons are strongly interacting particles. Muons do not interact strongly.

with a halflife of 1.8×10^{-8} sec. The π^0 decays very much faster into two photons by the electromagnetic rather than the weak interaction. The halflife of the π^0 is about 10^{-16} sec. The decay of a π^+ and also the decay of a μ^+ can be seen in the bubble chamber picture of Fig. 16-4. One of the two π^+ tracks actually comes to rest in the liquid hydrogen. Then it decays into a visible μ^+ track and invisible neutrino. After traveling 1.1 cm, the μ^+ comes to rest also and decays into a visible e^+ and two invisible neutrinos.

Actually the muon was discovered almost at the same time that Yukawa had predicted the pion. Since the muon mass is close to that of the pion, physicists thought for many years that the muon was indeed Yukawa's particle. Just before the pion was finally discovered, most physicists had given up on the muon (or else on Yukawa) because it never interacted strongly as Yukawa had predicted. The pion serves a useful purpose in explaining the strong, short-range nuclear force, but the only purpose seen for the muon is to give Nature one more chance to confuse man.

The K meson

A more recently established meson is the K meson. It is spin zero and has both positive and neutral charge (K^+ and K^0) with corresponding antiparticles K^- and $\overline{K}^0$. Because of its large mass (about half that of the proton), the K meson has a larger number of possibilities for decay by the weak interactions. The following decay modes of the K^+ have been observed:

$$K^+ \to \begin{cases} \pi^+ + \pi^0 & (\theta \text{ mode}) \\ \pi^+ + \pi^+ + \pi^- & (\tau \text{ mode}) \\ \pi^+ + 2\pi^0 \\ \mu^+ + \nu \\ \mu^+ + \nu + \pi^0 \\ e^+ + \nu + \pi^0 \end{cases}$$

The halflife of the charged K meson is 0.85×10^{-8} sec. The neutral K has similar decay modes with a halflife of 7.0×10^{-11} sec. Two contemporary problems in theoretical physics are why does the K^0 have a lifetime so much shorter than the K^+ and what is the relationship between K meson decay and the universal Fermi interaction? The production of K mesons is discussed in the next paragraph.

Q.6: Why can the K meson not decay into 4 pions?

TABLE 16-1. CATEGORIES OF ELEMENTARY PARTICLES

Hadrons (strongly interacting)
- Baryons
  - Hyperons Λ, Σ, Ξ, Ω (strange)
  - Nucleons N, P, N (nonstrange)
- Mesons
  - (strange) K
  - (nonstrange) π

Leptons (weakly interacting)
- Muon family μ, ν_μ
- Electron family e, ν_e

The Photon γ (electromagnetic interaction)

The hyperons (strange baryons)

Some of the baryons are called hyperons for reasons to be explained shortly. (See Table 16-1.) These hyperons decay by weak interaction and consequently live long enough to leave tracks in bubble chambers. There are four different kinds of elementary particles which are heavier than the proton and which have halflives long enough to leave tracks in bubble chambers. These are designated by the capital Greek letters Λ (lambda), Σ (sigma), Ξ (xi) and Ω (omega). They all decay by the weak interactions into nucleons, and thus obey the law of conservation of baryons. So far we have had two examples of what are called the strong interactions: (1) the nucleon-nucleon force and (2) the pion-nucleon interaction (pion production for example). Similarly, the interactions of hyperons and K mesons with nucleons and pions are examples of strong interactions. The production of hyperons and K mesons is one of the many examples of strong interactions. The most studied hyperon production interaction is

$$\pi^- + P \rightarrow \Lambda + K^0$$

A liquid hydrogen bubble chamber photograph of this is shown in Fig. 16-6. The necessary high-energy pion beams are easily produced in Bev proton accelerators. Note that the lambda is produced in association with a K meson. In fact, either a Λ, Σ, Ξ, or Ω is always produced in association with K mesons when the initial particle is a pion or nucleon. Because of this special property of associated production, these four baryons and the K meson were and still are called strange particles. The four strange particles which are baryons are called hyperons. What seemed strange about the strange particles was that they decay about 10^{14} times slower than was expected. Since they are produced easily by the strong interaction and they can decay into strongly interacting particles, it was expected that they should decay as quickly as they are produced which would be in about 10^{-23}

Fig. 16-6. Associated production of a lambda hyperon and K meson in a liquid hydrogen bubble chamber. A 1-Bev π^- from the Bevatron enters the chamber, hits a proton, and produces the Λ and K^0. (Courtesy liquid hydrogen bubble chamber group of the Lawrence Radiation Laboratory.) ▶

Ans. 6: Its rest mass is less than that of 4 pions.

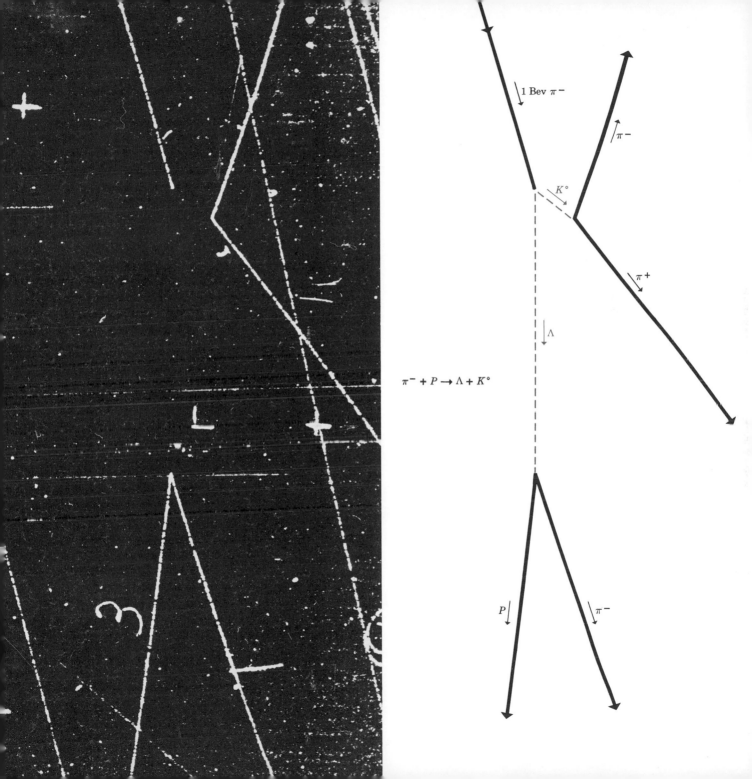

1 Bev $\pi-$

$K°$

$\pi-$

$\pi+$

Λ

$\pi- + P \rightarrow \Lambda + K°$

P

$\pi-$

sec. In order to explain this, it was necessary to invent one more conservation law which would slow down the decay of the strange particles by a factor of about 10^{14}. This new conservation law is called the conservation of strangeness. The K^+ and K^0 are assigned strangeness number $+1$, and the hyperons have the negative strangeness numbers shown in Fig. 16-8. The law of conservation of strangeness states that in a closed system the total strangeness number remains unchanged. This law is strictly obeyed by the strong and the electromagnetic interactions, but is completely ignored by the weak interaction. A strange particle decays by the weak interaction into particles having zero strangeness. And since the weak interaction is 10^{14} times weaker than the strong interaction, these strange particles will live 10^{14} times longer than normal.

The Λ is 37 Mev heavier than a proton plus a pion. It is neutral and has spin-$\frac{1}{2}$. Its principal decay modes are

$$\Lambda \quad \begin{matrix} \nearrow P + \pi^- \\ \text{or} \\ \searrow N + \pi^0 \end{matrix}$$

The Σ is 78 Mev heavier than the Λ and also has spin-$\frac{1}{2}$. However, the Σ can be positive, neutral, or negative. The principal decay modes are

$$\Sigma^+ \rightarrow \begin{cases} P + \pi^0 \\ N + \pi^+ \end{cases}$$
$$\Sigma^- \rightarrow N + \pi^-$$
$$\Sigma^0 \rightarrow \Lambda + \gamma$$

The Σ^0 decay is very much faster than the Σ^+ or Σ^- because it is an electromagnetic interaction that is about 10^{12} times stronger than the weak interaction.

The Ξ hyperon is 205 Mev heavier than a Λ and can be negative or neutral. The observed decay modes are

$$\Xi^- \rightarrow \Lambda + \pi^-$$
$$\Xi^0 \rightarrow \Lambda + \pi^0$$

For each hyperon there should be an antihyperon of opposite charge. So far the $\overline{\Lambda}$, $\overline{\Sigma}^+$, and the $\overline{\Sigma}^0$ are the only antihyperons that have been observed. However, physicists feel certain that the others must exist also. A Λ produced by the process $\overline{P} + P \rightarrow \overline{\Lambda} + \Lambda$ is shown in Fig. 16-7.

In increasing order of mass, the present (1967) list of long-lived elementary particles is: the photon γ; the leptons ν_e, ν_μ, e, and μ; the mesons π and K: the nucleons P and N; the hyperons Λ, Σ, Ξ, and Ω. If we take into account their different charges and their antiparticles, the grand total is 34. All of them are tabulated in Fig. 16-8. The particles are on the left and their corresponding antiparticles are "reflected" on the right.

The "resonances"

In the past few years quite a few new mesons and baryons have been discovered which decay directly by strong interaction. The first example, a baryon, was uncovered by Fermi in the early 1950's. It is called the N_1 and is 160 Mev heavier than a proton plus a pion. It decays in about 10^{-23} sec into a proton and a pion:

$$N_1 \rightarrow P + \pi$$

The strong interaction is so strong that when one strongly interacting particle touches another, there is a high probability of interaction. The time the particles will be in contact is the diameter of an elementary particle divided by the velocity: $t = D/v \approx 3 \times 10^{-13}$ cm$/3 \times 10^{10}$ cm $= 10^{-23}$ sec. Hence it takes about 10^{-23} sec for the strong interaction to act. Recently dozens of new hadrons have been discovered which have halflives ranging from 10^{-22} to 10^{-23} sec. These all decay by the strong interaction and for historical reasons are called resonances. There are a few others with halflives ranging from 10^{-21} to 10^{-16} sec which decay by the electromagnetic interaction. Most of these newly discovered particles fit into simple groups of 8 or 10 based on their quantum numbers. One of these groups of 8 contains the proton and neutron. Another group of 8 contains the three pions. These groups are sometimes referred to as the Eight-fold Way.

It is now felt that the proton or pion is no more elementary than any other hadron. Each hadron is just another state or energy level of strongly interaction matter. Over 190 "elementary" particles have been found thus far. New

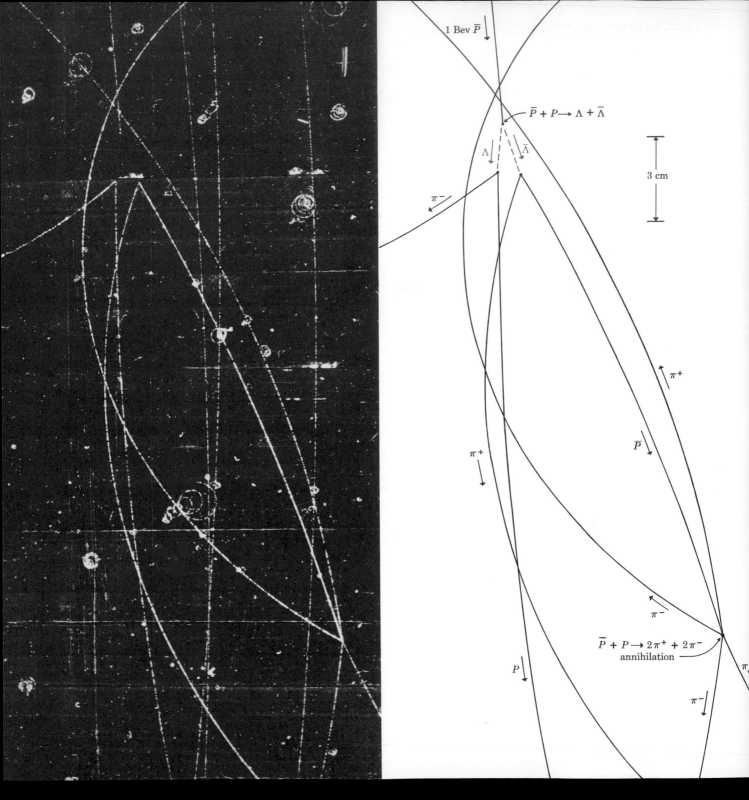

1 Bev $\overline{P}$

$\overline{P} + P \longrightarrow \Lambda + \overline{\Lambda}$

Λ $\overline{\Lambda}$

3 cm

π^-

π^+

π^+

$\overline{P}$

π^-

$\overline{P} + P \longrightarrow 2\pi^+ + 2\pi^-$
annihilation

P

π^-

π^-

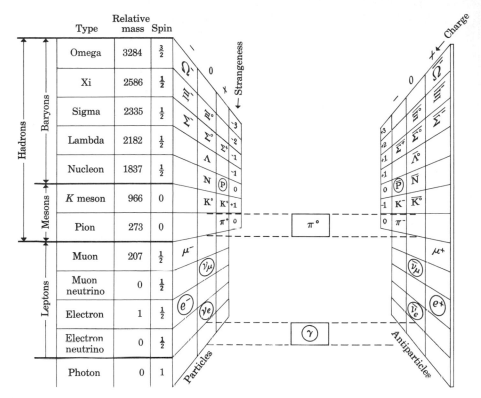

Fig. 16-8. Table of the elementary particles. The particles are tabulated on the left. The "reflections" on the right are the corresponding antiparticles. The stable particles are circled.

◀ **Fig. 16-7.** Production of an antilambda by the process $\bar{P} + P \rightarrow \bar{\Lambda} + \Lambda$. The antilambda decays into a π^+ and an antiproton. Then the antiproton comes to rest and annihilates a proton of the liquid hydrogen. (Courtesy 72-in. bubble chamber group of the Lawrence Radiation Laboratory.)

particles are being found almost each month. One of the current theoretical hopes is to explain all hadrons in terms of three or four basic building blocks called quarks, but so far no quarks have been found. In order to see quarks, if they exist, higher energy accelerators will probably be needed.

Since the weak interaction is about 10^{14} times weaker than the strong interaction, the lifetimes of particles which decay by weak interaction will be about 10^{14} times 10^{-23} sec or about 10^{-9} sec. All of the charged particles listed in Fig. 16-8 decay by weak interaction or else are stable. The reason why some hadrons decay by weak interaction rather than by strong interaction is that they are forbidden to decay by the strong interaction because of the conservation laws such as conservation of strangeness.

16-6 Nonconservation of Parity

"How would you like to live in a Looking-glass, Kitty? I wonder if they'd give you milk in there? Perhaps Looking-glass milk isn't good to drink." Alice

In this section we will learn that Lewis Carroll was right—looking-glass milk is indigestible if not poisonous. Also we shall see that not only is the law of conservation of parity violated by the weak interaction but so is the principle of antiparticle symmetry which is discussed in Section 16-4. The conservation of parity is the mathematical formalism of the symmetry principle called reflection invariance. The principle of reflection invariance states that the mirror image of any physical phenomenon itself is just as true a physical phenomenon. According to conservation of parity, if someone observed any physics experiment in a mirror and was not told he was looking in a mirror, there would be no way he could tell from the results whether he was looking in a mirror. Another way of saying it is that all the fundamental laws of physics should have the same mathematical form whether one is using a left-handed or a right-handed coordinate system. One consequence of the conservation of parity is that there is no way for an absent-minded professor to determine which is his right hand by performing experiments. It would be cheating for the professor to determine which side of his body his heart lies on. That would be equivalent to handing him a glove labeled "left."

Actually, the molecules of his body and, for that matter, all life on earth are equivalent to labeled gloves (see Chapter 14 opening). Biologically produced protein molecules are built up from amino acids—all of which are of a left-handed screw-type structure (except for antibiotics such as penicillin which contain a certain percentage of right-handed amino acids. This is thought to make these molds poisonous to bacteria and accounts for their use as antibiotics). On the other hand, right-handed proteins can be synthesized by chemists; and, as would be expected from conservation of parity, they have exactly the same chemical properties as the natural varieties. The only difference is that one is the mirror image of the other (see Fig. 16-9). The fact that biology on the earth always produces molecules of

Fig. 16-9. A right-handed screw and its mirror image.

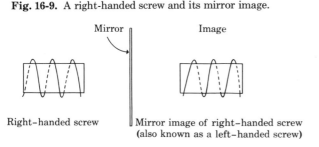

Right–handed screw

Mirror image of right–handed screw (also known as a left–handed screw)

Fig. 16-10. Photograph of T. D. Lee whose work in collaboration with C. N. Yang let to the downfall of parity conservation. Is this the true image, or the mirror image of Professor Lee?

one definite mirror symmetry, whereas the same molecules when synthesized by chemists are always produced as a fifty-fifty mixture of right- and left-handed varieties, may seem puzzling. An explanation may be that early forms of life on earth started out both left- and right-handed. Then the animals and plants of one form would be indigestible and probably poisonous to those of the other form. Finally one of the two forms of life would win out over the other in the battle for survival.

The photograph of T. D. Lee of Columbia University in Fig. 16-10 is presented as a challenge to the alert student. Is there any way you can tell for sure whether or not this picture was taken looking into a mirror? At first glance one would think that this is the mirror image of Dr. Lee. The symbols on the blackboard are reversed and Lee is using what appears to be his left hand. The alert student might suspect the picture was rigged—that Lee deliberately wrote the symbols backwards as a trick. More careful examination reveals that the buttons of Lee's jacket appear to be on his right side. On the basis of the rigid, world-wide custom that all men must have their buttons on the right side and all women must have theirs on the left, the alert student may correctly conclude that Fig. 16-10 is the true image of T. D. Lee. We did not go to the extreme and hire a tailor to manufacture a special jacket with bottons on the left and buttonholes on the right.

For reasons beyond the scope of this book, the conservation of parity forbids the K meson to have both two-pion and three-pion decay modes. The K meson is permitted to do one or the other, but not both. Both modes were observed experimentally. This led T. D. Lee and C. N. Yang in 1956 to question seriously the "self-evident truth" that nature should have no preference of right over left or vice versa. Lee and Yang proposed that the weak interactions do indeed violate the "sacred" principle of conservation of parity. They also proposed some specific experiments to test their hypothesis. We shall now discuss one of these experiments in detail. Our example shall be the decay of the π^+ which is a consequence of the weak interactions.

$$\pi^+ \rightarrow \mu^+ + \nu$$

Q.7: If an absent-minded professor knew which direction was north, could he then determine his right hand?

Fig. 16-11. (*a*) Decay of the π^+. (*b*) is the mirror image of the decay of the π^+.

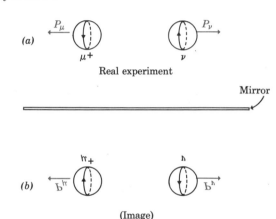

(*a*)

P_μ ← ⊕   ⊕ P_ν →
μ^+ ν

Real experiment

Mirror

(*b*)

Image

Lee and Yang suggested that the decay muons and neutrinos might have a preferred spin direction along their direction of motion.

We shall now proceed to show that if this is the case, reflection invariance would be violated. We shall represent spin schematically as the motion in the equatorial plane of the spinning particle. The particles will be drawn as spinning spheres. Lee and Yang proposed that the π^+ decay should always look as shown in Fig. 16-11*a*. The μ^+ and ν must be spinning in opposite directions as shown because their spins must add up to the initial pion spin which is zero. Figure 16-11*b* shows the mirror image of the decay particles of Fig. 16-11*a*. Note that in the mirror the spheres will appear as if they are spinning in the opposite sense. A "piece" of the equator of the μ^+ or ν would trace out a left-handed screw in the original view (*a*) and a right-handed screw in the mirror image (*b*). The situation shown in (*a*) was first observed in 1957 using cyclotron produced muons. The mirror image experiment shown in (*b*) never occurs in nature. Thus conservation of parity is clearly violated. The neutrinos always come off as left-handed screws. Nonconservation of parity was first observed in a beta decay experiment performed by C. S. Wu of Columbia University and a group at the National Bureau of Standards in Washington. It is now known that neutrinos are always spinning as left-handed screws, whereas antineutrinos are always spinning as right-handed screws.

It is still hard to believe that space has a built-in preference for left over right, but the evidence is so simple and clear that everyone almost immediately accepted it as true. Thus all an absent-minded professor need do to determine his left hand is to look at any neutrino or to a μ^+ coming from a π^+ decay. Here is a clear case of a law of nature that is not symmetrical. One reason for all the excitement is that this is the first time that violation of a basic symmetry principle has been found.

Violation of antiparticle symmetry

If we charge conjugate Fig. 16-11*a* we obtain the decay as shown in Fig. 16-12*a*. Note that this would make the antineutrino left-handed. But we now know from experiment that the antineutrino is always right-handed as shown in

Ans. 7: Yes. By watching the stars or sun he can point his fingers in the direction of the earth's rotation. Then if his thumb points north, it is his right hand.

Fig. 16-12. (*a*) is obtained from the π^+ decay by replacing particles with their antiparticles. (*b*) is the observed experimental result.

(a) Charge conjugate of π^+ decay

(b) π^- decay: experimental result

Q.8: What resulting experiment is obtained if Fig. 16-11*a* is reflected in a mirror and then charge conjugated? (This is the reverse order of the operations in Fig. 16-12.)

Fig. 16-12*b*. Figure 16-12*b* is what the π^- decay looks like experimentally. Thus here is a case where changing particles to their corresponding antiparticles predicts results that do not occur in nature. This is a clear violation of antiparticle symmetry.

Note that if we reflect π^+ decay (Fig. 16-11*a*) in a mirror and also change particles to their antiparticles, we arrive at the correct answer for π^- decay (Fig. 16-12*b*). Thus an over-all symmetry is preserved. There still is no way to tell an absent-minded professor in a remote galaxy which is his right hand because we do not know whether his galaxy is made of antimatter or not, and we would have no way to distinguish a μ^+ from a μ^- for him. Conversely, there is no way to tell him whether his atoms are made of orbital electrons or positrons unless he knows right from left. This over-all symmetry is called CP invariance by the theoretical physicists and it is still being checked. At the time of this writing (1960) it appears that the weak interactions do observe CP invariance. This possibility of such an over-all symmetry is philosophically satisfying and makes it easier to accept the loss of conservation of parity and antiparticle symmetry.

The preceding paragraph was intentionally left untouched for this second edition in order to emphasize what has happened since then. In 1965, a group working at the Brookhaven AGS did indeed find a small violation of CP invariance by carefully measuring decay modes in the decay of the K°. Up until their experiment it had been believed that the decay curve for K° mesons would be exactly the same as the decay curve of $\overline{K}^\circ$ mesons. Now experiments show that these two decay curves should have some small differences and that it should be possible in principle to distinguish a K° beam from a $\overline{K}^\circ$ beam. So finally we have an effect with which to determine our right hand. Now we ask the professor in a remote galaxy to build an AGS, produce what he thinks are K° mesons, and measure their decay curve. Depending on which curve he obtains, we can tell him whether or not he and his AGS are built of antimatter. At present intensive searches are underway to detect possible CP violations in other reactions and to test whether there is any CP violation in the electromagnetic or the strong interaction.

Prohibitions

The basic conservation laws of physics are in a sense stronger than other "laws." This is because anything that can happen does happen, unless it is forbidden by a conservation law. We cannot tell how often something will happen from the conservation laws—although the relative strengths of the four basic interactions are a good guideline. We recall that electromagnetic interactions are about 10^{-2} as probable as the strong interaction, and weak interactions are about 10^{-14} as probable as the strong interaction. The important point is that any reaction or decay mode that you can think of will occur in nature provided it is not prohibited by any of the following conservation laws.

Some of the following 13 conservation laws are stated as symmetry principles. However, in quantum mechanics it is always possible to find a mathematically equivalent conservation law corresponding to each symmetry principle. The sudden and recent loss of conservation of parity and antiparticle symmetry serves as an additional warning to scientists and philosophers that other sacred laws of physics may also not be correct. For example, one can never prove in a foolproof way that the law of conservation of energy is true. However, if a single clear-cut violation of this law were ever found, this would be an absolute proof that the law of conservation of energy as now stated is incorrect. With this warning we list our final collection of conservation laws. For completeness, some of the laws not covered in the text are also listed.

1. Conservation of total energy. Rest mass must be included.
2. Conservation of total linear momentum.
3. Conservation of total angular momentum.
4. Conservation of charge.
5. Conservation of baryons. The nucleons and hyperons have baryon number $+1$. Their antiparticles have baryon number -1. This law says that the total baryon number must remain unchanged.
6. Conservation of leptons. This law may be thought of as the light particle counterpart of the above law. The

Ans. 8: One still ends up with the experimentally observed π^- decay as shown in Fig. 16-12*b*.

leptons ν_e and e^- have electron lepton number $+1$ and their antiparticles have lepton number -1. According to this law the total electron lepton number must remain unchanged before and after any interaction. There is a corresponding, but independent, law of conservation of muon leptons.

7. Charge independence (also called conservation of isotopic spin). This law holds only for the strong interactions. Because of the electromagnetic interactions its predictions are only accurate to within about 1%. Charge independence makes such predictions as that the proton-proton force should be the same as the proton-neutron force.

8. Conservation of strangeness (associated production of strange particles). This laws holds for all strong and electromagnetic interactions but not for the weak interactions. It is because of this law that hyperons and K mesons decay slowly enough for us to see their tracks.

9. Antiparticle symmetry. This law holds also for all strong and electromagnetic interactions but not for the weak interactions.

10. Conservation of parity. This law also holds for all strong and electromagnetic interactions but not for the weak interactions.

11. Overall antiparticle-parity symmetry (CP invariance). This law says that if all the particles in the mirror image of any experiment are changed to their corresponding antiparticles, this new experiment is also a legitimate experiment. This law appears to hold for strong and electromagnetic interactions, but a small violation has been observed in the decay of the neutral K meson.

12. CPT invariance. Overall antiparticle-parity-time reversal symmetry. This law says that if all the particles in the mirror image of any experiment are changed to their corresponding antiparticles, and all velocities and rotations are reversed, the new experiment so obtained is a legitimate experiment. This law is believed to hold for all interactions.

13. Time reversal invariance. This law says that if the velocities and rotations of all the particles of any experiment are reversed, this new experiment is also a legitimate experiment. It appears to hold for strong and electromagnetic interactions, but must be slightly violated for the weak interaction because CP is slightly violated and CPT is not.

Q.9: Are there any conservation laws which apply to the weak interaction, but are violated by the strong interaction?

16-8 Problems for the Future

Only the beginning

The use of the term "elementary" begins to look ridiculous when we start talking of over 190 elementary particles. It might be reasonable to think in terms of just four different kinds of particles: the photon, electron leptons, muon leptons, and hadrons. A hopeful prediction for the future is that there may be a reduction in the number of truly elementary particles. The fact that all these elementary particles can transform into one another at will (consistent with the conservation laws) raises the hope that there might be one grand field of which these particles are different "quantum states." Such a grand unified theory would have to predict the masses of the existing "elementary" particles. Also with such an ultimate theory we should be able to calculate the charge of the electron and all other physical constants. At present the physical constants such as c, e, h, m_e, m_P, etc., are completely independent. In general, as we get closer to the ultimate truth we should be able to calculate some of these constants in terms of the others. For example, the binding energy of the hydrogen atom can now be calculated from e, h, and m_e. Another example is that the recent theory of the universal Fermi interaction permits us to calculate the muon lifetime in terms of the neutron lifetime.

The ultimate theory should not only give us the way to calculate the charge of the electron (or strength of the electromagnetic interaction), but it also should explain the strong, weak, and gravitational interactions. The gravitational interaction is even much weaker than the weak interaction. There have been unsuccessful attempts to explain gravity in terms of neutrinos. Perhaps someday gravity will be explained in terms of other apparently unrelated phenomena. Other unsolved problems are the origin, size, and evolution of the universe. Is matter being created out of nothing? Are there just as many galaxies made out of antimatter as ordinary matter?

Our understanding of the physical world has come a long way from the days of Aristotle when everything was explained in terms of the four basic elements: Fire, Water, Air, and Earth. We now have a thorough and satisfying explana-

Ans. 9: No.

tion of the structure of ordinary matter in terms of quantum electrodynamics. However, we are just scratching the surface in our attempt to understand what we believe to be truly fundamental, the multitude of elementary particles and their interactions.

References:

Ford, K. W. *Elementary Particles,* Blaisdell Publishing Company, New York.

Frisch, D. H. and A. M. Thorndike, *Elementary Particles,* D. Van Nostrand Company, Princeton, New Jersey.

Feynman, R. P. *Lectures on Physics,* Vol. I, Ch. 52, Addison Wesley, Reading, Mass.

Swartz, C. E. *The Fundamental Particles,* Addison Wesley, Reading, Mass.

Problems

1. The Ξ^- hyperon decays as follows:

$$\Xi^- \rightarrow \pi^- + \Lambda^0$$

   (a) What would be the decay products of the anti-Ξ^- hyperon?
   (b) What is the charge of the anti-Ξ^- hyperon?

2. What is the antiparticle of the antineutron?

3. What is the halflife of the antineutron? What are the decay products?

4. In the $\pi - \mu$ decay $\pi \rightarrow \mu + \nu_\mu$ is the total amount of mass immediately after decay less than that before decay? (Remember that a moving particle has more mass than when stationary.)

5. The following decay modes are forbidden. For each decay mode list the conservation laws that would be violated.
$$\Lambda \rightarrow \pi^+ + \pi^-$$
$$K^+ \rightarrow \pi^+ + \pi^- + \pi^0$$
$$\overline{N} \rightarrow e^- + P + \overline{\nu}$$
$$P \rightarrow N + e^+ + \nu$$
$$N \rightarrow e^- + e^+ + \nu$$

6. What is the charge of the $\overline{\Sigma}^+$?

7. For each of the following reactions state whether or not it is forbidden. If it is forbidden, state a conservation law that is violated.
   (a) $\Lambda^0 \rightarrow P + \pi^0$
   (b) $\overline{P} + P \rightarrow \mu^+ + e^-$
   (c) $N \rightarrow P + e^- + \nu_e$
   (d) $P \rightarrow N + e^+ + \nu_e$
   (e) $\Sigma^+ \rightarrow \Lambda^0 + \pi^+$

8. The U^{238} alpha decay is followed by two successive beta (e^-) decays. What is the Z and A of this great-granddaughter of U^{238}?

9. A Pu^{239} $(Z = 94)$ nucleus alpha decays. The daughter nucleus beta decays (emits e^-) into the granddaughter nucleus which also beta decays (emits e^-). Then this great-granddaughter is bombarded by neutrons and absorbs 4 neutrons. What is the Z and A of the final product?

10. Of the four kinds of neutrinos $(\nu_e, \overline{\nu}_e, \nu_\mu, \overline{\nu}_\mu)$ which can give rise to the reaction

    (a) $(?) + P \rightarrow N + e^+$
    (b) $(?) + N \rightarrow P + \mu^-$
    (c) $(?) + N \rightarrow P + e^-$

11. Assume that none of the annihilation products in Fig. 16-4 has further interactions, but that they all decay. Assume the charge pions decay into muons, and then the muons decay. The final product will all be e^-, e^+, ν, $\overline{\nu}$, and photons. How many of each will there be?

12. The image of a right-handed screw is projected onto a frosted glass screen and appears as a right-handed screw. When the screen is viewed from the other side, is the image a right-handed or left-handed screw?

13. An antiproton comes to rest and annihilates with a proton. They produce three pions of equal energy. What is the kinetic energy in Mev of each pion?

14. An "atom" of an electron bound to a positron can exist for a short time ($\sim 10^{-6}$ sec) before the eventual annihilation. Such an atomic system is called positronium. What is the charge conjugate of positronium?

15. A right-handed screw is being screwed into a threaded hole. As viewed from the hole, does it appear as a right-handed or left-handed screw?

16. In Fig. 16-11a pretend the mirror is vertical at the right side rather than horizontal at the bottom. Draw a picture of the image in this mirror. Is the μ^+ in the image right-handed or left-handed?

17. In μ^- capture by the proton the two can convert by the universal Fermi interaction to a neutron plus another particle. What is this other particle? (It must satisfy conservation of leptons.)

18. Draw a sketch of what a bubble chamber picture of $K^- + P \rightarrow \Sigma^- + \pi^+$ would look like. Assume the K^- comes to rest and the Σ^- decays in flight.

19. When Co 60 nuclei are lined up with their spins pointing up, more beta rays are observed in the down direction than the up. Draw the mirror image of this.

    (a) When the mirror is horizontal, what is the spin direction of

Prob. 19

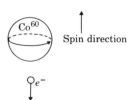

Spin direction

e^-

the reflected Co^{60}? Use the same convention for spin direction as shown in the figure.

(b) When the mirror is vertical, what is the spin direction of the reflected Co^{60}?

(c) In each of the above mirror images state whether the electron decay is parallel or antiparallel to the spin direction.

20. Experimentally when the left-handed positive muons come to rest, their decay positrons come off predominantly in the backward direction. Consider the case of a positron coming off exactly backwards and the ν and $\bar{\nu}$ going exactly forward. Is the positron left-handed or right-handed?

21. In the decay $\pi^+ \rightarrow \mu^+ + \nu$ the neutrino energy cP plus the muon kinetic energy $(P^2/2M_\mu)$ must be supplied by the mass difference. Find P and the muon kinetic energy in Mev.

22. What is the maximum electron energy in the decay $\mu^- \rightarrow e^- + \nu + \bar{\nu}$? Assume the electron is so relativistic that its momentum $P = W/c$.

Tables

TABLE I USEFUL CONVERSIONS

1 in.	$= 2.54$ cm
1 mi	$= 1.61$ km or 5280 ft
1 m	$= 39.37$ in.
1 mph	$= 44.7$ cm/sec
60 mph	$= 88$ ft/sec
1 kg	$= 2.204$ lb
1 newton	$= 10^5$ dynes
1 joule	$= 10^7$ ergs
1 calorie	$= 4.18$ joules
1 coulomb	$= 3 \times 10^9$ statcoulombs
1 statvolt	$= 300$ volts
1 ev	$= 1.6 \times 10^{-12}$ ergs
1 Mev	$= 1.6 \times 10^{-6}$ ergs
1 Å	$= 10^{-8}$ cm

TABLE II SOME PHYSICAL CONSTANTS

$c \quad = 3 \times 10^{10}$ cm/sec   speed of light

$e \quad = 4.8 \times 10^{-10}$ statcoulomb   charge of the electron

$g \quad = 980$ cm/sec^2 acceleration due to gravity at earth's surface

$G \quad = 6.67 \times 10^{-8}$ cm^3/gm-sec^2 gravitational constant

$h \quad = 6.62 \times 10^{-27}$ ergs-sec   Planck's constant

$k \quad = 1.38 \times 10^{-16}$ ergs/°C   Boltzmann constant

$m \quad = 9.11 \times 10^{-28}$ gm   mass of the electron

$M_P \quad = 1.67 \times 10^{-24}$ gm   mass of the proton $M_P/m = 1837$

$mc^2 \quad = 0.51$ Mev   rest energy of the electron

$M_P c^2 = 938$ Mev   rest energy of the proton

For other masses see Fig. 16-8

$N_o \quad = 6.02 \times 10^{23}$ Avogadro's number

$P_o \quad = 1.01 \times 10^6$ dynes/cm^2 atmospheric pressure

$T \quad = 0°$K $= -273°$C   absolute zero

Diameter of earth	$= 7918$ mi
Distance earth to moon	$= 240,000$ mi
Distance earth to sun	$= 93$ million mi

TABLE III THE ELEMENTS

Element	Symbol	Atomic number	Average atomic mass
Actinium	Ac	89	227
Aluminum	Al	13	26.99
Americium	Am	95	[243]
Antimony	Sb	51	121.79
Argon	A	18	39.955
Arsenic	As	33	74.93
Astatine	At	85	210
Barium	Ba	56	137.40
Berkelium	Bk	97	[249]
Beryllium	Be	4	9.015
Bismuth	Bi	83	209.06
Boron	B	5	10.82
Bromine	Br	35	79.938
Cadmium	Cd	48	112.44
Calcium	Ca	20	40.09
Californium	Cf	98	[249]
Carbon	C	6	12.014
Cerium	Ce	58	140.17
Cesium	Cs	55	132.95
Chlorine	Cl	17	35.467
Chromium	Cr	24	52.02
Cobalt	Co	27	58.96
Copper	Cu	29	63.56
Curium	Cm	96	[245]
Dysprosium	Dy	66	162.55
Einsteinium	E	99	[255]
Erbium	Er	68	167.32
Europium	Eu	63	152.0
Fermium	Fm	100	[255]
Fluorine	F	9	19.01
Francium	Fr	87	223
Gadolinium	Gd	64	157.30
Gallium	Ga	31	69.74
Germanium	Ge	32	72.62
Gold	Au	79	197.1
Hafnium	Hf	72	178.55
Helium	He	2	4.004
Holmium	Ho	67	164.98
Hydrogen	H	1	1.0083
Indium	In	49	114.85
Iodine	I	53	126.94
Iridium	Ir	77	192.2
Iron	Fe	26	55.87
Krypton	Kr	36	83.82
Lanthanum	La	57	138.96
Lawrencium	Lw	103	[?]
Lead	Pb	82	207.27
Lithium	Li	3	6.942
Lutetium	Lu	71	175.04
Magnesium	Mg	12	24.33
Manganese	Mn	25	54.95
Mendeleevium	Mv	101	[256]
Mercury	Hg	80	200.66
Molybdenum	Mo	42	95.98
Neodymium	Nd	60	144.31
Neon	Ne	10	20.188
Neptunium	Np	93	[237]
Nickel	Ni	28	58.73
Niobium	Nb	41	92.94
Nitrogen	N	7	14.012
Nobelium	No	102	[253]
Osmium	Os	76	190.3
Oxygen	O	8	16.0044
Palladium	Pd	46	106.4
Phosphorus	P	15	30.983
Platinum	Pt	78	195.14
Plutonium	Pu	94	[242]
Polonium	Po	84	210
Potassium	K	19	39.111
Praseodymium	Pr	59	140.96
Promethium	Pm	61	[145]
Protactinium	Pa	91	231
Radium	Ra	88	226.11
Radon	Rn	86	222
Rhenium	Re	75	186.27
Rhodium	Rh	45	102.94
Rubidium	Rb	37	85.50
Ruthenium	Ru	44	101.7
Samarium	Sm	62	150.39
Scandium	Sc	21	44.97
Selenium	Se	34	78.98
Silicon	Si	14	28.10
Silver	Ag	47	107.909
Sodium	Na	11	22.997
Strontium	Sr	38	87.65
Sulfur	S	16	32.075
Tantalum	Ta	73	181.00
Technetium	Tc	43	[99]
Tellurium	Te	52	127.64
Terbium	Tb	65	158.97
Thallium	Tl	81	204.45
Thorium	Th	90	232.11
Thulium	Tm	69	168.99
Tin	Sn	50	118.73
Titanium	Ti	22	47.91
Tungsten	W	74	183.91
Uranium	U	92	238.13
Vanadium	V	23	50.96
Xenon	Xe	54	131.34
Ytterbium	Yb	70	173.09
Yttrium	Y	39	88.94
Zinc	Zn	30	65.40
Zirconium	Zr	40	91.24

TABLE IV PERIODIC TABLE OF THE ELEMENTS (after Longuet-Higgins)

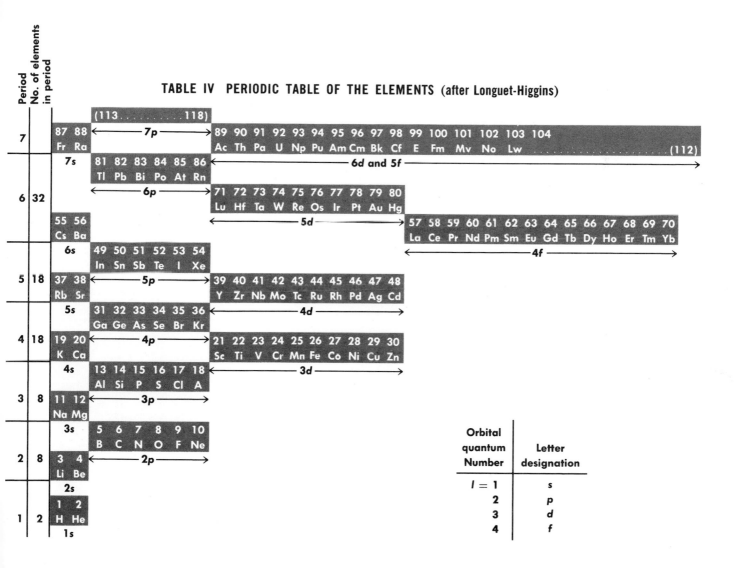

Orbital quantum Number	Letter designation
$l = 1$	s
2	p
3	d
4	f

Answers
to
even-numbered
problems

Chapter 1

2. $\dfrac{AB}{A + B}$

4. 1.67×10^6

6. $\dfrac{1 + a}{1 - a}$

8. $2 \times 10^{-11}, 5 \times 10^{-10}$

10. 6.65×10^{-24} gm

12. $\dfrac{v^2}{2s}$

14. 1.66×10^{-24} gm

16. x, 1.

18. $v = c \sqrt{1 - \left(\dfrac{M_0}{M}\right)^2}$

20. $(-W_0 + U_0), + 5.$

22. 1.293, 4 significant figures.

24. 1.75 m

26. $y = \dfrac{v_y}{v_x} x - \dfrac{g}{2v_x{}^2} x^2$

28. $x = \dfrac{x' - \beta ct'}{\sqrt{1 - \beta^2}}$ and $ct = \dfrac{ct' - \beta x'}{\sqrt{1 - \beta^2}}.$

Chapter 2

2. $\bar{v} = \dfrac{0 + v}{2}$

$\left(\dfrac{s}{t}\right) = \dfrac{v}{2}$

$v = \dfrac{2s}{t}$

4. $\bar{v} = \dfrac{v_0 + v}{2}$

$\left(\dfrac{s}{t}\right) = \dfrac{v_0 + v}{2}$

$v = \dfrac{2s}{t} - v_0$

6. $(a)\ \bar{v} = \dfrac{x_2 - x_1}{t_2 - t_1},\ (b)\ \bar{a} = \dfrac{v_2 - v_1}{t_2 - t_1}.$

8. The highest point.

10. 32 ft/sec, 16 ft.

12. 204 sec.

14. B − A, X − Y.

16. 2 mi/sec, 330 sec.

18. 121 ft.

20. 181.8 sec, 5810 ft/sec.

22. $a = 388$ ft/sec$^2 = 12.1g$

24. (*a*) 4 hr, (*b*) 4.27 hr, (*c*) 4.13 hr.

26. 2.15×10^8 m/sec after 1 yr 2.985×10^8 m/sec ($v/c = 0.995$) after 10 yr.

28. It has an acceleration of g which is about 60^2 times that of the previous question.

30. (*c*) 20 ft/sec, (*d*) 20 ft.

32. 50.4°

34. $\frac{1}{6}$

Chapter 3

2. Take any object from the pocket and throw it out.

4. gm/sec^2.

6. Newton's third law only requires that the force of the plow against the ground equal the force of the ground on the plow. The force pulling the plow can be different from this.

8. (*a*) 5.86 newtons, (*b*) 7.07 newtons, 4.71 newtons.

10. (*a*) Zero, (*b*) $\frac{F}{3}$, (*c*) $\frac{F}{3M}$, (*d*) $\frac{2}{3}F$.

12. 2.36×10^8 cm/sec, 7.51×10^{15} revolutions/sec.

14. $a = 333$ cm/sec^2. Tension in string between cars is 3.33×10^3 dynes. Tension in string pulled by child is 1.41×10^4 dynes. Force of floor on 20 gm car is 1.96×10^4 dynes.

16. 19,600 dynes, 29,400 dynes, 14,700 dynes.

18. 4π/sec.

20. $T_1 = 2.4$ nt, $T_2 = 2$ nt, $T_3 = 1.2$ nt. The inner string will break first.

22. When $F = 200$ nt, $F_{net} = 100$ nt in the horizontal direction. When $F = 800$ nt, $F_{net} = 500$ nt.

24. (*a*) $(T_1 - M_1g), (M_3g - T_2)$; (*b*) zero, $T_1 = M_1g$; (*c*) $T_1 = M_3g$; (*d*) $a = \dfrac{M_3 - M_1}{M_1 + M_2 + M_3} g$.

Chapter 4

2. No. No. Yes.

4. (*a*) Increased, (*b*) increase, (*c*) remain the same.

6. The same as now.

8. 881 nt.   588 nt.   Zero.

10. 1765 nt.

12. 9.9%

14. (*a*) 468 min, (*b*) $\frac{1}{36}g$, (*c*) $\frac{1}{3600}g$.

16. 23,700 mi from the moon.

18. The acceleration due to gravity at a height of 100 mi is $0.952g$.

20. See p. 80.

22. (*a*) $G\dfrac{M_E m}{(R_E + h)^2}$, (*b*) $\dfrac{mv^2}{R_E + h}$, (*c*) same as (*a*) or (*b*), (*d*) $v = \sqrt{\dfrac{GM_E}{R_E + h}}$, (*e*) increase.

24. Force of the earth = 272 dynes. Force of the sun = 599 dynes.

Chapter 5

2. Conservation of heavy particles.

4. To the earth.

6. 66.7 cm

8. 1.8 kg cm^2/sec^2

10. $v_{max} = 6.26$ m/sec, $a_{max} = 4g$.

12. (*a*) 4.9×10^5 ergs. (*b*) 7.0×10^5 ergs. (*c*) 200 cm/sec.

14. (*a*) $\sqrt{\dfrac{gR}{2}}$, (*b*) $\dfrac{R}{4}$.

16. 408 nt.

18. 6.23×10^5 ergs of KE were lost and converted into heating up the bullet and the wood.

20. Initial KE = 284 joules. Final KE = 10,220 joules. The student did work in pulling the dumbbells toward him. The KE of the system was increased by this amount of work.

22. (*a*) 500 nt, (*b*) 0.5 m/sec^2, (*c*) 196 m.

24. 1.99 m

26. The other fission product is Rb^{95} (37 protons and 58 neutrons).

28. (*a*) Mgh, (*b*) $\sqrt{gh}$, (*c*) $\dfrac{gh}{2L}$.

Chapter 6

2. (*a*) 546°C, (*b*) 3.12×10^5 cm/sec.

4. 39 gm

6. No.

8. (*a*) 0.4 nt, (*b*) 0.4 cm.

10. (a) 3 gm/cm³, (b) 3.06×10^4 cm³.

12. 0.1 cm

14. Helium

16. (a) 2×10^5 nt/m², (b) 6.7×10^4 nt.

18. (a) 2.18×10^6 dynes/cm², (b) 12 meters.

20. 3.54×10^3 molecules/cm³

22. $N = \dfrac{PV}{kT}$, $D = \dfrac{Pm}{kT}$.

24. 600°K or 327°C

26. (a) One, (b) $\dfrac{T_2}{T_1} = \dfrac{1}{2}$.

Chapter 7

2. (a) -36 dynes. (b) 200π.

4. (b) 20π

6. 50 dynes

8. (a) $+56.6$ statvolts. (b) Zero. (c) Zero.

10. (a) 1.6×10^{-6} erg. (b) Positive. (c) 1.44×10^{-13} cm.

12. Negative, positive.

14. (a) 1.02×10^7 dynes. (b) 1.535×10^{-6} ergs.

16. 1.24×10^{36}

18. 40 statvolts

20. 1.6×10^{-6} ergs

22. (a) -1.29×10^8 statcoul. (b) 2.58×10^8 dynes/statcoul.

24. (a) 3 ergs. (b) Zero. (c) -4 ergs. (d) -1 erg.

28. (a) $\frac{1}{3}M_0$. (b) $\frac{1}{2}g$. (c) $\frac{1}{2}g$.

30. $C = C_1 + C_2$

32. (a) $\dfrac{2(\rho_1 + \rho_2)}{r}$. (b) $\dfrac{2\rho_1}{r}$.

34. $V = \dfrac{QL}{r^2 - (L^2/4)}$

36. (a) $-\dfrac{Ze^2}{R^2}$. (b) $A_c = \dfrac{4\pi}{T^2}R$. (c) $T = 2\pi \sqrt{\dfrac{mR^3}{Ze^2}}$.

Chapter 8

2. (a) Zero. (b) 0.2 gauss.

4. emf $= 10^{-8} \dfrac{\Delta N_B}{\Delta t}$

6. 0.3π statvolts

8. 0.6 gauss

10. (a) Parabola; (b) up, $\dfrac{eE}{m}$; (c) $t = \sqrt{\dfrac{2m(d - h)}{eE}}$

12. (a) 0.4 gauss. (b) Zero.

14. (a) $\dfrac{2I}{cr}$. (b) Zero.

16. (a) $\dfrac{1}{240}$ of a second. (b) 0.8 statvolts. (c) 0.08 statvolts.

18. (a) 5.35×10^{-26} gm. (b) 59 times the rest mass.

20. 43.2 m

22. (a) 7.95×10^7 statamps/cm. (b) 1.67×10^{-2} gauss.

Chapter 9

2. The light bulb.

4. 222 m; 19,600

6. 18,000 volts

8. 216

10. $\dfrac{R}{3}$

12. (a) 3 volts. (b) 6 volts.

14. It will be a sinewave with the negative halves clipped off.

16. $x = \dfrac{R_2 R_3}{R_1}$

18. (a) 240 cm. (b) 13,900 gauss.

20. $M = \dfrac{eBD}{2vc}$

Chapter 10

2. (a) 6. (b) $2\frac{1}{2}$. (c) $\frac{1}{3}$ vibration per sec.

4. One-half wavelength.

8. 1048 cps.

10. Not at any time.

12. (a) $D_1 - D_2 = (N + \frac{1}{4})\lambda$. (b) $D_1 - D_2 = (N + \frac{3}{4})\lambda$

14. 4600 A

16. $D_1 - D_2 = N\lambda$

18. $D_1 - D_2 = \dfrac{N\lambda}{2}$ for max; $D_1 - D_2 = \frac{1}{2}(N + \frac{1}{2})\lambda$ for min.

20. (b) Yes. (c) Yes.

22. At the same spot.

24. The image would be erect and larger than the object.

26. $\sin \theta = \dfrac{\lambda}{d}$

28. 30 cm

30. (a) $d \cos \theta$. (b) $d \cos \theta = N\lambda$.

Chapter 11

2. New matter is continuously formed to replace the matter that is drifting away.

4. Away.

6. Slower.

8. 35 micrograms.

10. (a) 117 Mev. (b) 147 Mev.

12. Three to one.

14. 0.75 c.

16. 1.25.

18. (a) 825 Mev. (b) 8.81×10^{-25} gm. (c) $v = 0.8\,c$. (d) 1.67. (e) 1.67×10^{-8} sec.

20. 0.235 of a fringe shift.

22. (a) 625 cm. (b) B fired before A. (c) 400 cm.

24. (a) $f' = \dfrac{1 - (v/c)}{\sqrt{1 - (v^2/c^2)}}\, f$. (b) $f' = \dfrac{1 + (v/c)}{\sqrt{1 - (v^2/c^2)}}\, f$.

Chapter 12

2. 0.0124 A.

4. hf

6. 3×10^{-5} gm cm/sec, 2.21×10^{-22} cm.

8. The electron.

10. $\lambda_0 = 2880$ A.

12. 5.16 photons/sec.

14. (a) 167 counts/sec. (b) 26.7 counts/sec.

16. 1.47 A.

18. (a) 1.06×10^{-18} gm cm/sec. (b) 6.15×10^{-6} erg or 3.84 ev. (c) 5×10^{-8} cm. (d) 1.32×10^{-19} gm cm/sec. (e) 6.25×10^{-9} cm.

20. (a) $hf = mv - \dfrac{hf'}{c}$. (b) $hf = \frac{1}{2} mv^2 + hf'$.

Chapter 13

2. 0.265×10^{-8} cm.

4. 2×10^{-8} cm, 10^{-8} cm, $\frac{2}{3} \times 10^{-8}$ cm.

6. (a) 13.6 ev. (b) -27.2 ev.

8. 10

10. 14

12. 0.82 ev.

14. $hf = 13.6\left(1 - \dfrac{1}{N^2}\right)$. The wavelengths are those of the Lyman series.

16. (a) 54.4 ev. (b) 24.6 ev.

18. (a) -9 ev. (b) 1 ev, 2 ev, 3 ev.

20. (a) $-\dfrac{qQ}{R^2}$. (b) $U = -mv^2$. (c) $-\frac{1}{2}mv^2$. (d) $Rmv = N\dfrac{h}{2\pi}$. (e) $R = \dfrac{N^2 h^2}{4\pi^2 Qqm}$. (f) $W = -\dfrac{2\pi^2 m Q^2 q^2}{h^2 N^2}$.

22. (a) 2.82×10^3 ev. (b) 2.57×10^{-11} cm. (c) 2.11×10^3 ev.

24. (a) -3.2 ev. (b) 1.2, 1.8, 2.2, 6.1, and 0.4 ev.

26. (a) $\dfrac{mv^2}{R}$. (b) $\frac{1}{2}G\dfrac{Mm}{R}$. (c) $-G\dfrac{Mm}{R}$. (d) $-\frac{1}{2}G\dfrac{Mm}{R}$. (e) $Rmv = N\dfrac{h}{2\pi}$. (f) $R = \dfrac{h^2}{4\pi^2 m^2 GM} = 1.2 \times 10^{31}$ cm.

Chapter 14

2. Decrease

4. 40 min.

6. 2.62×10^{22} conduction electrons per gm of sodium. 3.3×10^{16} conduction electrons per gm for the germanium sample.

8. 5 ev.

10. 10.3×10^4 cal.

12. 5.12×10^{-11} cm.

14. (a) 3 volts. (b) negative.

16. $U_0 = 7.95$ ev.

18. (a) $P_x = N_x \dfrac{h}{2L}$, $P_y = N_y \dfrac{h}{2L}$, $P_z = N_z \dfrac{h}{2L}$. (b) $\text{KE} = \dfrac{h^2}{8mL^2}(N_x{}^2 + N_y{}^2 + N_z{}^2)$.

20. (a) $P_x = \dfrac{N_x h}{2L_x}$, $P_y = N_y \dfrac{h}{2L_y}$, $P_z = N_z \dfrac{h}{2L_z}$. (b) $\text{KE} = \dfrac{h^2}{8m}\left(\dfrac{N_x{}^2}{L_x{}^2} + \dfrac{N_y{}^2}{L_y{}^2} + \dfrac{N_z{}^2}{L_z{}^2}\right)$.

Chapter 15

2. $\frac{1}{2}, \frac{1}{2}$.

4. 1 min.

6. $\left(1 - \dfrac{1}{e}\right) = 63.2\%$, $\left(1 - \dfrac{1}{e^2}\right) = 86.5\%$.

8. 707.

10. 15.8 mi.

12. (*a*) 190/128 Mev. (*b*) 16 Mev.

14. 1.93×10^8.

16. (*a*) 32 Mev. (*b*) 2 Mev. (*c*) −40 Mev. (*d*) 8 Mev. (*e*) 10^{-6} sec. (*f*) 7 Mev. (*g*) Greater. (*h*) 8 Mev.

Chapter 16

2. The neutron.

4. No.

6. Negative.

8. U^{234} ($Z = 92, A = 234$).

10. (*a*) $\bar{v}_e$. (*b*) v_μ. (*c*) v_e.

12. Left-handed.

14. Positronium also.

16. Right-handed.

20. Right-handed positron.

22. 53 Mev.

Index